DATE DUE

Aquatic
Oligochaete
Biology

International Symposium on Aquatic Oligochaete Biology (1st: 1979: Sidney, B.C.)

Aquatic Oligochaete Biology

Edited by

Ralph O. Brinkhurst

Institute of Ocean Sciences
Sidney, British Columbia, Canada

and

David G. Cook

Fisheries and Oceans Canada
Ottawa, Ontario, Canada

Plenum Press · New York and London

Library of Congress Cataloging in Publication Data

International Symposium on Aquatic Oligochaete Biology, 1st, Sidney, B.C., 1979.
  Aquatic oligochaete biology.

  "Proceedings of the first International Symposium on Aquatic Oligochaete Biology,
held in Sidney, British Columbia, Canada, May 1–4, 1979."
  Includes index.
  1. Oligochaeta–Congresses. 2. Aquatic invertebrates–Congresses. I. Brinkhurst,
Ralph O. II. Cook, David G., 1941-   III. Title.
QL391.A6I57   1979   595.1'46   79-28164
ISBN 0-306-40338-2

Proceedings of the First International Symposium on Aquatic Oligochaete
Biology, held in Sidney, British Columbia, Canada, May 1–4, 1979.

© 1980 Plenum Press, New York
A Division of Plenum Publishing Corporation
227 West 17th Street, New York, N.Y. 10011

Printed in the United States of America

FOREWORD

     After some conversations with Professor Dr. H. Caspers and
other participants at the triennial congress of the International
Association of Theoretical and Applied Limnology (S.I.L.) held in
Copenhagen, Denmark (1977), the senior editor approached the
international delegates at the business meeting for approval of
the concept of holding the First International Symposium on
Aquatic Oligochaete Biology at Sidney (near Victoria) British
Columbia on May 1-4, 1979. The S.I.L. agreed to sponsor such a
meeting, and this sponsorship in turn led to the provision of
space and technical facilities at the Institute of Ocean Sciences
Patricia Bay, the Pacific Regional headquarters of the Ocean and
Aquatic Sciences component of the federal Fisheries and Oceans
Department. The National Research Council of Canada provided
travel support for a number of non-Canadian participants.

     Invitations were sent to as many active workers in the field
as the senior editor could name, and in addition two representatives
of closely allied fields were invited - Dr. V. Standen who works
with closely related but terrestrial species, and Dr. J. Grassle
who works with polychaetes but especially with *Capitella capitata*
which is much like an oligochaete in some aspects of its biology -
so much so that the senior editor in his salad days succumbed to
some ill-founded advice and described *Capitella* as a new marine
tubificid! The addition of these "outside" influences prevented
the group from making a number of unfounded assumptions during
discussions, and provided valuable cross-linkages.

     As the first meeting was such a spectacular success in terms
of the personal contacts made, the group has decided to hold
meeting at three-year intervals and the indication is that our
parent body will again give us its nominal sponsorship. It may
be that, at the next meeting, some of the outstanding taxonomic
problems might be addressed in a workshop setting, so that some
group decisions might be taken. The senior editor has taken it
upon himself to add an appendix to this volume, simply listing
current differences in nomenclature that show up in the manuscripts
but those names cited in a historical context are also listed with

v

their usually-accepted current names. No attempt has been made
to impose editorial control over the names used in the text. The
reader should consult the list of alternate names in the Appendix
where no attempt has been made to justify the use of any particular
alternative where two or more names are in current use. The taxonomic literature contains the various arguments pro and con each
name used. At the end of the lists the senior editor has attempted
to present, in as unbiased a fashion as possible, some of the basic
questions that remain unresolved and which lead to the instability
of the nomenclature. Many of these involve primarily the Tubificidae, the family in which most ecological and physiological interest has been taken in that they are a regular component of lake
and river benthic associations, have well-known relationships to
pollution, and have recently been found to extend to the sea-bed
well beyond the intertidal. This Appendix is presented not only
for the reader, but as a challenge to the second symposium.

I wish to thank Mrs. Lorena Magsalin Quay and the staff of
Prime Personnel, who carried out the typing, Technigraphics who
provided the final versions of the illustrations, and the staff
of Plenum Press for making our task so easy, and my colleague,
Dr. D.G. Cook for beginning the editorial work. Due to his unfor-
tunate illness at the critical time I have had to take on the bulk
of the work, and so those errors and omissions that have arisen
due to my more amateur approach to editing should be charged to
me rather than to Dr. Cook. The various language problems involved
often led to an extensive re-writing on my part but I trust I have
been able to preserve the author's intentions while rendering the
papers somewhat more comprehensible to readers. Time did not
permit all the articles to be re-checked, as the editors were
obliged to retype every paper to bring them into the required
format so far as this proved possible.

Ralph O. Brinkhurst
July 13, 1979

CONTENTS

SPECIFIC AND GENERIC CRITERIA IN FRESHWATER OLIGOCHAETA, WITH SPECIAL EMPHASIS ON NAIDIDAE

W.J. Harman

Department of Zoology
Louisiana State University
Baton Rouge, Louisiana, U.S.A. 70803

ABSTRACT

Definitions of the species, especially those that are applicable to uniparental species, are reviewed. Characters used to distinguish species may be qualitative or quantitative: the former are possibly more useful than the latter. Among the purely sexual taxa, species determinations are likely to be based upon the genital apparatus; while in the Naididae, the genus is likely to have as a primary character the gonadal placement. At the slightly higher subfamily, or in some cases family, we strongly consider the number of segments formed in anterior regeneration.

Examples of ecologically induced variation in Naididae are cited and their taxonomic implications discussed.

SPECIFIC CHARACTERS

In assessing the magnitude of the implications of the title of my presentation, I first went to the standard references to renew my concept of the definitions of "species" and "genus". Among the most common, and least acceptable, definitions of species for the Oligochaeta is that one which imposes criteria of "... interbreeding natural populations that are reproductively isolated from other such groups" (Mayr, 1969). This supposes a universality of syngamic reproduction to which not all oligochaetes subscribe.

While I recognize that many of you work with the Enchytraeidae, Lumbriculidae, and the Tubificidae where biparental reproduction is predominant, there are those of us who work with the Naididae and species in other families where sexual reproduction may be seasonal or sporadic and for which asexual reproduction assumes the predominant role. Generations of individuals may come and go with no breeding taking place, and the extreme application of that definition would require that the individual be synonymous with the species. No one is, fortunately, following that concept, or working with that definition.

Another concept of the species is all those individuals which have a common inheritance back to a point where the ancestors differed in enough features to be considered as distinct (Blackwelder, 1967). While this kind of a definition of a species offers some improvement in that it does not impose the criterion of interbreeding, or reproductive isolation except by inference, it will require a long term knowledge of a species and will work best for those who study taxa that have left a fossil record. We are all painfully aware that oligochaetes have left no such record, and that work with oligochaetes is cast entirely in the framework of the present.

In a very practical sense, a species consists of all those individuals which are considered to be members of a single kind as shown by the evidence that they are as alike as their offspring or their hereditary relatives over the span of a very few generations (Blackwelder, 1967). This is a much more plausible concept for uniparental forms and seems not to complicate the role of those who work with biparental species. It places upon the taxonomist only the responsibility of assessing similarity (and difference) and omits all illusion to breeding as a criterion.

In the latter two definitions of a species we are asked to assess similarity or differences, but we are given no yardstick or quantified criteria by which to make these assessments. And this is the point at which the greatest difficulty arises. Subjectivity cannot be avoided, and there is more subjectivity than I would like, but there is still objectivity involved. Each of us makes these assessments of similarity and differences each time we look at a specimen, either consciously or subconsciously. And it is the summation of these subjective and objective comparisons that culminates in a taxonomic decision.

Taxonomists in general, and oligochaete taxonomists in particular, are basically honest, conscientious individuals who earnestly seek uniformity in their collective decisions. But there are the lumpers and the splitters. Each is still acting in a

conscientious manner. It is their assessment of the quantifications
that differ, and some are more demanding than others. So it has
always been and so it will continue. We are not suffering from a
contemporary difference of opinion. It has been with us through
the years and will remain with us as long as there are people
concerned with these definitions translated into practical
application.

Most of us work with a "population" concept of the species
today. This is a far step away from the older concept of "the type"
where no variation was allowed or even considered. If two specimens
varied one from the other in any way, they could not possibly be in
the same species. Parasitologists even went through the "one host,
one parasite" phase in which it was considered that a parasite was
adapted to life in or on only one host species, and any deviation
from that was considered another species. If you doubt that
oligochaetes have been treated to no small degree of this "type"
concept, you have but to look at the list of synonyms in any
monograph. But before condemning some earlier effort let us try
to think of a species without letting any knowledge of genes,
chromosomes, or heredity enter into the process. Try to view a
specimen with a primitive microscope and a candle or the sun as the
only source of illumination. I dare say that your respect for these
older workers will go up, not down, under those limitations.

With the advent of genetics and the knowledge of the mutability
of the gene, most of us are willing to allow some variation within
the parameters of our concept of the species. Some more than others.
But we are just now on the threshold of recognizing a phenomenon
that botanists have recognized for a long time. That phenomenon is
ecologically induced variation.

Botanists have long known that a plant grown in the sun or in
the shade will have different growth patterns, and that these had
an infinite pattern of variability as one adds nutrients, soil type,
pH, and other factors as variables or in combinations. Zoologists
have remained relatively smug in the concept that animal characteristics were either divinely endowed, or more recently, genetically
endowed. This is not always the case.

I had a graduate student a few years ago who was able to put
spination on the shaft of a sponge spicule by growing the animal
in alkaline water high in dissolved solids. Conversely, he could
prevent this spination in the next generation of spicules by
removing the sponge to acid waters low in dissolved solids. By
several of these transfers of an actively growing sponge, he could
produce alternate rings of sponge spicules in a single specimen
that would correspond to the limitations of different species

(Poirier, 1974). If you have ever worked with freshwater sponges, you can imagine the confusion which this creates in the field.

Lest you offer all your pity to sponge taxonomists, I advise you to retain a small portion for self application. A current worker in my laboratory, Dr. Michael Loden, has been able to remove and then replace the intermediate teeth in a naidid needle seta by similar manipulation. Almost all of us will agree that intermediate teeth in a needle is close to being a single character quantification of what constitutes enough difference to recognize a distinct species. But this is not all.

Those who have worked with the Naididae have seen *Pristina aequiseta, Pristina foreli,* and/or *Pristina evelinae*. They are all, in my opinion, a single species recognizable in a static sense as distinct species, influenced by environmental factors, complicated by aging, and these grade from one into another, none of which do we understand.

If you place a specimen of *Pristina aequiseta* (recognized by the giant seta) in an artificial pond water or dilute habitat water, the posterior zooid is recognizable as *Pristina foreli*. These posterior zooids never develop the giant seta of *Pristina aequiseta* although they had been budded asexually from an anterior zooid of that species. If the posterior zooid is cultured in dilute pond water or in dilute artificial pond water, the characteristics remain that of *Pristina foreli*. When this zooid of *Pristina foreli* is returned to full strength habitat water or to artificial pond water with added sodium chloride, the posterior zooid that is produced under the new regimen will be recognizable as *Pristina aequiseta*. Dr. Loden's very interesting data on this manipulation has now been presented for publication (Loden and Harman, in press).

If we go back at this point to the definition of a species based on reproductive isolation from all related populations, we are in even greater difficulty. In a north-south gradient, or even in an east-west gradient, there will be differences in reproductive strategies. *Nais pardalis,* for example, reaches sexual maturity in North Carolina in October, but the same species in Louisiana is not known to mature sexually at any season. In northern latitudes, it is likely that the onset of winter is the seasonal stress that initiates maturity. In Louisiana, it is likely to be the onset of summer that acts as this stimulus. So we have the same species having different reproductive strategies according to their geography.

A similar phenomenon is known in a planarian, *Dugesia tigrina*. Kenk (1944) reports that collections "...from different localities have different reproductive habits, and that in certain localities *Dugesia tigrina* propagates exclusively by fission." Longest (1966)

reports populations of the same species in Louisiana which are
sexual and some which are entirely asexual. The latter never
become sexual even under experimental conditions. Workers with
planarians have not had difficulties in working with a single
species that has different reproductive strategies in different
habitats or in different parts of its distribution. I submit
that workers with oligochaetes also have no difficulty in working
with this same set of conditions because we have been doing it.
And the thing that we have going for us is that we are actually
working with the concept of distinguishable populations, alike
their immediate ancestors as well as their progeny, instead of the
interbreeding concept of the species.

Among the Naididae the principal characters used in specific
assessment are both external and somatic. The latter includes
things like the degree of symmetry of the dorsal vessels, the
segment in which the digestive tube is dilated, commissural vessels
in the anterior segments. And there are those characteristics that
are neither internal nor external, like swimming behavior.

At the species level we work with qualitative and quantitative
differences. Some examples of qualitative differences are:
>   serrated setae vs. nonserrated setae
>   bifid needle vs. simple pointed needle
>   pectinate needle vs. spatulate needle
>   presence of a proboscis vs. absence of a proboscis
>   fragmentation vs. zooid formation
>   swimming vs. non-swimming.

Some quantitative differences might be:
>   tooth proportion
>   nodular position
>   length of proboscis.
I give greater weight to the former than to the latter.

While there are populations of oligochaetes that show very
stable genomes and therefore stable phenotypes, there are many
which show variation to a greater degree than any of us would like.
Several species in *Dero* and *Nais* will illustrate the point. Perhaps
these are in reality complexes of species which we have not been
able to sort out. I feel particularly sensitive to this possibility
when dealing with *Dero digitata-obtusa-nivea*. Sometimes complexes
are created through negative characteristics: if the specimen
possesses a characteristic, it is Species A; if it does not possess
the characteristic, the specimen is Species B. Such use of negative
characteristics always favors the lumper and may, in fact, cast
into the same taxon several species. Any taxonomist will feel more
comfortable with species identifications that are qualitative,
clear cut, and non-overlapping. but what we would like to work
with and what we get to work with are frequently different.

GENERIC CHARACTERS

The recognition of a species, then, is that population or group of populations similar enough to be recognized as a single kind, and different enough from related forms to have that difference recognized by all workers familiar with the biology of the taxon. We like to think of the species as the only finite unit in taxonomy, and certainly no taxonomist can make any claim that the genus is other than an artificial concept for the convenience of the biologist and required by the Linnaean system under which we work.

When we transfer these deliberation from the species level to that of genus, we are on even more shaky ground. The definition of a genus ranges from the rather loose and inaccurate group of related species to the more plausible taxon "...containing a single species, or a monophyletic group of species, which is separated from other taxa of the same rank by a decided gap." (Mayr, 1969). Again, it is up to the taxonomist to decide what are related species, how closely they are related, and the width of the gap that exists between one genus and another. It is also recommended that the size of the gap be in inverse proportion to the size of the genera it separates. That is, large genera may have smaller gaps, and small genera should have large gaps. The obvious advantage of this ratio is the elimination of burdensome numbers of monotypic genera.

As workers with the oligochaetes increase in number - and this number has increased in the last decade - I rather suspect that these gaps will narrow as a function of our increased knowledge of the group. In well worked groups such as some of the insect taxa, the quantification of the differences between species and genera will appear frightfully small to us.

Immediately we are back to the job of comparison of similarities and differences. Species contained within a genus will presumably be of common phyletic origin, have many characteristics in common, but will differ one from another, ever so slightly. Again, we are not given yardsticks for the assessment of similarity required for congeneric placement just as there are no criteria for the separation of species.

It will do us no good to look at common characteristics. Having nuclei in our cells does not ally us to the Protozoa. Finding an oligochaete with a ventral nerve cord tells us very little, and finding segmentation in an annelid is of no practical value if we already know what phylum of animals we are dealing with, but the loss of these stable characteristics could tell us a great deal about specialization and comparative age of an two taxa in question.

Lacking assistance generally among the somatic characteristics which are notoriously stable, we must look for characteristics that appear to be more rapidly evolving, and this seems to be the reproductive structures. Such a consensus is also held by workers with Arthropoda, Platyhelminthes, and other major taxa.

Among Naididae, setal characteristics are frequently used for the recognition of species, but it would take a phenomenal setal difference for that to be the sole or even the primary characteristic for the recognition of a genus. In general, we rely upon reproductive characteristics, gonadal placement, spermathecal number, whether spermathecae are single or paired, number of segments possessing gonads, or whether the gonadal segments are separated by one or more segments without gonads of any kind.

In a poorly known family like the Opistocystidae, the genus *Opistocysta* has testes in XXI and ovaries in XXII; in *Trieminentia* testes are in XIV or XV with ovaries in XV or XVI; in *Crustipellis* testes are in XI and ovaries in XII. (Harman and Loden, 1978). Such displacement is, I think, recognizable to most of us as sufficiently great to be recognized at the generic level. Differences need not be so great.

In Aulodrilinae (Tubificidae) species of *Aulodrilus* have the gonadal segments moved forward as far as V, while other genera retain the more traditional segment X (♂) and XI (♀).

SUPRAGENERIC CHARACTERS

At a slightly higher taxonomic level, say the subfamily, we rely heavily upon the method of asexual reproduction. For example, whether there is zooid formation or fragmentation as a primary method of reproduction and the number of segments regenerated anteriorly following such asexual reproduction.

Among the Lumbriculidae and the Pristininae, the number of segments regenerated following asexual reproduction is 7; among the Tubificidae and the Naidinae the number is 5; and in the Enchytraeidae the number is 3. In some taxa, like the Naidinae, those regenerated segments possess no dorsal setae.

Ivanov (1937) believes that the original (primitive) number of these regenerated segments is 7 for all oligochaetes. This tendency toward an evolutionary reduction in specialization is not only common among the oligochaetes, but is also widespread in animal phylogeny. It tends to hold very well also among plants, but can be violated in the case of hybrids and polyploids which botanists recognize as species, but zoologists do not.

REFERENCES

Blackwelder, R.E., 1967, Taxonomy. John Wiley & Sons, Inc.,
     New York, 698pp.
Harman, W.J. and M.S. Loden, 1978, A re-evaluation of the
     Opistocystidae (Oligochaeta) with descriptions of two new
     genera. Proc. Biol. Soc. Wash., 91:453-462.
Ivanov, P.P., 1937, General and Comparative Embryology. Moscow.
Kenk, R., 1944, Freshwater triclads of Michigan. Misc. Publ.
     Mus. Zool., Univ. of Mich., 60:1-44.
Loden, M.S. and W.J. Harman, 1980, Ecomorphic variation in the
     setae of Naididae (Oligochaeta). In "Aquatic Oligochaete
     Biology", R.O. Brinkhurst and D.G. Cook (eds.) Plenum Press,
     New York.
Longest, W.D., 1966, The freshwater Tricladida of the Florida
     parishes of Louisiana. Ph.D. Dissertation. Louisiana State
     University.
Mayr, E., 1969, Principles of Systematic Zoology. McGraw-Hill
     Book Company, New York.
Poirrier, M.A., 1974; Ecomorphic variation in gemmoscleres of
     Ephydatia fluriatilis L. (Porifera, Spongillidae) with
     comments upon its systematics and ecology. Hydrobiologia,
     44:337-347.

SPECIFIC AND GENERIC CRITERIA IN MARINE OLIGOCHAETA, WITH

SPECIAL EMPHASIS ON TUBIFICIDAE

Christer Erséus

Department of Zoology
University of Gothenburg
Fack S-400 33 Göteborg, Sweden

ABSTRACT

The enchytraeids *Marionina* and *Lumbricillus*, each containing many marine, littoral species, are in great need of revision; their generic and specific characteristics require reconsideration. The taxonomic problems due to morphological differentiation between geographically separated populations of the exclusively marine genus *Grania* are elucidated.

The supra-specific classification of the Tubificidae is, by tradition,based upon the principal organization of the male duct and its accessories, besides the presence or absence of coelomo-cytes, gut diverticulae and spermatozeugmata. Specific criteria vary slightly between tubificid genera. Generally, somatic characters provide little or no species-distinguishing information. Much more fundamental are the sexual characters: proportions in the male genitalia; number and appearance of the genital setae (when present); and morphology of the spermathecae. The importance of using histological techniques when studying the morphology of male genital organs is emphasized.

A thorough taxonomic analysis of the sub-family Phallodrilinae has provided evidence of radiative evolution among marine Tubifici-dae. Striking morphological modifications, such as enlarged somatic and genital setae, absence of alimentary canal and flattened body shape, are noted for some aberrant phallodriline species.

9

INTRODUCTION

The marine Oligochaeta are being increasingly studied, and their significance as members of the marine benthos is becoming more and more evident (Lasserre, 1971a; Cook and Brinkhurst, 1973; Giere, 1975). Marine species are largely found among the Naididae, Enchytraeidae and Tubificidae. The characters principally used for their identification are the same as those used for the non-marine forms of the respective families.

The Naididae are reviewed separately by Harman (1979). The marine naidids are very few, and they are, therefore, not further treated here. However, the marine enchytraeids and especially the tubificids do deserve further attention. I will give a brief account of some selected features and problems, some of which are not purely taxonomic, but are of importance for the understanding of the biology and phylogeny of marine oligochaetes.

ENCHYTRAEIDAE

An enchytraeid (Fig. 1) is characterized by having testes and ovaries in segments XI and XII, respectively. Its male pores are found in XII, and its spermathecae are located in segment V. Typical for an enchytraeid are also the so called septal glands in some of the anterior segments.

Nielsen and Christensen (1959) published their comprehensive revision of the Enchytraeidae. Their work was not complete regarding the non-European species, and, as shown by Brinkhurst (in Brinkhurst and Jamieson, 1971), it did not always follow the rulings of the International Code of Zoological Nomenclature. However, Nielsen and Christensen's generic division of the family has been applied by most subsequent workers.

Enchytraeid genera are distinguished by means of several characters, most of which are somatic: shape and arrangements of setae; organization of septal glands and seminal vesicles; absence and presence of head pores, dorsal pores, oesophagal diverticula, so called peptonephridia, etc. (Nielsen and Christensen, 1959). It is important to note that sexual characters play a minor role at the generic level in enchytraeid classification. This is quite the contrary of the situation in the Tubificidae (cf. below).

Enchytraeids are numerous in marine littoral habitats (Lasserre, 1971a; Giere, 1975), especially in temperate and polar regions of the world. The most diverse and abundant genera here are *Lumbricillus* and *Marionina*. A key to the European littoral

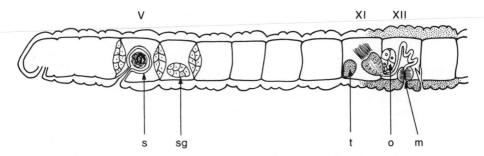

Figure 1. Lateral view of an enchytraeid, anterior end. m, male
pore; o, ovary; s, spermatheca; sg, septal gland;
t, testis.

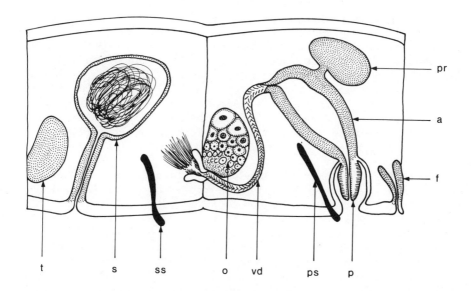

Figure 2. Lateral view of genital organs in a tubificid, segments
X and Xi. a, atrium; f, female funnel; o, ovary;
p, penis; pr, prostate gland; ps, penial seta; s,
spermatheca; ss, spermathecal seta; t, testis; v,
vas deferens.

Enchytraeidae was published by Tynen and Nurminen (1969), and a
checklist and bibliography of the North American Enchytraeidae,
including the marine forms,* was compiled by Tynen (1975).

   Lumbricillus Orsted, 1844 contains many marine species (cf.
e.g. Nielsen and Christensen, 1959; Nurminen, 1965, 1970; Tynen,
1966, 1969; Lasserre, 1971b; Shurova, 1974, 1977; Lasserre and
Erséus, 1976; Erséus, 1976a, 1977), but includes also a few limnic
and terrestrial forms (cf. Nielsen and Christensen, 1959). One of
the most important generic characters of *Lumbricillus* is the subdivision of the seminal vesicles into several small lobes containing
ripening spermatozoa (cf. e.g. Erséus, 1977, fig. 2). The worms
are often yellowish, reddish or greenish, due to the color of their
blood. About 70 species of *Lumbricillus* are known. Important
specific characters are the color, the number of segments, the
number of setae within the bundles, the appearances of the spermathecae, the shape of the sperm funnels, and the position of the
so called copulatory glands.

   Many *Lumbricillus* species are very similar to each other.
Because of the very great intra-specific variability so characteristic of enchytraeids, one could even suspect that many species
are synonymous. A careful re-examination of all species within
this genus is, therefore, badly needed. One case of suspected
synonymous species *L. pagenstecheri* (Ratzel, 1869), *L. kalatdlitus*
(Nurminen, 1970) and *L. ritteri* (Eisen, 1904), was discussed by
Erséus (1976a). It should be noted, however, that sibling species
of *Lumbricillus (L. rivalis* Levinsen, 1883) have recently been
revealed by electrophoretic studies and breeding experiments
(Christensen and Jelnes, 1976).

   Most of the meiobenthic species of littoral and sublittoral
marine enchytraeids are found within *Marionina* Michaelsen, 1889
(Nielsen and Christensen, 1959; Lasserre, 1964, 1966, 1971b; Finogenova, 1972, 1973; Giere, 1974; Erséus, 1976b; Lasserre and Erséus,
1976). The whole genus comprises about 60 species, more than half
of which are non-marine. Lasserre (1971a, fig. 7) has given an
instructive key to marine species of *Marionina*. Many of the
species show peculiar and species-specific reductions in the distribution of the setae. Very essential are the spermathecae, which
are of many different types. Other characters used are the arrangement of the septal glands and the number of the segments. However,
Marionina, as it is presently defined, is somewhat heterogeneous.
For instance, some species possess, others lack seminal vesicles.
Future work on the group might entail a division of *Marionina* into

---

* See also Coates, K.A. 1979, Revision of the taxonomy of intertidal Enchytraeidae (Annelida, Oligochaeta) of British Columbia,
University of Victoria, MSc. Thesis, 396 pp. Edit.

two or more genera. Furthermore, the species themselves need
reconsideration, not only because of the great intra-specific
variability, but also considering the great number of forms that
are still to be described from warmer seas (Erséus, unpublished).

        The exclusively marine genus *Grania* Southern, 1913, is a group
of small, transparent and very slender enchytraeids, which predominantly inhabit coarse, subtidal sediments. The genus is cosmopolitan and morphologically very closely related forms are found
in widely separated geographical areas. The taxonomy of *Grania*
was studied by Erséus and Lasserre (1976, 1977). The forms of the
genus principally differ from each other in their setal pattern
and in the morphology of their spermathecae (Erseus and Lasserre,
1976, table I). All forms show reductions in the distribution of
setae, and some species lack dorsal setae completely. The spermathecae are of a few different kinds, but a common feature is the
arrangement of sperm in rings or balls in the wall of the spermathecal ampulla. Two polytypic species, *G. macrochaeta* (Pierantoni,
1901) and *G. postclitellochaeta* (Knöllner, 1935), both amphi-
Atlantic in their distribution, have been recognized (Erséus and
Lasserre, 1976). However, the genus is now known from many more
parts of the world, and the classification of *Grania* has to be
further revised and supplemented. For instance, forms that are
more or less identical to *G. macrochaeta* have been found in the
South Atlantic, the Indian Ocean, and both North and South Pacific
(Jamieson, 1977; Erséus, in preparation). In addition, new evidence
supports the splitting of *G. postclitellochaeta* into two distinct
species (Erséus, 1977).

TUBIFICIDAE

        The tubificids are well represented in the sea. Ten years
ago only a little more than 40 tubificid species were known from
marine or brackish waters (cf. e.g. Brinkhurst and Jamieson, 1971).
Today, almost 120 marine species are described (a number of new
species descriptions that are in press are included), and I anti-
cipate that many more are still to be discovered. The increased
knowledge of this diversity has allowed revisions of several species and genera, but still we are only beginning to formulate an
ultimate classification of the marine Tubificidae.

        A tubificid is principally defined by the location of its
genital organs (Fig. 2). The male pores are generally in segment
XI, and the spermathecae generally open on segment X. The male
efferent ducts are taxonomically very important: they provide
both generic and specific criteria. A sperm funnel on the septum
between X and XI leads to the vas deferens, which terminates in
the atrium. The atrium is glandular, sometimes muscular, and often
it bears one or more external prostate glands. Often the male duct
ends in a penial apparatus of some kind. Modified, genital setae
(so called penial and spermathecal setae) may be present.

The supra-specific classification of the Tubificidae is, by
tradition, based upon the principal organization of the male duct
and its accessories, besides the presence or absence of coelomo-
cytes, gut diverticula and the so called spermatozeugmata (the kind
of internal spermatophores found in, for instance, the sub-family
Tubificinae (Brinkhurst and Jamieson, 1971:33). The phylogenetic
relations between marine and freshwater tubificids are still not
fully understood, but the marine Tubificidae are not just one sister
group of the limnic ones. Marine genera are found among the subfamilies Tubificinae, Rhyacodrilinae and Aulodrilinae, as these
are presently defined (Brinkhurst and Jamieson, 1971; Brinkhurst
and Baker, in press). However, fully marine and freshwater species
generally do not belong to the same genera. The Phallodrilinae is
the only most exclusively marine sub-family.

Generic Criteria

The problems of defining genera of marine tubificids are not
of the same kind throughout the different sub-families. However,
the principal appearance of the genital organs always plays an
important role.

The sub-family Tubificinae Eisen, 1885 comprises species in
which the prostate glands are stalked and spermatozeugmata are
developed. There are many marine species within the Tubificinae.
Most of these were long known as members of the genus *Peloscolex*
Leidy, 1850. However, *Peloscolex* was recently revised by Holmquist
(1978, 1979), who divided it into a number of smaller genera, most
of which are purely limnic. In her revision, Holmquist studied a
few marine species. One of these, *Peloscolex amplivasatus* Erséus
1975, exhibits a very prominent histological tripartition in its
atrium (cf. Erséus, 1976b, fig. 2), a character that was used by
Holmquist to define the marine genus *Tubificoides* Lastochkin, 1937.
The tripartite atrium is found also in *T. swirencowi* Jaroschenko,
1948 (Holmquist, 1978; fig. 5C), which is the type species of
Tubificoides. Most marine species of the former *Peloscolex* complex,
it seems, have a characteristic ental cap-like portion of the atrium
(Holmquist, 1978: figs. 5B-C, 6). This "cap" has never been found
in any freshwater Tubificinae. The well-known marine species
Peloscolex benedeni (d'Udekem, 1855) has such a "cap", but the rest
of its atrium is histologically not further divided (Holmquist,
1978: fig. 5B) as in *T. amplivasatus* and *T. swirencowi*. Therefore,
Holmquist assigned *benedeni* to a new genus, *Edukemius,* as *E. benedii*
(d'Udekem, 1855) (*benedii* being the original spelling of the species).

Unfortunately, most of the marine species of the former *Peloscolex* are not adequately described, and the available material is
not good enough, to fit these species into Holmquist's system.
Regarding the bipartite or tripartite atria, intermediates might

occur, and this would perhaps support a union of the two marine genera into one. This has been proposed by Brinkhurst and Baker (in press), but the matter requires further studies.

Limnodriloides Pierantoni, 1903 belongs to the sub-family Aulodrilinae Brinkhurst, 1971. Its prostate glands are broadly attached to the atria, and true spermatozeugmata are not developed. Although only about ten species are described at present (Cook, 1974; Jamieson, 1977), the genus is very diverse (Erséus, in preparation). A pair of oesophagal diverticula found in segment VIII or IX has long been used as a most important diagnostic feature of Limnodriloides. However, it is possible that the diverticula could become a suprageneric character, when more species are described. Hrabê (1971) proposed splitting the genus into Bohadschia Hrabê, 1971 for those species possessing true penes, and Limnodriloides for those provided with pseudopenes. However, Cook (1974) has demonstrated that this particular division is difficult to uphold, since intermediates do occur.

The sub-family Rhyacodrilinae Hrabê, 1963, defined by diffuse prostates, presence of numerous coelomocytes, but absence of spermatozeugmata, includes a few marine forms. For instance, many brackish-water species are found within the rhyacodriline genus Monopylephorus Levinsen, 1884 (Brinkhurst and Jamieson, 1971; Brinkhurst and Baker, in press).

The sub-family Phallodrilinae Brinkhurst, 1971 comprises at least eleven genera, all of which are marine. The basic feature of phallodriline genera is the possession of two pairs of distinct and more or less stalked prostate glands associated with the male ducts (Fig. 3). However, there are deviations from this.

In the type genus Phallodrilus Pierantoni, 1902, each atrium bears two prostates, one at each end of the atrium. The vas deferens enters the ental part of the atrium. Penial setae are generally present. Phallodrilus contains about 25 species at present (Erséus, in press b; Giere, in press), and is the largest genus within the Phallodrilinae. Even if it is not possible to say whether or not Phallodrilus is the most conservative of the different genera, it is possible to interpret the homologies in the male ducts of others phallodriline genera from the arrangement found in Phallodrilus.

Bathydrilus Cook, 1970, which includes many deep-sea species, is very reminiscent of Phallodrilus, but its vas deferens enters the ectal part of the atrium, not the ental end. Intermediate forms do exist, which supports the idea of a close phylogenetic relationship between Bathydrilus and Phallodrilus (Erséus, in press a).

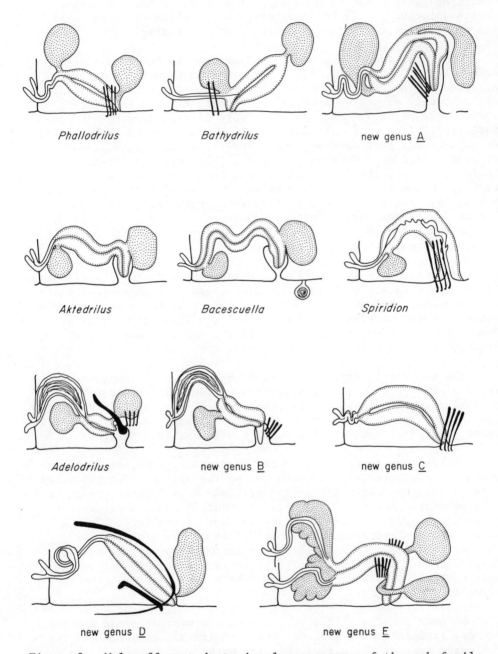

Figure 3.   Male efferent ducts in eleven genera of the sub-family
            Phallodrilinae.

Aktedrilus Knöllner, 1935 has two prostates, but the penial setae are missing. Instead a penis is developed. Furthermore, all *Aktedrilus* species (Erséus, in preparation) have an unpaired, mid-dorsal spermatheca, not two lateral ones, which is the normal case for most other genera. *Bacescuella* Hrabě, 1973 is very closely related to *Aktedrilus*. However, *Bacescuella* generally lacks the spermathecae completely. The worms transfer their sperm by means of external spermatophores of a kind not known elsewhere in the Tubificidae (Erséus, 1978b).

Spiridion Knöllner, 1935 lacks the posterior prostate, but it has a peculiar ectal portion of the atrium, which can be everted and form a so called pseudopenis (Erséus, 1979b).

In *Adelodrilus* Cook, 1969, the vas deferens is not ciliated, but it is dilated and contains a large bundle of sperm. The bundle of penial setae generally consists of one giant scoop or spoonshaped seta followed by a row of hooked smaller ones (Erséus, 1978a). In a new, monotypic genus from Bermuda (new genus B in Fig. 3), which apparently is closely related to *Adelodrilus,* the posterior prostate is missing, and instead of the giant penial seta a cuticularized penis is found (Erséus, in press e). However, the dilated sperm-storing vas deferens and the hooked smaller penial setae are still present.

A few other monotypic genera have been discovered recently. One (new genus A in Fig. 3) is equipped with a prominent penis, but unlike *Aktedrilus* the genus possesses penial setae and paired spermathecae (Baker and Erséus, in press). In another species (new genus C), prostates are completely lacking, but the atrium is histologically very similar to those of other phallodriline genera (Erséus, 1979a). In a form from Florida (new genus D) the anterior prostate is missing, and the penial setae are highly modified. In addition, the spermatheca of genus D is unpaired and located mid-dorsally (Erséus, in press c). The new genus E, finally, has the atria of the two sides united, forming a single median structure, though the sperm ducts and the prostates are still paired. The only known species of genus E has an unpaired, mid-ventral spermatheca with its opening in segment IX, which is a very unusual location for a spermathecal pore in the Tubificidae (Erséus, in press d).

Specific Criteria

The specific criteria in marine tubificids are of many different kinds, and the combination used differs from genus to genus. Somatic characters alone are seldom reliable for species identification, but generally they support the distinction of a species, and they should of course never be neglected.

Setal shapes and arrangements sometimes provide valuable key characteristics (Fig. 4). The Tubificinae, for instance, exhibit a great variety of setal types and different combinations and distributions of these. However, in other groups, such as the Phallodrilinae, most species possess plain bifid setae, and only a few species have aberrant setal types.

The sexual characters of a tubificid worm provide the basis for its identification as a distinct species. The following main features are generally used: (1) proportions and histological subdivision of the male efferent ducts (including prostate glands); (2) morphology of penes (when present); (3) morphology of genital setae (when present); and (4) morphology of spermathecae. Obviously, identification is facilitated when penes or genital setae (or both) are present, when compared to genera in which neither of these structures is present.

An example of species-distinguishing characteristics is shown in Fig. 5 for two species of *Phallodrilus*. *P. coeloprostatus* Cook, 1969 (Fig. 5A) has a long and narrow vas deferens, which enters the apical, inner end of a cylindrical and non-muscular atrium. The prostates are relatively large. The spermathecal ampulla is pear-shaped and completely filled with sperm. In *P. prostatus* (Knöllner, 1935)(Fig. 5B), however, the vas deferens is short and wide, and it enters the very heavily muscular and pear-shaped atrium sub-apically. The prostates are relatively small. The spermathecal ampulla is divided into two parts by a constriction, sperm generally being present only in the innermost, small and spherical part of the ampulla. A key to 23 species of *Phallodrilus* is found in Erséus (in press b).

The specific shapes of penes will be demonstrated from *Aktedrilus* by Erséus (in preparation). Some *Aktedrilus* species have short, conical and non-cuticular penes, while others have elongated and cuticularized copulatory organs.

The species-specific penial setae are shown for three species of *Adelodrilus* in Fig. 6.

Aberrant Marine Tubificids

A thorough taxonomic and morphological analysis of the subfamily Phallodrilinae (Erséus, 1978a, b, 1979 a,b, in press a-e; Baker and Erséus, in press; Giere, in press) has provided evidence of a radiative evolution among marine Tubificidae. Perhaps the most peculiar phenomenon is the total absence of a normal alimentary system recently discovered in a number of new species in warmer seas (Erséus, in press b, c; Giere, in press). These worms are in most cases clear members of *Phallodrilus*, and they have all

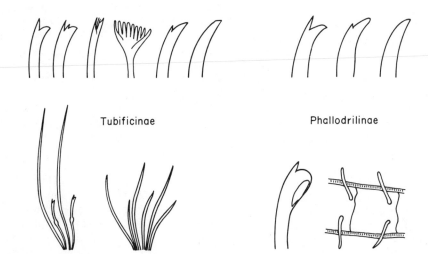

Figure 4.   Different types of somatic setae found among marine
            Tubificinae and Phallodrilinae.

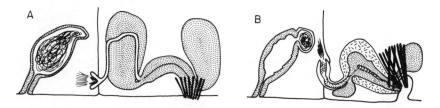

Figure 5.   Spermathecae and male ducts of two species of Phallo-
            drilus. (A) *P. coeloprostatus* Cook. (B) *P. prostatus.*

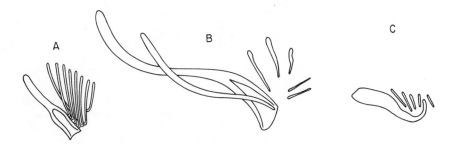

Figure 6.   Bundles of penial setae of three species of *Adelodrilus.*
            (A) *A. pusillus* Erséus (B) *A. cooki* Erséus (C) *A. vora-
            ginus* (Cook).

been found in coral sand from various coral reef areas. They are
very slender and conspicuously white or, in a few cases, completely
transparent. The body wall is very thick and granulated, but its
prospective role in the uptake of dissolved organic matter is not
yet established. Normal blood vessels and chloragogen cells are
well developed, but the gut is completely, or, at least in one
species, partly reduced. Both mouth and anus are absent. Two of
the gutless species are dorso-ventrally flattened, probably as a
modification to increase the body wall area in relation to volume;
this would have an adaptive value if the worms do take up dissolved
organics. The flat species also possess long posterior processes
(Erséus, in press b).

Giant somatic and genital setae are found in a few phallodriline species. For instance, in *Bathydrilus rarisetis* (Erséus, 1975)
most of the somatic setal bundles are represented by large individual setae, which are very thick and single-pointed (Fig. 4, lower
right corner). The species inhabits coarse sand and gravel bottoms
in Norwegian fjords (Erséus, 1975, 1976b, in press a). It is noteworthy that it lives together with species of the enchytraeid genus
Grania, which is also characterized by having one large seta in each
bundle (Erséus and Lasserre, 1976). The large size of these setae
is apparently correlated with large sediment particles, and is
probably a modification to facilitate locomotion in the interstitial
spaces of these coarse bottoms.

Very much enlarged genital setae are found 'in genera other than
Adelodrilus (Fig. 6). A new species of *Phallodrilus* (described in
Erséus, in press b) from Florida has a pair of very long ribbon-

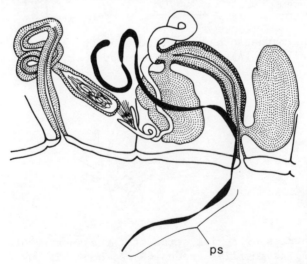

Figure 7.   Genitalia of *Phallodrilus* n. sp. (ps, penial seta).

shaped and slightly spiral penial setae (Fig. 7) which are protruded through muscular muffs (not shown in Fig. 7) near the male pores. The setae are apparently inserted into the very long spermathecal ducts of the mate during copulation. Similar, very long penial setae are found in the new genus D̲ (Fig. 3).

CONCLUSIONS

     We are still very far from a full knowledge of the diversity of marine oligochaetes. Hundreds of new species, specially of Tubificidae, are undoubtedly still to be discovered, and we will probably have many reasons to revise the generic and specific criteria used today. The importance of intra-specific variation has to be studied further and quantitatively analysed. The complexity of many marine oligochaete groups demands great accuracy and detail in descriptions of species. For instance, the histology of the male ducts in the Tubificidae must always be carefully considered. Good preparations of types and other reference material have to be deposited and made accessible to subsequent workers. It is also important to encourage the approach of taxonomical methods other than the traditional morphological ones. For instance, both cytological (e.g. Christensen, 1960, 1961; Christensen and Nielsen, 1955; Nielsen and Christensen, 1959; Jelinek, 1974; Sharma et al., 1976; McGee and French, 1977) and electrophoretic, protein taxonomic studies (Melbrink and Nyman, 1973; Christensen and Jelnes, 1976; Christensen et al., 1976, 1978) have proved to be useful in microdrile taxonomy.

REFERENCES

Baker, H. R., and C. Erséus, in press, *Peosidrilus biprostatus* n.g., n.sp., a marine tubificid (Oligochaeta) from the Eastern United States, Proc. Biol. Soc. Wash.,
Brinkhurst, R. O., and H.R. Baker, in press, The marine Tubificidae (Oligochaeta) of North America, Can. J. Zool., 57:1553-1569.
Brinkhurst, R. O., and B. G. M. Jamieson, 1971, "Aquatic Oligochaeta of the World", Oliver and Boyd, Edinburgh.
Christensen, B., 1960, A comparative cytological investigation of the reproductive cycle of an amphimictic diploid and a parthenogenetic triploid form of *Lumbricillus lineatus* (O. F. M.) (Oligochaeta, Enchytraeidae), Chromosoma, 11:365-379.
Christensen, B., 1961, Studies on cyto-taxonomy and reproduction in the Enchytraeidae, Hereditas, 47:387-450.
Christensen, B., U. Berg, and J. Jelnes, 1976, A comparative study on enzyme polymorphism in sympatric diploid and triploid forms of *Lumbricillus lineatus* (Enchytraeidae, Oligochaeta), Hereditas, 84:41-48.
Christensen, B., and J. Jelnes, 1976, Sibling species in the oligochaete worm *Lumbricillus rivalis* (Enchytraeidae) revealed by enzyme polymorphisms and breeding experiments, Hereditas, 83:237-244.

Christensen, B., J. Jelnes and U. Berg, 1978, Long-term isoenzyme
    variation in parthenogenetic polyploid forms of *Lumbricillus*
    lineatus (Enchytraeidae, Oligochaeta) in recently established
    environments, Hereditas, 88:65-73.
Christensen, B. and C. O. Nielsen, 1955, Studies on Enchytraeidae
    IV. Preliminary report on chromosome numbers of 7 Danish
    genera, Chromosoma, 7:460-468.
Cook, D. G., 1974, The systematics and distribution of marine Tubi-
    ficidae (Annelida: Oligochaeta) in the Bahia de San Quintin,
    Baja California, with descriptions of five new species, Bull.
    Sth. Calif. Acad. Sci., 73:126-140.
Cook, D. G., and R. O. Brinkhurst, 1973, Marine flora and fauna of
    the northeastern United States. Annelida: Oligochaeta, NOAA
    Techn. Report NMFS Circ., 374.
Erséus, C., 1975, *Peloscoles amplivasatus* sp.n. and *Macroseta*
    rarisetis gen. et sp.n. (Oligochaeta, Tubificidae) from the
    west coast of Norway, Sarsia, 58:1-8
Erséus, C., 1976a, Littoral Oligochaeta (Annelida) from Eyjafjördur,
    north coast of Iceland, Zool. Scr., 5:5-11.
Erséus, C., 1976b, Marine subtidal Tubificidae and Enchytraeidae
    (Oligochaeta) of the Bergen area, western Norway, Sarsia,
    62:25-48.
Erséus, C., 1977, Marine Oligochaeta from the Koster area, west
    coast of Sweden, with descriptions of two new enchytraeid
    species, Zool. Scr., 6:293-298.
Erséus, C., 1978a, New species of *Adelodrilus* and a revision of
    the genera *Adelodrilus* and *Adelodriloides* (Oligochaeta,
    Tubificidae), Sarsia, 63:135-144.
Erséus, C., 1978b, Two new species of the little-known genus
    Bacescuella Hrabe (Oligochaeta, Tubificidae) from the north
    Atlantic, Zool. Scr., 7:263-267.
Erséus, C., in press a, Taxonomic revision of the marine genera
    Bathydrilus Cook and *Macroseta* Erseus (Oligochaeta, Tubifici-
    dae), with descriptions of six new species and subspecies,
    Zool. Scr., 8:
Erséus, C., in press b, Taxonomic revision of the marine genus
    Phallodrilus Pierantoni (Oligochaeta, Tubificidae), with
    descriptions of thirteen new species, Zool. Scr., 8:
Erséus, C., in press c, *Inanidrilus bulbosus* gen. et sp.n., a
    marine tubificid (Oligochaeta) from Florida, USA, Zool.Scr., 8:
Erséus, C., in press d, *Uniporodrilus granulothecus* n.g., n.sp.,
    a marine tubificid (Oligochaeta) from eastern United States,
    Trans.Amer. Micros. Soc., 98:
Erséus, C., in press e, *Bermudrilus peniatus* n.g., n.sp. (Oligo-
    chaeta, Tubificidae) and two new species of *Adelodrilus*
    from the Northwest Atlantic, Trans. Amer. Micros. Soc., 98:
Erséus, C., 1979a, *Coralliodrilus leviatriatus* gen. et sp.n., a
    marine tubificid (Oligochaeta) from Bermuda, Sarsia, 64:179-
    182.

Erséus, C., 1979b, Re-examination of the marine genus *Spiridion* Knöllner (Oligochaeta, Tubificidae), Sarsia, 64:183-187.

Erséus, C., and P. Lasserre, 1976, Taxonomic status and geographic variation of the marine enchytraeid genus *Grania* Southern (Oligochaeta), Zool. Scr., 5:121-132.

Erséus, C., and P. Lasserre, 1977, Redescription of *Grania monochaeta* (Michaelsen), a marine enchytraeid (Oligochaeta) from South Georgia (SW Atlantic), Zool. Scr., 6:299-300.

Finogenova, N. P., 1972, New species of Oligochaeta from Dniepr and Bug Firth and Black Sea and revision of some species (in Russian), Trudy Zool. Inst., Leningrad, 52:95-116.

Finogenova, N. P., 1973, New species of Oligochaeta from the Caspian Sea (in Russian), Zool. Zh., 52:121-124.

Giere, O., 1974, *Marionina istriae* n.sp., ein mariner Enchytraeidae (Oligochaeta) aus dem mediterranen Hygropsammal, Helgoländer wiss. Meeresunters., 26:359-369.

Giere, O., 1975, Population structure, food relations and ecological role of marine oligochaetes, with special reference to meiobenthic species, Mar. Biol., 31:139-156.

Giere, O., in press, Studies on marine Oligochaeta from Bermuda, with emphasis on new *Phallodrilus*-species (Tubificidae), Cah. Mar. Biol.,

Harman, W. J., 1979, Specific and generic criteria in freshwater Oligochaeta, with special emphasis on Naididae. in: "Aquatic Oligochaete Biology", R. O. Brinkhurst and D.G. Cook, eds., Plenum Press, New York.

Holmquist, C., 1978, Revision of the genus *Peloscolex* (Oligochaeta, Tubificidae). 1. Morphological and anatomical scrutiny; with discussion on the generic level, Zool. Scr., 7:187-208.

Holmquist, C., 1979, Revision of the genus *Peloscolex* (Oligochaeta, Tubificidae). 2. Scrutiny of the species, Zool. Scr., 8:37-60.

Hrabê, S., 1971, On new marine Tubificidae of the Adriatic Sea, Scr. Fac. Sci. Nat. Univ. Brno., 1:215-226.

Jamieson, B. G. M., 1977, Marine meiobenthic Oligochaeta from Heron and Wistari Reefs (Great Barrier Reef) of the genera *Clitellio*, *Limnodriloides* and *Phallodrilus* (Tubificidae) and *Grania* (Enchytraeidae), Zool. J. Linn. Soc., 61:329-349.

Jelinek, H., 1974, Untersuchungen zur Karyologie von vier naidomorphen Oligochaeten, Mitt. Hamburg. Zool. Mus. Inst., 71:135-145.

Lasserre, P., 1964, Note sur quelques oligochètes Enchytraeidae présents dans les plages du bassin d'Arcachon, P. V. Soc. Linn. Bordeaux, 101:87-91.

Lasserre, P., 1966, Oligochetes marins des cotes de France. I. Bassin d'Arcachon: Systematique, Cah. Mar. Biol., 7:295-317.

Lasserre, P., 1971a, Oligochaeta from the marine meiobenthos: taxonomy and ecology, Smithson Contr. Zool., 76:71-86.

Lasserre, P., 1971b, The marine Enchytraeidae (Annelida, Oligochaeta) of the eastern coast of North America with notes on their geographical distribution and habitat, Biol. Bull. Mar. Biol. Lab., Woods Hole, 140:440-460.

Lasserre, P., and C. Erseus, 1976, Oligochetes marins des Bermudes.
    Nouvelle especes et remarques sur la distribution geographique
    de quelques Tubificidae et Enchytraeidae, Cah. Biol. Mar.,
    17:447-462.
McGee, D. R., and W. L. French, 1977, Chromosomes of some fresh-
    water oligochaetes, Trans. Amer. Micros. Soc., 96:535-537.
Milbrink, G., and L. Nyman, 1973, On the protein taxonomy of
    aquatic oligochaetes, Zoon, 1:29-35.
Nielsen, C. O., and B., Christensen, 1959, The enchytraeidae,
    Critical revision and taxonomy of European species, Natura
    jutl., 8-9:1-160.
Nurminen, M., 1965, Enchytraeid and Lumbricid records (Oligochaeta)
    from Spitsbergen, Annls zool. fenn., 2:1-10.
Nurminen, M., 1970, Records of Enchytraeidae (Oligochaeta) from the
    west coast of Greenland, Annls zool. fenn., 7:199-209.
Sharma, G. P., S. M. Handa and Komal Sahi, 1976, Analysis of
    chromosomes in five species of aquatic oligochaetes, Nucleus,
    19:102-105.
Shurova, N. M., 1974, Enchytraeidae of the genus Lumbricillus
    (Oligochaeta) from the intertidal zone of the Kurile Islands
    (in Russian), in: "Plant and Animal World of the Littoral Zone
    of the Kuril Islands." Nauka, Novosibirsk, pp. 128-136.
Shurova, N. M., 1977, New species of littoral oligochaetes of the
    genus Lumbricillus (in Russian), Biologiya Morya, 1:57-62.
Tynen, M.M., 1966, A new species of Lumbricillus with a revised
    check-list of the British Enchytraeidae (Oligochaeta),
    J. mar. biol. Ass. U. K., 46:89-96.
Tynen, M. J., 1969, New Enchytraeidae (Oligochaeta) from the east
    coast of Vancouver Island, Can. J. Zool., 47:387-393.
Tynen, M.J., 1975, A checklist and bibliography of the North
    American Enchytraeidae (Annelida: Oligochaeta), Syllogeus,
    9:1-14.
Tynen, M. J. and M. Nurminen, 1969, A key to the European littoral
    Enchytraeidae (Oligochaeta), Annls zool. fenn., 6:150-155.

POLYCHAETE SIBLING SPECIES

Judith Grassle

Marine Biological Laboratory
Woods Hole
Massachusetts 02543

ABSTRACT

Capitella capitata was formerly regarded as an excellent cosmopolitan indicator species for marine pollution or environmental disturbance. Following an oil spill in West Falmouth, Massachusetts in September 1969, when most of the benthic marine fauna was killed, the subsequent responses of a number of polychaete and other invertebrate species allowed us to rank species in order of decreasing opportunism: 1. *Capitella capitata* 2. *Polydora ligni* 3. *Syllides verrilli* 4. *Micropthalmus aberrans* 5. *Streblospio benedicti* 6. *Mediomastus ambiseta*. These polychaete species all showed the ability to increase rapidly, large population size, early maturation and high mortality that are characteristics of opportunists. Preliminary electrophoretic studies on two Mdh (malate dehydrogenase) loci in *Capitella* collected in the oil spill area and in adjacent control areas indicated short-term selection at the Mdh-2 locus (Grassle and Grassle, 1974).

Subsequent electrophoretic and life history studies on *Capitella* collected from 5 populations in the vicinity of Woods Hole, and from two populations in Gloucester, Massachusetts, indicated that there were in fact six sibling species, difficult or impossible to distinguish on morphological grounds, but showing marked differences in life histories and reproductive modes, and with virtually no alleles in common between the species at eight enzyme loci that were identified electrophoretically (Grassle and Grassle, 1976).

The life history characteristics of species that have been
shown to occur sympatrically vary from those of species IIIa (egg
diameter 250μ; 30-50 eggs/brood; no planktonic larval stage) to
those of species II (egg diameter 230μ; 30-400 eggs/brood; a planktonic larval stage lasting up to 2 days depending on temperature)
and those of species III (egg diameter 50μ; 200-1000 eggs/brood;
a planktonic larval stage lasting up to two weeks depending on
temperature). That is, the dispersal capability of the larval
stages differs greatly from species to species. Moreover some of
the species (e.g. Ia and III) have relatively short breeding
seasons limited to the winter and early spring, while others (e.g.
I and II) breed throughout the year. The differences between the
species in colonizing ability reflect differences in the environmental variability of their respective habitats. Those species
with the greatest powers of dispersal (Ia and III) occur in the
less variable subtidal habitats and have a limited breeding season.
The species with the most rapid response to disturbed habitats
(i.e. the most opportunistic) is species I. Although it has relatively limited dispersal ability (a planktonic stage lasting several
hours) this is the species with the largest rate of increase,
largest maximum population size and highest mortality (Grassle and
Grassle, 1974, 1977). This species may be uniquely adapted to
exploit transitory high concentrations of readily assimilated
organic matter. It has been demonstrated experimentally that
Capitella species I has the ability to increase its incorporation
of a detritus source, that is rapidly decomposed by bacteria,
exponentially with a linear increase in the level of added organic
nitrogen (Tenore et al., 1979).

The distinct differences in temporal adaptations among the
sibling species of *Capitella* are reflected in the relative abun-
dances of the species collected in different habitats and at
different times of the year (Table 1).

There also appears to be a relationship between the fraction
of enzyme loci polymorphic and the dispersal ability, measured as
length of larval life, and the time it takes for the species to
respond to disturbance by colonizing the disturbed habitat.
Species Ia, III and II with relatively good dispersal have 12.5-
25% of their enzyme loci polymorphic, species I and IIIa with
limited or no dispersal, have 50% of their loci polymorphic
(Grassle and Grassle, 1978). The *Capitella* species showing the
highest genetic variability so far seen (5 out of 8 loci polymorphic, or 62.5%) is a deep-sea species, not yet described, collected from wood panels placed on the bottom at 3644 m for nine
months (Turner, 1973). Other deep-sea experiments have indicated
that the same species, or a closely related sibling species, has a
rate of colonization of azoic substrate measured in years (Grassle,
1977). It may be possible then for the species that are the most
genetically variable and have the least efficient dispersal, to

Table 1. Per cent occurrence of *Capitella* sibling species in field populations collected at different times of the year. The March samples from Gloucester were collected on the same day at stations approximately 1 km apart.

| Species | Buzzard's Bay | | Vinyard Sound | | | | Gloucester, Mass. (March) | |
	Wild Harbor Intertidal	Sippewissett Intertidal	Woods Hole Outfall Jan	Sept.	Falmouth Harbour Subtidal	Bourne Pond Estuarine/ Subtidal	Captain's Cour. Subtidal	Fish Pier West Subtidal
I	0-90%	70-100%	0-5%	0-5%	3-48%	>95%	---	---
Ia	---	0-30%	<5%	<5%	---	---	5%	>90%
II	<5%	0-5%	59%	94%	52-97%	---	5%	<10%
IIa	---	---	<5%	---	---	---	---	---
III	10-100%	0-20%	---	<1%	---	<5%	---	---
IIIa	---	---	30%	<1%	---	---	90%	---

respond to selection over several generations under certain environmental conditions.

How many sibling species of *Capitella* are there that were formerly lumped together as *Capitella capitata*? To date it has only been possible to obtain small collections of live *Capitella* from lócalities outside the immediate vicinity of Woods Hole through the courtesy of colleagues, but it appears that there may be at least four additional species that can be distinguished electrophoretically (Table 2).

Capitella species that have no planktonic stage, or a relatively brief one, also have a short generation time and are easily cultured in the laboratory. Over the last two years I have established a number of lines of species I, II, IIIa and Fort Pierce (a) in culture. In considering the relative potentials for dispersal in the sibling species it seemed likely that the different species might respond differently to the effects of close inbreeding. I have therefore carried out sibling matings in succeeding generations and have followed the effects of such close inbreeding on development, mortality, time to maturity, sex ratio and fecundity (measured as the number of eggs produced in the first brood). In species I some lines have been inbred in this fashion for 15 generations. We expect to compare these inbred lines with outcrossed lines in their response to certain environmental stresses to establish the levels of genetic variability·in wild *Capitella* in such multigenic, quantitative characters.

In *Capitella* species I in the first few generations after the inbreeding regime was established a sharp drop in fecundity was noted, and a high indicence of rather specific abnormalities including twinning in the metatrochophore larvae. As inbreeding continued marked changes in the sex ratios in these lines soon appeared. Most lines tended toward a preponderance of ♂ and ⚥ individuals, some lines having 100% ♂ and ⚥ individuals. Other lines showed a preponderance of ♀ 's with approximately 20% ♂ 's. In species I *Capitella*, hermaphrodite individuals are rarely seen in natural populations of high density. Experiments with one line of inbred *Capitella* species I with a high incidence of ⚥ 's, have indicated that the ⚥ 's produce significantly more eggs than ♀ 's in the same line, although ⚥'s in a highly inbred line produce fewer eggs that ⚥'s from a less highly inbred line. In some cases hermaphrodite individuals isolated early on in development, and then paired with females have produced viable offspring, indicating that they are capable of self-fertilization. It seems probable that the existence of hermaphrodites in this species is promoted under conditions that facilitate inbreeding, and that these hermaphrodites are capable of self-fertilization and production of viable offspring. This may be of considerable importance for a highly opportunistic species under conditions in

Table 2. *Capitella* species identified electrophoretically in addition to the six *Capitella* sibling species described from the Woods Hole area (Grassle and Grassle, 1976). Provisional letter designations have been assigned pending further life history data. Species (d) from the deep Atlantic may represent two species.

COLLECTION SITE	SPECIES	LENGTH OF PLANKTONIC LARVAL STAGE	REFERENCE
Deep-sea Atlantic (wood panels and azoic mud boxes)	species (d)	---	Grassle and Grassle, 1978
Fort Pierce (Fla.)	species (a)	1 day	unpublished
Fraser River estuary (B.C., Canada)	species (b)	Long-lived, on order of several weeks	unpublished
	species (c)	---	
Monterey Bay (Cal.)	species (b)	long-lived, on order of several weeks	unpublished
Los Angeles (Cal.) White's Point sewer outfall	species (b)	long-lived on order of several weeks	unpublished
Boat Harbor	species I	several hours	Grassle and Grassle, 1976

which it is present in low densities, for example following a
population crash.

Other *Capitella* species have a completely different reproductive mode. For example species IIIa is a gonochoristic; hermaphrodites have never been observed in extensive field studies
and under close inbreeding in the laboratory. In species III
however, all individuals are protandrous simultaneous hermaphrodites.

How common are groups of sibling species likely to be in the
Polychaeta? Sibling species in the genus *Ophryotrocha* (F. Dorvi-
lleidae) have been the object of close study by Åkesson (1973,
1978) and others. In this genus the reproductive modes of the
species seem to form a pattern which corresponds in some degree
to the diploid chromosome number. Other examples of paired sibling
species may be signalled by reports of two different kinds of
larvae in a single species (e.g. *Spio setosa*, Simon, 1968) or by
a combination of small morphological differences together with
some electrophoretic evidence (e.g. *Scolelepis fuliginosa*,
Kerambrun and Guérin, 1978). Rice (in press) has examined
populations of *Polydora ligni*, the species that ranked second
in our examination of opportunism in polychaete species (Grassle
and Grassle, 1974), from Tampa Bay, Fort Pierce and Los Angeles.
He found evidence for marked genetic differentiation in one Tampa
Bay population, and some indication of infertility between Tampa
Bay and Los Angeles individuals.

All of these polychaetes may be considered relatively opportunistic. Since such species frequently form a dominant element
in benthic studies of disturbed marine environments it is of
considerable importance to develop techniques for characterizing
and identifying such sibling species. Moreover, such groups of
sibling species, particularly those in the genera *Capitella* and
Ophrytrocha, provide a rich source of material for studying basic
genetic mechanisms in species that are closely related but that
differ sharply in their life history characteristics.

SUMMARY

Recent work on the characterization of life history features
and genetic variability in sibling species in the polychaete genus
Capitella is reviewed and compared with studies on sibling species
in other polychaete genera.

ACKNOWLEDGMENTS

This research was supported by N.S.F. grant DEB 76-20336.
Thanks are due to the following colleagues who collected *Capitella*
for this study: P. Chapman, K. Eckelbarger, J. Gilfillan, J.G.
Grassle, G. Hampson, C.A. Hannan, L. Harris, M. Moore, D. Reish,
D. Schnieder, R. Turner and J. Word. Excellent technical assis-
tance in culturing and electrophoresing the *Capitella* was provided
by: A.M. Benfatto, F. Birtwistle, L. Carey, A. Fugère, R. Cross,
C. Lanyon-Duncan and M. Philbin-Munson.

REFERENCES

Akkeson, B., 1973, Reproduction and larval morphology of five
    Ophryotrocha species (Polychaeta, Dorvilleidae). Zool. Scr.,
    2:145-155.
Akesson, B., 1978, A new *Ophryotrocha* species of the *labronica*
    group (Polychaeta, Dorvilleidae) revealed in crossbreeding
    experiments. In: Marine Organisms, B. Battaglia and J.
    Beardmore, eds., Plenum Publishing Corporation, New York,
    pp 573-590.
Grassle, J.F., 1977, Slow recolonization of deep-sea sediment.
    Nature,265:618-619.
Grassle, J.F., and J.P. Grassle, 1974, Opportunistic life
    histories and genetic systems in marine benthic polychaetes.
    J. Mar. Res., 32:253-284.
Grassle, J.P., and J.F. Grassle, 1976, Sibling species in the
    marine pollution indicator *Capitella* (Polychaeta). Science,
    192:567-569.
Grassle, J.F., and J.P. Grassle, 1977, Temporal adaptations in
    sibling species of *Capitella*. In: Ecology of Marine Benthos,
    B.C. Coull, ed., Bell Baruch Library in Marine Science No. 6,
    University of South Carolina Press, Columbia, S.C., pp 177-189.
Grassle, J.F., and J.P. Grassle, 1978, Life histories and genetic
    variation in marine invertebrates. In: Marine Organisms,
    B. Battaglia and J. Beardmore, eds., Plenum Publishing
    Corporation, New York, pp 347-364.
Kerambrun, P., and J.-M. Pérès, 1978, Mise en évidence, par
    electrofocalisation, de l'hétérogénéité de la population de
    Scolelepis (Malacoceros) fuliginosa de Cortiou (Bouches-du-
    Rhône). C.R. Acad. Sc. Paris,286:1207-1210.
Rice, S.A., and J.C. Simon, In Press, Intraspecific variation in
    the pollution indicator species *Polydora ligni* (Spionidae).
    Ophelia.
Simon, J.L., 1968, Occurrence of pelagic larvae in *Spio setosa*
    Verrill, 1873 (Polychaeta: Spionidae). Biol. Bull., 134:
    503-515.

Tenore, K.R., R.B. Hanson, B.E. Dornseif, and C.N. Wiederhold,
    1979, The effect of organic nitrogen supplement on the
    utilization of different sources of detritus. Limnol.
    Oceanogr.,24:350-355.
Turner, R.D., 1973, Wood-boring bivalves, opportunistic species
    in the deep sea. Science,180:1377-1379.

ECOPHENOTYPIC VARIATION IN SETAE

OF NAIDIDAE (OLIGOCHAETA)

M.S. Loden
   and
W.J. Harman

Department of Zoology
Louisiana State University
Baton Rouge, Louisiana 70803

ABSTRACT

Specimens from a clone of *Pristina aequiseta* Bourne were found to lose and regain giant setae characteristic of the species when cultured in varying ionic concentrations of an artificial pond water. *Pristina foreli* (Piguet) is declared a junior synonym of *P. aequiseta*. Species of *Dero, Nais,* and *Pristina* were found to produce morphologically different setae in response to environmental changes in field and laboratory situations.

INTRODUCTION

Among the various families of the Oligochaeta, chaetotaxy reaches its greatest importance in the Naididae. Throughout the other oligochaete families, specific and generic differences are based on genitalia; with the uncommon occurrence of sexual maturity among species of Naididae, diversity in setal morphology has dictated increased emphasis on setal characters in classification.

It has long been accepted that the setal configuration on naidid species is relatively stable. This has led to the acceptance of such characters as the segment on which dorsal setae commence, the number of setae per bundle, relative sizes of the teeth of bifid setae, and setal lengths for major, as well as

minor, taxonomic considerations. Reliance on chaetotaxy has
resulted in considerable subjectivity among authors; Pop (1973)
discussed this subjectivity in his description of *Pristina napo-
censis*, one of a group of species whose differences are based
almost entirely on rather fine aspects of setal form.

Intraspecific variation of setal characters was suspected of
several species by some authors (e.g. Sperber, 1948; Brinkhurst,
1971), but they were unable to demonstrate its occurrence.

Pristina aequiseta Bourne is characterized by setae in the
ventral bundles of segments IV, V, or both, that are thicker and
longer than other ventral setae. This species differs from *P.
foreli* (Piguet) primarily in that *P. foreli* lacks the enlarged
setae. These enlarged, or "giant" setae (Sperber, 1948) vary in
form from only slightly thicker and longer than setae in adjacent
segments to much thicker, and with the proximal tooth reduced or
absent. Harman (1974) synonomized *P. evelinae* Marcus with *P.
aequiseta* because of overlapping characters associated with giant
setae of the two species.

LABORATORY INVESTIGATIONS

Specimens of *P. aequiseta* were collected from a local lake.
They were severed in mid-body; the anterior end was fixed and
identified, and the posterior end was allowed to regenerate in a
medium of artificial pond water (Dietz and Alvaredo, 1970). The
specimens were allowed to remain in culture conditions for two
weeks, after which they were fixed and examined. None of the
cultured individuals had the giant setae characteristic of the
anterior ventral bundles of *P. aequiseta,* although giant setae
had been present in the original field-collected specimens. The
cultured specimens were identical to *P. foreli*.

A clone of *P. aequiseta* was established in artificial pond
water through asexual reproduction. The anterior zooid of a
budding individual specimen was fixed, and its identity verified;
the posterior zooid underwent somatic maturation and formed the
clone by budding.

A non-budding individual from the clone was placed in each
of a series of Petri dishes to which were added concentrations of
artificial pond water that had been diluted with distilled water.
Additional specimens were placed in a series of Petri dishes with
pond water that had the ionic concentration increased with sodium
chloride. A small quantity (ca. 0.5 mm^3) of commercial dry cat
food was placed in each culture as a source of food for the worms.
While it was recognized that the cat food would add ions to the
media, adding approximately the same amount to each dish should
have negated its effects.

The *P. aequiseta* specimens were maintained in culture for 14 days, after which they were examined for thickened setae. Individuals from solutions with a specific conductance of less than 500 μmho/cm remained identical to those in artificial pond water: they were indistinguishable from *P. foreli*. Those in concentrations with a specific conductance of 500 μmho/cm or greater had thicker, longer setae ventrally in segment IV (Table 1, Figure 1).

A comparison of individuals from the solutions with greater ionic concentrations revealed a relationship between the conductance and the degree of exaggeration of the giant setae. With progressively larger and thicker (Figure 1b-f). Even with a concentration of 1300 μmho/cm conductance, the setae did not regain the highly exaggerated characteristics of freshly-collected specimens. This indicates that while sodium chloride can account for a partial change in setal form, unknown factors in the natural habitat provide the stimulus for a greater modification.

Individuals from the clone were placed in habitat water from the collecting locality (conductance 310 μmho/cm). After 30 days the specimens were examined; approximately 80% had regained the giant setae typical of freshly-collected specimens (Figure 1g). A control group in artificial pond water continued to have setae typical of *P. foreli*.

DISCUSSION

The ease at which specimens of *P. aequiseta* underwent asexual reproduction in culture conditions provided an unusual opportunity to investigate ecophenotypic variation. Because all the individuals were descended from a single specimen, variation in the appearance of the anterior ventral setae was due solely to environmental factors. Genetic variation among individuals could not affect the outcome of the experiments.

Because of the variation in the giant setae of *P. aequiseta*, and because laboratory investigations demonstrated that environmental factors can alter the setae to the extent that *P. aequiseta* is indistinguishable from *P. foreli,* the taxonomic status of *P. foreli* is in doubt. Two alternatives are possible: either *P. foreli* is a sibling species of *P. aequiseta* or *P. foreli* represents an ecomorph of *P. aequiseta*. If they are sibling species, studies of life-history patterns, microhabitat preference, or other ecological studies delineating their niches will be necessary for their separation. An occurrence of specimens of both species at the same locality would not aid in separating the two; it is only after an undetermined time following budding and the separation of zooids that specimens of *P. aequiseta* develop first enlarged setae, then giant setae (Figure 1h). In the intervening period *P. aequiseta* specimens remain inseparable from *P. foreli*.

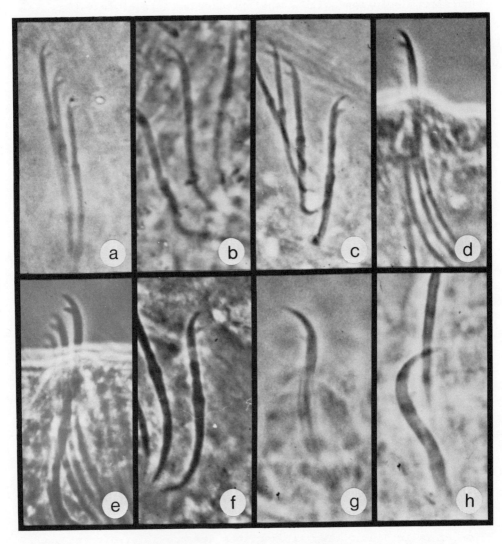

Figure 1. *Pristina aequiseta*: ventral setae of segment IV. a,
cultured in artificial pond water (90 µmho/cm); b,
500 µmho/cm; c, 700 µmho/cm; d, 900 µmho/cm; e,
1100 µmho/cm; f, 1300 µmho/cm; g, cultured in habitat
water; h, generation of giant seta in a bundle
previously lacking giant setae.

Table 1. Form variations of ventral setae of segment IV of
          Pristina aequiseta Bourne in cultures of artificial
          pond water (specific conductance 90 μmho/cm),
          dilutions with distilled water, and pond water
          with sodium chloride added to increase ionic
          concentration.

Conductance (μmho/cm)	Setal Modification
0	None (specimens died)
1.8	None (specimens died)
18	Unmodified (Figure 1a)
45	Unmodified (Figure 1a)
90	Unmodified (Figure 1a)
310	Unmodified (Figure 1a)
500	Figure 1b
700	Figure 1c
900	Figure 1d
1100	Figure 1e
1300	Figure 1f

We prefer the alternative that *P. foreli* is an ecomorph of
P. aequiseta, and we hereby declare *P. foreli* to be a junior
synonym of *P. aequiseta*. We feel that the differences in setal
structure described for the two species are the result of variation
induced by environmental conditions.

When setae are the primary characters used in identification,
subjectivity can create situations where it is difficult to ascribe
specific names. The common North American species of *Dero*, *D.
digitata* (Muller), *D. obtusa* d'Udekem, and *D. nivea* Aiyer have
setae that are somewhat similar: unless the branchial fossae of
individuals are particularly well-preserved, it is frequently
difficult to distinguish among species. There are several species
of *Nais* whose differences, when setae only are considered, appear
trivial; *N. communis* Piguet and *N. variabilis* Piguet are frequently
difficult to distinguish when certain slide preparation techniques
obliterate virtually all taxonomic characters except the setae.
Nais variabilis, *N. pardalis* Piguet, and *N. bretscheri* Michaelsen
differ in the shape of the ventral setae. The ventral setae of
N. pardalis are longer and thicker than those of *N. variabilis*,
and the teeth are more exaggerated. The ventral setae of *N.
bretscheri* are longer, thicker, and with more exxagerated teeth
than those of *N. pardalis*.

Individuals of *N. pardalis* and *N. variabilis* were placed in
the same culture dish to examine possible ecomorphic variation.
If the progeny remained different, genetic factors would have

maintained the differences. If the progeny of the cultured
specimens were identical, differences in field-collected specimens
would have been due to environmental factors. Specimens of the
two species were easily separable in the culture dish: *N. pardalis*
specimens were distinctly redder than the *N. variabilis* individuals,
and the rate of asexual reproduction of *N. variabilis* was faster;
setae of the worms remained distinctive after 30 days of culture,
indicating that recognition of both forms as species is desirable.

The presence of intermediate teeth in needle setae has been
used on numerous occasions as a major taxonomic character to
justify the description of a new species. In the LSU collection
there are specimens of *Pristina osborni* (Walton) and *Nais communis*
in which the needle setae of the anterior zooid are bifid and those
of the posterior zooid are trific or pectinate. This is apparently
a condition triggered by differing environmental conditions whereby
the generation of the older, anterior setae was influenced by one
set of environmental conditions that had become altered when the
younger, posterior segments were formed. Freshly collected specimens of an undescribed species of *Dero* had needle setae with an
intermediate tooth, but when individuals were cultured, the trifid
needles were shed, and bifid setae were formed.

The discovery of pectinate setae in *P. osborni* is likely to
result in a misidentification of specimens as *P. sima* (Marcus) if
current taxonomic keys are used. While some investigators may
favor the synonomizing of the two species, we feel that the illustration accompanying the original description of *P. sima* (Marcus,
1944), in which the proximal tooth is shorter than the distal,
indicates a relationship of *P. sima* to *P. rosea* (Piguet) and other
similar species. In *P. osborni* the teeth are equally long.

Brinkhurst (1966) described *Nais africana*, which differs from
other species of the genus largely because of its pectinate needle
teeth. As *N. communis* may also have pectinate needles, the status
of *N. africana* is in doubt. It may be necessary to declare *N.
africana* a species dubium until a reexamination of the type series
can be made.

Examinations of other species that differ from one another in
fine details of setal morphology are likely to demonstrate additional instances of ecophenotypic variation. Such discoveries
could alter the systematics of several of the genera.

Naidid setae are apparently not the stable phenotypic structures that differ only among species. Taxonomic conclusions based
solely on setal morphology may prove to be only as reliable as the
morphological stability of the setae, which is, in turn, dependent
on the stability of the environment.

REFERENCES

Brinkhurst, R.O., 1966, A contribution towards a revision of
    the aquatic Oligochaeta of Africa. Zool. Africana, 2:131-166.
Brinkhurst, R.O., 1971, Family Naididae. In R.O. Brinkhurst
    and B.G.M. Jamieson, Aquatic Oligochaeta of the World. Univ.
    Toronto Press. Toronto.
Dietz, T.H. and R.H. Alvaredo, 1970, Osmotic and ionic regulation
    in Lumbricus terrestris L. Biol. Bull., 138:247-261.
Harman, W.J., 1974, The Naididae (Oligochaeta) of Surinam.
    Zool. Verh., No. 133. 36 pp.
Marcus, E., 1944, Sombre Oligochaeta limnicos do Brasil. Bolm.
    Fac. Filos., Cienc., Let. Univ. Sao Paulo, 43:5-135.
Pop, V., 1973, Deux especes nouvelles de Pristina (Naididae,
    Oligochaeta) et leurs affinites. Stud. Univ. Babes-Bolyai
    Ser. Biol., 1:79-89.
Sperber, C., 1948, A taxonomical study of the Naididae. Zool.
    Bidr. Uppsala, 28:1-196.

ON THE QUESTION OF HYBRIDIZATION AND VARIATION

IN THE OLIGOCHAETE GENUS *LIMNODRILUS*

Michael Barbour

Ecological Analysts, Inc.
8600 LaSalle Road
Towson, Maryland 21204

David G. Cook

Scientific Information and Publication Branch
Fisheries and Oceans Canada
240 Sparks Street
Ottawa, Canada K1A 0E6

Robin S. Pomerantz

Ecological Analysts, Inc.
8600 LaSalle Road
Towson, Maryland 21204

ABSTRACT

Variant forms of *Limnodrilus* are common in many ecosystems in North America. The question of whether these variant forms are ecomorphs or hybrids of two local species has not been substantiated. In an attempt to determine the relationship of variant *Limnodrilus* to typical species, the length of the penis tube, minimum diameter of the shaft, and maximum diameter of the shaft were measured on numerous individuals of *L. hoffmeisteri*, *L. claparedeianus*, *L. spiralis*, and *L. hoffmeisteri* (variant) from several geographical areas in North America. The *Limnodrilus* were collected from Lake Okanagan, Lake Winnipeg, Lake Superior,

Lake Huron, Lake Ontario, and the Potomac River. These measurements were used to develop frequency histograms and bivariate
scatter plots to illustrate the relationship of the selected
species of *Limnodrilus*. Multivariate discriminant analysis was
also used to further test the separation of *L. spiralis* and
L. hoffmeisteri (variant) from the typical forms. Results
indicate that *L. hoffmeisteri* (variant) is indeed an ecomorph of
L. hoffmeisteri, and *L. spiralis* exhibits characteristics of both
L. hoffmeisteri and *L. claparedeianus* but is also most likely an
ecomorph of *L. hoffmeisteri*.

INTRODUCTION

    Species of *Limnodrilus* are distinguished primarily by the
shape of the penis tube and the configuration of the head at the
distal end of the tube. For instance, the head of the penis tube
of *L. hoffmeisteri* is variable in shape but typically resembles
a hood draped over one side of the tube (Figure 1a). *L. claparedeianus* has a noticeably longer tube and a triangular or pear-
shaped head which is attached to the tube at a slant (Figure 1b).

    Variant forms of *Limnodrilus* are common in many ecosystems in
North America. The question of whether these variant forms are
ecomorphs or hybrids of two local species has not been substantiated. Kennedy (1969) discussed the variability of the distal
ends of *Limnodrilus hoffmeisteri* and *L. claparedeianus* among
others. Kennedy stated that the length : breadth ratio is more
useful as an additional or confirmatory character to the shape
of the penial sheath which proved to be the most important
character.

    The shape of the penial sheath and complex distal end may be
variable in *L. hoffmeisteri* such that distinct populations of
variant forms can be found in various geographical areas of North
America. A commonly found variant form of *L. hoffmeisteri* is
distinguished by having a scalloped or lobate distal head of the
penis tube (Figure 1c). This form has been reported from Lake
Michigan (Hiltunen 1967; Stimpson et al. 1975) and Western United
States (Brinkhurst and Jamieson 1971).

    In addition to typical *L. hoffmeisteri* and the variant form
having a lobate distal end of the penial sheath, Hiltunen (1967,
1969) found a second form in Lake Michigan and Lake Erie which
he illustrated and photographed but included it under *L. hoffmeisteri* because of uncertainty of its proper placement (Figure
1d). Hiltunen (1969) described this form which had a relatively
long straight penis tube with a more or less flat, rounded head,
perpendicular to the long axis of the tube, as resembling Eisen's
(1885) *Camptodrilus (=Limnodrilus) spiralis*. Stimpson et al.
(1975) found the same form in Lake Michigan and Howmiller and

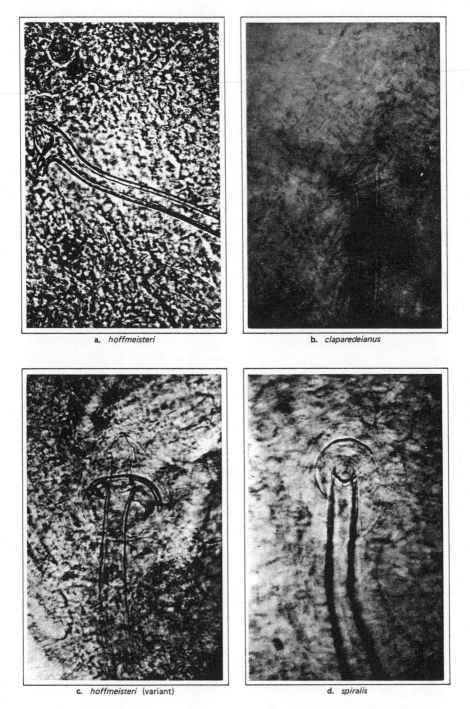

a. *hoffmeisteri*

b. *claparedeianus*

c. *hoffmeisteri* (variant)

d. *spiralis*

Figure 1. Penial sheath configurations of *Limnodrilus*.

Loden (1976) in Wisconsin, and the binomial *L. spiralis* for individuals of this type was retained in both studies. Hiltunen (1967)
first noted that the *spiralis* form had characters intermediate to
hoffmeisteri and *claparedeianus*. Cook and Johnson (1974) suggested
that *spiralis* may be a hybrid between *L. hoffmeisteri* and *L. claparedeianus* based on unpublished data from locations in Canada.

The primary purpose of this study was to describe variation
patterns of penial characters and to ascertain whether these
patterns would elucidate the taxonomic rank of the various taxa;
secondarily, because taxonomic separation involved subjective
judgements based on experience, we sought an objective, measureable
character that would separate taxa reliably. This presentation
reports some preliminary findings.

METHODS

In an attempt to determine the relationship of variant
Limnodrilus to typical species, the length of the penis tube,
minimum diameter of the shaft, and maximum diameter of the shaft
were measured on numerous individuals of *L. hoffmeisteri, L. claparedeianus, L. hoffmeisteri* (variant), and *L. spiralis* from
several geographical areas in North America. The *Limnodrilus*
were collected from Lake Okanagan, Lake Winnipeg, Lake Superior,
Lake Huron, Lake Ontario, and the Potomac River (Figure 2). The
minimum diameter measurements were taken invariably at the neck
region of the tube and the maximum diameter was ,at the base.
The length included the distal end of the tube. All individuals
measured had spermatozeugmata present in the spermatheca.

PENIS SIZE DISTRIBUTION

Frequency versus each variable and ratios of the variables
were plotted for *hoffmeisteri, spiralis, hoffmeisteri* (variant),
and *claparedeianus*. The data were combined from all sites.
Results of the length-frequency histograms show similar ranges
of length for *hoffmeisteri, spiralis,* and *hoffmeisteri* (variant)
(Figure 3). Some overlap occurred between the greater lengths
of *spiralis* and the lesser lengths of *claparedeianus*.

Frequency maximum diameter did not differ greatly among the
four species groupings and therefore is not presented. The range
and frequency of minimum diameter was the same for *hoffmeisteri,
spiralis,* and *claparedeianus*. The minimum diameter of *hoffmeisteri*
(variant) was within the same range but at the greater end of the
scale (Figure 4).

Frequency plots for length : minimum diameter ratio indicated
that *hoffmeisteri* (variant) had the smallest length : minimum

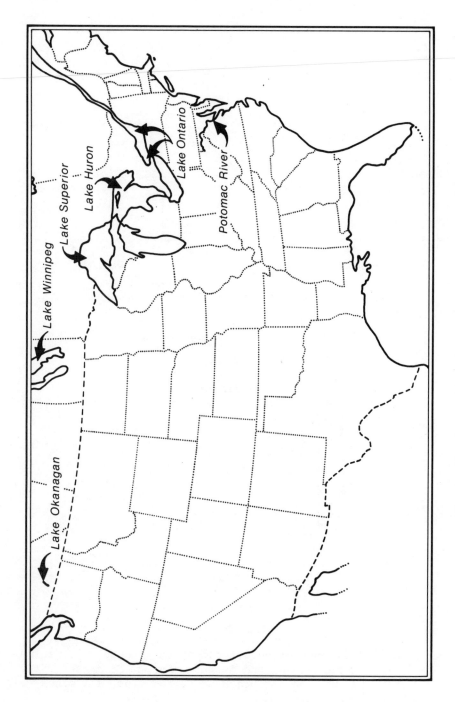

Figure 2. Geographical locations of collections.

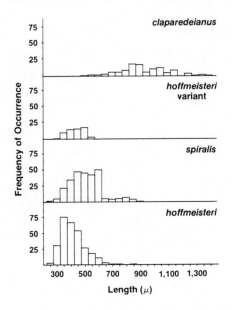

Fig. 3

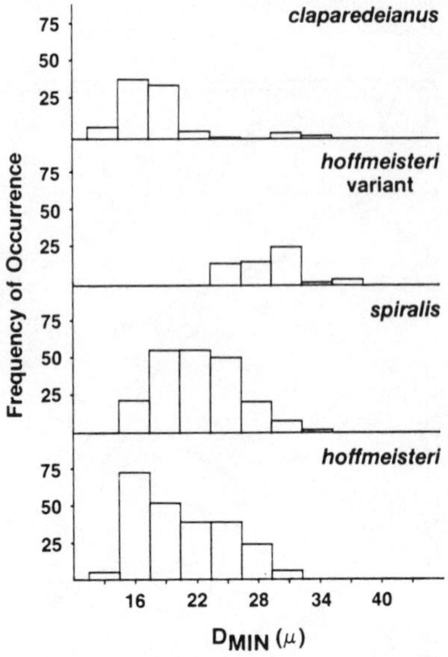

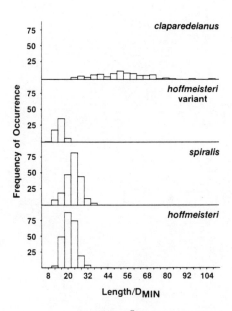

Fig. 4

Fig. 5

Figures 3,4,5.   Frequency of occurrence of the length (3), minimum diameter (4), and length: minimum diamenter ratio (5) of the penial sheath of *Limnodrilus*, all sites combined.

diameter ratio because of the broader neck region of the penial shaft (Figure 5). Length : minimum diameter reflects the most distinct separation between *claparedeianus* and the other species groupings.

TAXONOMIC DISCRIMINATION

A multivariate discriminant analysis (MDA) was used to determine what linear combinations of the variates would give the greatest separation between group means and aid in the determination of the extent of intergradation among *hoffmeisteri, spiralis,* and *claparedeianus*. Atchley et al. (1976) showed that multivariate statistical procedures are greatly affected when data upon which the analyses are based include ratios or proportions. Patterns of covariance generally derived from linear combinations are biased if it is composed of ratio variables in which many have the same denominator or numerator. The variables used in the MDA in this study were restricted to length, maximum diameter, and minimum diameter. Those sites having sufficient data for all three species groupings were treated individually, and the results are presented in Table 1.

The "jackknifed" classification matrix is a summary of cases (measured individuals) classified into the different species groupings according to the classification functions computed based on the variables (or variates) found to provide the maximum separation from the stepwise MDA. In the case of Western Lake Ontario all three variates were found to be useful for detecting separation; length and maximum diameter were important at Lake Okanagan; length and minimum diameter at Lake Winnipeg; and only length was the key variate at Eastern Lake Ontario.

The classification matrix is computed by treating each case individually with its variables being entered into the classification function. The result is that it is assigned to one of the species groupings based on the computed function. This process is repeated until all cases have been assigned. The result of the classification matrix applied to data from Lake Winnipeg indicate that only 49.1% of *hoffmeisteri* and 50.9% of *spiralis* were classified correctly. The variables of those two species groupings were so similar in dimensions that there was considerable overlap among the assigned cases.

92.3% of the *claparedeianus* were classified correctly; but those that were incorrect were assigned to *spiralis*. *L. clapa-redeianus* from Lake Okanagan and Eastern Lake Ontario were all classified correctly, as were most of the *hoffmeisteri*. The *spiralis*, however, were classified 75.7% correctly at Lake Okanagan and 84% at Eastern Lake Ontario. Eastern Lake Ontario was the only site where some *spiralis* were classified as

Table 1.   "Jackknifed" classification matrix resulting from
multivariate discriminant analyses.

LAKE WINNIPEG   Key Variates: Length, Min. Diam.

	_%_	_hoffmeisteri_	_spiralis_	_claparedeianus_
hoffmeisteri	49.1	28	29	0
spiralis	50.9	27	28	0
claparedeianus	92.3	0	5	60

LAKE OKANAGAN   Key Variates: Length, Max. Diam.

	_%_	_hoffmeisteri_	_spiralis_	_claparedeianus_
hoffmeisteri	98.2	15	2	0
spiralis	75.7	9	28	0
claparedeianus	100.0	0	0	5

EASTERN LAKE ONTARIO   Key Variates: Length

	_%_	_hoffmeisteri_	_spiralis_	_claparedeianus_
hoffmeisteri	92.9	13	1	0
spiralis	84.0	2	21	2
claparedeianus	100.0	0	0	4

WESTERN LAKE ONTARIO   Key Variates: Length, Min. Diam., Max. Diam.

	_%_	_hoffmeisteri_	_spiralis_	_claparedeianus_
hoffmeisteri	88.0	73	9	1
spiralis	64.9	26	48	0
claparedeianus	81.0	0	4	17

claparedeianus. The greatest amount of inter-classification among all species groupings occurred at Western Lake Ontario. However, the most inter-classification at all sites occurred between *hoffmeisteri* and *spiralis* with minimal overlap between *spiralis* and *claparedeianus*.

COVARIATION OF PENIAL LENGTH AND DIAMETER

Bivariate scatter plots were constructed for each of the major sites using the two foremost variates determined by the MDA. Results at all sites were similar in that *claparedeianus* was mostly separated from the mixture of *hoffmeisteri* and *spiralis* reflecting the greater length of *claparedeianus*. At Lake Winnipeg, *hoffmeisteri* and *spiralis* were indistinguishable from each other, complementing the results of the MDA for that site (Figure 6). Similar covariation of *hoffmeisteri* exhibited generally lesser length and maximum diameter than did *spiralis* (Figure 7). Few *claparedeianus* were measured from Eastern Lake Ontario, and plots of minimum diameter versus length of those individuals exhibited a wide range (Figure 8). Some cases of *spiralis* overlapped with the *claparedeianus,* indicating highly variable lengths and minimum diameter of the penial sheaths of both species groupings. Both *hoffmeisteri* and *spiralis* were similar in dimensions at Western Lake Ontario which included an extended range of lengths and minimum diameters (Figure 9). Cases of *claparedeianus* were clumped into a tight pattern, except for three isolated cases found at the top end of the minimum-diameter range.

Bivariate scatter plots were also constructed on the data from the Potomac River which included several individuals of *hoffmeisteri* (variant) as well as typical *hoffmeisteri* and *spiralis*. Length was plotted against minimum diameter (Figure 10) as well as maximum diameter (Figure 11). Cases of three species groupings were entirely interspersed, indicating that no separation of characters was determined.

CONCLUSIONS

Further study of *Limnodrilus* should include detailed measurement and description of the vas deferens and spermatozeugmata, which have been noted to differ among species of Tubificidae. In this study hybridization has not been shown to exist between *L. hoffmeisteri* and *L. claparedeianus* based on length, minimum diameter, and maximum diameter of the penial sheath. The *spiralis* form does not represent a distinct separation of characters from *hoffmeisteri,* although it does exhibit tendencies towards *claparedeianus* at Eastern and Western Lake Ontario. The *spiralis* form is most likely an ecomorph of *hoffmeisteri*, as is the *hoffmeisteri* (variant).

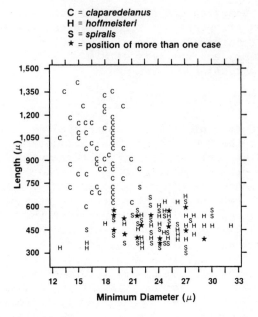

Figure 6.   Bivariate scatter plot minimum diameter versus length
Lake Winnipeg.

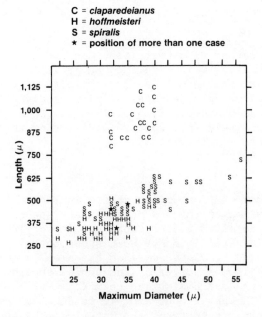

Figure 7.   Bivariate scatter plot maximum diameter versus length
Lake Okanagan.

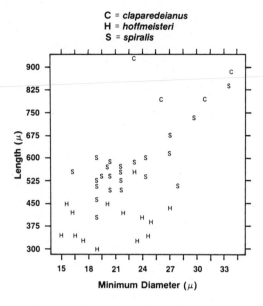

Figure 8. Bivariate scatter plot minimum diameter versus length eastern Lake Ontario.

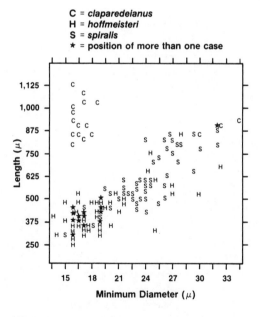

Figure 9. Bivariate scatter plot minimum diameter versus length western Lake Ontario.

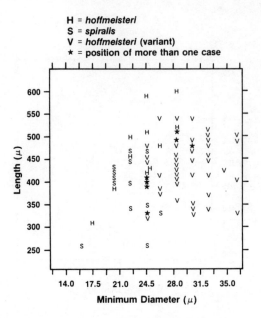

Figure 10.   Bivariate scatter plot minimum diameter versus length Potomac River.

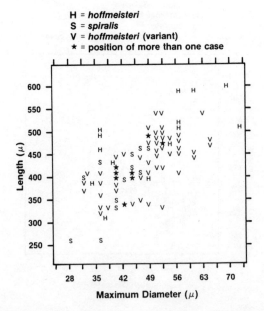

Figure 11.   Bivariate scatter plot maximum diameter versus length Potomac River.

ACKNOWLEDGEMENTS

The majority of these data were obtained from studies being
conducted by the Canada Centre for Inland Waters (CCIW), Burlington,
Ontario and by the Freshwater Institute, Winnipeg, Manitoba. Data
were also compiled from studies conducted by Ecological Analysts,
Inc., Towson, Maryland. We acknowledge Peter Sly as Senior
Investigator for CCIW and thank him for his approval of the use
of his data.

We acknowledge Ecological Analysts for the use of their
computer and report production facilities to analyze and compile
the data. Special thanks are due to Ingrid and Thomas Farnam
for constructing and photographing the graphs and table and to
Anna Klein for photographing the penial sheaths.

REFERENCES

Atchley, W.R., C.T. Gaskin, and D. Anderson, 1976, Statistical
    Properties of Ratios. I. Empirical Results. Systematic
    Zoology, Vol. 25:137-148.
Brinkhurst, R.O. and B.G.M. Jamieson, 1971, Aquatic Oligochaeta
    of the World. U. of Toronto Press. Toronto, Ontario. 860 pp.
Cook, D.G. and M.G. Johnson, 1974, Benthic Macroinvertebrates
    of the St. Lawrence Great Lakes. J. Fish. Res. Board Can.,
    31:763-782.
Eisen, G., 1885, Oligochaetological researches. Dept. U.S. Comm.
    Fish. Fisheries for 1883, Part XI, Appendix D, pp.879-964.
Hiltunen, J.K., 1969, Distribution of Oligochaetes in Western
    Lake Erie, 1961. Limnol. Oceanog., 14:260-264.
Howmiller, R. and M.S. Loden, 1976, Identification of Wisconsin
    Tubificidae and Naididae. Wis. Acad. Sci., Arts Letters,
    64:184-197.
Kennedy, C.R., 1969, The variability of some characters used for
    species recognition in the genus Limnodrilus Claparede
    (Oligochaeta: Tubificidae). J. Nat. Hist., 3:53-60.
Stimpson, K.S., J.R. Brice, M.T. Barbour and P. Howe, 1975,
    Distribution and Abundance of Inshore oligochates in Lake
    Michigan. Trans. Amer. Micro. Soc., 94:384-394.

DISTRIBUTION OF AQUATIC OLIGOCHAETES

T. Timm

Institute of Zoology and Botany of the
Academy of Sciences of the Estonian S.S.R.
Vortsjärv Limnological Station
202454 Rannu, Estonian S.S.R.

ABSTRACT

About 700 limicolous and 100 true marine species are known
so far. The passive ways of their distribution are more important than active migration. Watersheds and the current are the
main mechanical barriers while salinity, oxygen content and
temperature of the water are the ecological ones. Most of the
species need hibernation in cold for successful sexual reproduction and therefore cannot colonize the torrid zone. Only 4.5%
of the species (most of them antropochorous) are cosmopolitan
while 72% are distributed in one region or subregion only, and
42% are known from a single locality. Nine of the 15 families
surely originate from the northern temperate zone. Probably
the class arose from its polychaete ancestors in marsh biotopes
and differentiated during the dry and cold Perm period, which
may be reflected in the thermophobe character of their sexual
reproduction. After the dissociation of Pangaea, the ranges of
families were distributed between the new continents. The evolution of some groups became quicker in the Pleistocene again.

The inland waters are divided by the author into 6 zoogeographical regions, some of them into subregions. There are some
subfamilies, genera and separate species secondarily and inde-
pendently adapted to marine life. An attempt is made to describe
the changes in the limicolous oligochaete fauna of the north-
western part of the Soviet Union in the Holocene.

About 1,300 species of aquatic oligochaetes in the wider
meaning of the word (including Aeolosomatidae and Branchiobdellidae) have been observed. Discarding amphibious forms and
those casually fallen into water or those populating only the
sea littoral (from among the Enchytraeidae and Megadrili), about
700 limicolous and 100 genuine marine species remain. During
the recent decades, abundant data from all over the world became
available on their distribution. Contours of regional peculiarities are appearing. Therefore it is time to sum up the data
on the zoogeography of aquatic oligochaetes.

The material of the present report includes the distribution
charts of all known aquatic oligochaetes drawn on the basis of
data from literature and the samples collected by the author from
the water bodies of the north-western part of the Soviet Union.

HISTORICAL SURVEY

Earthworms became an object of interest in zoogeography
rather a long time ago (Michaelsen, 1903, etc.). As for the smaller
and less known aquatic oligochaetes, only few habitats were known
for a long time, and even those were often rather far away from
each other. Of zoogeographical problems, possible ways of distribution were mainly discussed in the literature (Zacharias, 1889;
Bretscher, 1903, Benham, 1903, etc.).

Even Stephenson (1930) in his monograph mainly confines
himself to the treatment of the ways of distribution of aquatic
forms. He is of the opinion that, due to the easy distribution,
some genera or even species are very widely distributed, being
cosmopolitan as, for instance, rotifers or infusoria. The
problems of distribution barriers, peregrinity, regional differences of fauna, etc. are observed only in the case of earthworms,
but he did note Lake Baikal as a nest of numerous endemic aquatic
oligochaetes.

Cekanovskaja (1962) supposed that there is no principal
difference between the zoogeography of terrestrial and aquatic
oligochaetes, although her material was insufficient for gene-
ralizations.

The further increase of the data on distribution from different continents, especially thanks to R.O. Brinkhurst's activities
in the 1960's enabled Brinkhurst and Jamieson (1971) to treat
many problems of the zoogeography of aquatic oligochaetes on the
universal level. Some of their viewpoints seem to be debatable
now, but the majority of them, are accepted by the author of the
present report.

DISTRIBUTION MECHANISMS

The active migration of the species populating bottom sediments is very slow. *Isochaetides newaensis*, a big and strong psammophilous tubificid, experimentally introduced into Lake Vortsjarv, has extended its population area no more than 10 metres a year. The swimming forms, especially Naididae, spread more quickly. For instance, *Stylaria lacustris* is often found in the pelagial of small lakes where it cannot live permanently. In 1959 in Lake Peipsi the author could observe how the quickly propagating *Nais barbata* and *Stylaria lacustris* from a narrow coastal strip spread towards the open lake with the speed of 1 km per month.

Still the passive ways of distribution are of greater importance. Especially rapid and abundant is the drift in rivers downstream: the swimming forms by water, the bottom-dwelling ones rolling along the bottom or together with floating plant wisps, logs, ice, etc. Distribution against the stream is very slow in rivers, and is practically impossible across waterfalls and dams. In such cases, larger aquatic animals such as fishes, crayfish, waterfowl, mammals, etc. may possibly be of some help. In lakes and in the sea, worms may also be carried by currents. Migration from one water body to another proceeds along water-ways: either along permanent (rivers, canals, straigs, phreatic water, karst, etc.) or temporary connections (floodings). During a longer period changes in the form and size of the water body as well as shifts of watersheds should also be considered.

The passive distribution over dry land through air is rare but possible. Dryness and frost-resistant stages of some forms (cysts of Aeolosomatidae, hibernating cocoons of Naididae) can be spread together with dust by wind. Such worms can most often be found in ephemeral pools. There have often been suppositions in the literature as regards the possibility of the distribution of aquatic oligochaetes by the waterfowl, flying aquatic insects and even mammals dirty with mud. Unfortunately, no direct proof is available. As oligochaetes themselves (and in most cases their cocoons as well) do not stand dryness, such air transport is possible over short distances and only during favourable (moist and cool) weather. Benham's (1903) supposition that birds could carry oligochaetes over oceans seems to be extremely improbable. Rare but effective carriers of aquatic animals over dry land are waterspouts.

Man himself changes nature powerfully. Involuntarily he carries oligochaetes over oceans and dry land: in the ballast of ships and in their periphyton, together with live plants and fishes, etc. In addition to direct distribution man also influences them by damming and draining rivers and lakes, joining water bodies by canals and tubing, polluting water with sewage and heat. Conscious distribution of oligochaetes, mostly for the feed of fishes, is rare.

Eurytopic worms distribute relatively more successfully. So do the ones having many ova in their cocoons or those reproducing asexually or parthenogenetically - a single specimen which finds itself in a new place may form a new population.

DISTRIBUTION BARRIERS

Distribution barriers of aquatic oligochaetes may convention- ally be divided into mechanical and ecological. Mechanical barriers are, first of all, any kind of dry land or watershed. These barriers can be crossed under exceptional circumstances only, for instance, in the case of flooding, or with the help of land animals or by human interference. The current may be another barrier. In several Estonian rivers *Isochaetides newaensis, Pota- mothrix moldaviensis* and other species carried in by ships can be found only in the lower reaches, before the first dam or waterfall. For limnobionts, distribution from one lake to another along a river against the stream without the help of a man is in no case easier than over the dry land. Torrential upper reaches of bit rivers in mountains are mostly poorer in species than their middle and lower reaches.

A most essential ecological barrier is salinity. The sea is a barrier for fresh-water forms, and vice versa. In both cases euryhaline species are in more preferable conditions.

For many rheophilous and oxyphilous species of oligochaetes, lakes and parts of rivers with slow or no current may become dis- tribution barriers. Not the current itself but the oxygen content in water seems to be essential here. Many species commonly known as rheophilous may be found in masses in the littoral, some of them (e.g. *Stylodrilus heringianus*) even in the profundal of northern oligotrophic lakes rich in oxygen.

Although there are few examples in the literature, tempera- ture conditions may also be a distribution barrier. It is a known fact that *Branchiura sowerbyi* which stems from the tropics, populates mainly the waste-water of thermal power plants in Europe and North America, while in natural water bodies it is either absent or has dwarf size. Relatively cold water is probably the barrier to the distribution of some tropical Naididae (e.g. genus *Allonais* and most of the species of *Aulophorus*) to the temperate zone. In general, the common temperate zone species of the Tubificidae, Naididae, Lumbriculidae and Enchytraeidae success- fully populate spring brooks or bottoms of northern lakes where the temperature of water never exceeds +4° C. Many species of

Haplotaxidae, Lumbriculidae and even Tubificidae are specific for permanently cool phreatic water. The giant Lake Baikal, in which the surface never warms up over 10 or 12° C, is populated by an oligochaete fauna rich in species.

The author has bred several common Estonian species in aquaria and at different temperatures. It appeared that their sexual reproduction is a relatively stenotherm process requiring lower temperatures than other vital functions. Of the micropopulations kept at 25° C only few reproduce sexually: first of all *Tubifex tubifex* (cocoons were got even at 30° C), also *Limnodrilus hoffmeisteri, Ilyodrilus templetoni* and *Potamothrix moldaviensis*. Kept at the room temperature (18–20° C) for the whole year round *Psammoryctides barbatus* also reproduces easily. The other species of Tubificidae and Lumbriculidae (*Potamothrix hammoniensis, Peloscolex ferox, Isochaetides newaensis, Rhyacodrilus coccineus, Stylodrilus heringianus*) become sexually mature at the room temperature but either do not reproduce or lay only single cocoons. The regular sexual reproduction occurs only if they can spend at least a part of the year (winter) at temperatures not exceeding 10–15°C. In the case of *Potamothrix bedoti* architomy (fragmentation) was observed even at 25° C while the sexual reproduction occurred only in the temperature range of 10–18° C. Of the studied species *Rhynchelmis tetratheca* and *Lamprodrilus isoporus* were most stenotherm. They become mature and reproduce only at temperature below 10° C. A constantly low temperature of water seems to be no barrier for the maturing or reproduction of the above mentioned species although it may decelerate the process.

Probably the aquatic families of the oligochaetes, possibly the whole class, originate from a cool climate — a fact "remembered" by their sexual reproduction. Only a few eurythermal forms, especially when the asexual reproduction alternates with the sexual one or dominates over it (as Aeolosomatidae, Naididae, *Branchiura, Bothrioneurum,* and *Aulodrilus*), have been able to adapt themselves to the permanently warm tropic waters. An analogous phenomenon can be observed among the extraordinary thermophobe family of Lumbriculidae populating mainly the phreatic water, brooks, springs and cool lakes: only one species, mainly reproducing by architomy, *Lumbriculus variegatus* is widely spread in warmer surface waters.

The inability of terrestrial earthworms of the family Lumbricidae to populate the tropics was stated earlier by Gates (1958).

To what extent biotic barriers restrain the distribution of aquatic oligochaetes, is not known so far. They may have a certain role in the fact that the fauna of Lake Baikal does not get mixed with the fauna of surrounding water bodies. The partial vicariate and extinction of one species by another was observed in Branchiobdella (Järvekülg, 1957).

DISTRIBUTION RANGES OF SPECIES

The distribution range or distribution area is due to the antagonism of spreading and extinction, formation and perishing of new populations in change, in dynamic balance. We can observe these changes directly very rarely. Usually we can reconstruct the way of formation of a range by presumption, on the basis of the present form of the range and development of the geographic environment. Sometimes the ranges of relative species are of some use in defining the centre of origin of the group.

As the possibilities for the distribution of aquatic oligo-chaetes are rather limited, few cosmopolitans are found among them. Distribution charts reveal that only 15 species have been found on all the six continents (excluding the Antarctic Conti-nent) so far: *Aeolosoma hemprichi*, *Slavina appendiculata*, *Dero digitata*, *D. nivea*, *Aulophorus furcatus*, *Nais elinguis*, *Chaeto-gaster limnaei*, *Pristina longiseta*, *Tubifex tubifex*, *Limnodrilus hoffmeisteri*, *L. udekemianus*, *Bothrioneurum vejdovskyanum*, *Bran-chiura sowerbyi*, *Aulodrilus pigueti* and *Eiseniella tetraedra*. Another 17 species have been found on 5 continents: *Aeolosoma niveum*, *A. quaternarium*, *A. variegatum*, *Dero obtusa*, *Allonais paraguayensis*, *A. inaequalis*, *Nais communis*, *N. variabilis*, *Hae-monais waldvogeli*, *Chaetogaster langi*, *Pristina aequiseta*, *P. jenkinae*, *P. proboscidea*, *Limnodrilus claparedeanus*, *Rhyacodrilus coccineus*, *Aulodrilus limnobius*, and *Lumbriculus variegatus*. Thus, of the 709 inland water species of oligochaetes only 32 or 4.5% (among them 4 species of Aeolosomatidae, 17 Naididae, 9 Tubifici-dae, 1 Lumbriculidae and 1 Lumbricidae) are very widely distributed and can conventionally be called cosmopolitan. Most of them have obviously been distributed by man and are antropochorous or pere-grine species (surely most of the above checked Tubificidae, also *Lumbriculus* and *Eiseniella*) whose earlier natural area of distri-bution was more limited. Perhaps some species of Aeolosomatidae and Naididae are distributed all over the world naturally. Twenty six out of the 32 cosmopolitan species can reproduce asexually (by paratomy or architomy), a fact which facilitates their dis-tribution.

Unlike the low number of cosmopolitans, 568 inland water species (80% of the total number) are distributed in one zoogeo-graphical region only, of them 301 only in the Holarctic. 507 species (72%) are distributed only in one subregion. It is difficult to find a species, the range of which would exactly cover a whole zoogeographical region or subregion. In most cases the ranges are either considerably more limited or reach even the neighbouring regions. For instance, transholarctic species can, as a rule, also be found in the Sino-Indian region (e.g. *Stylaria lacustris*, *Uncinais uncinata*, *Chaetogaster diaphanus*) or even in Central America (*Limnodrilus profundicola*). Many European species

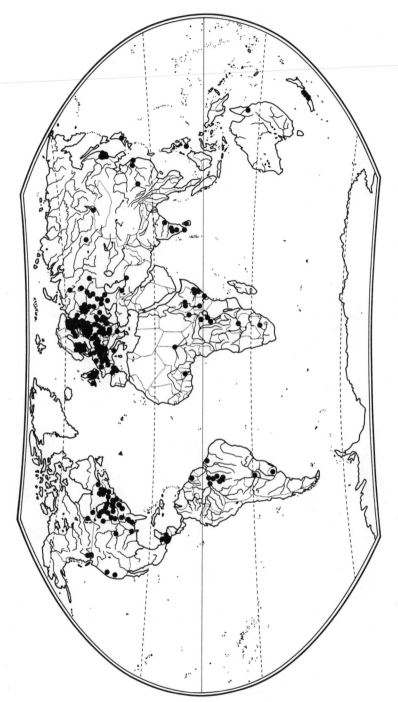

Figure 1. The distribution of *Dero digitata*, a cosmopolitan species.

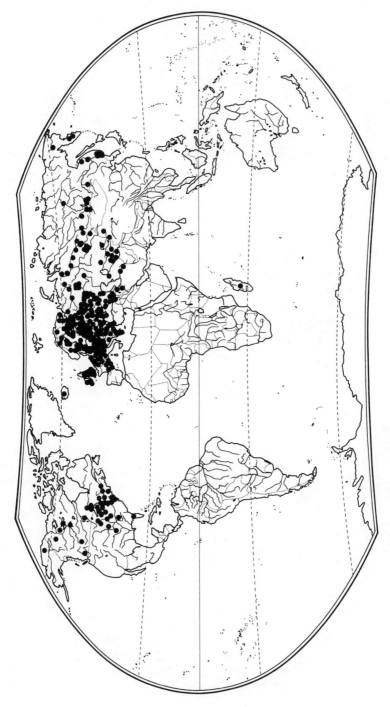

Figure 2.   The distribution of *Stylaria lacustris*, a trans-holarctic species.

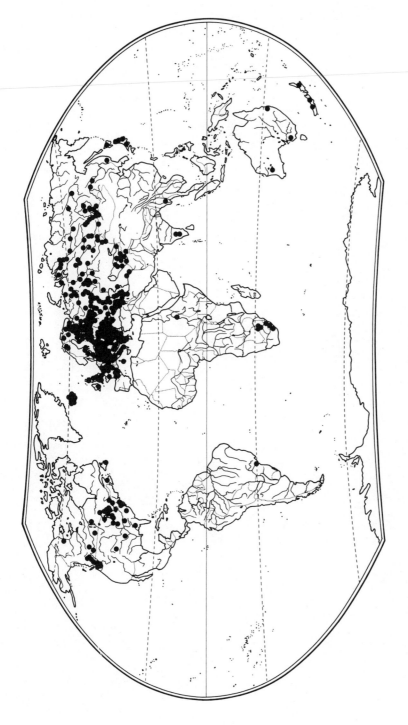

Figure 3. The distribution of *Tubifex tubifex* (s.s.), a peregrine species originating from the Holarctic.

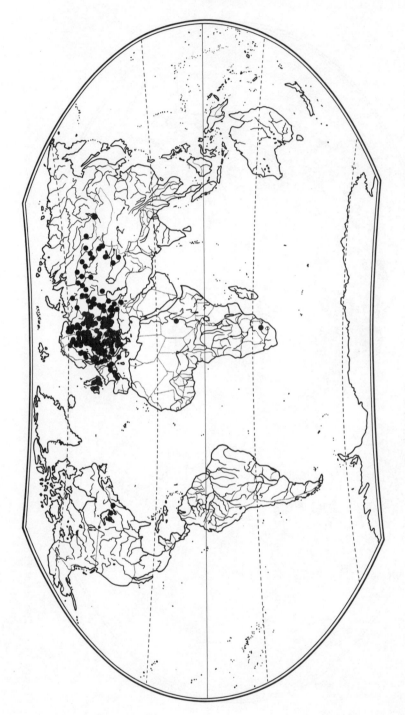

Figure 4. The distribution of *Potamothrix hammoniensis*, an Euro-Siberian species.

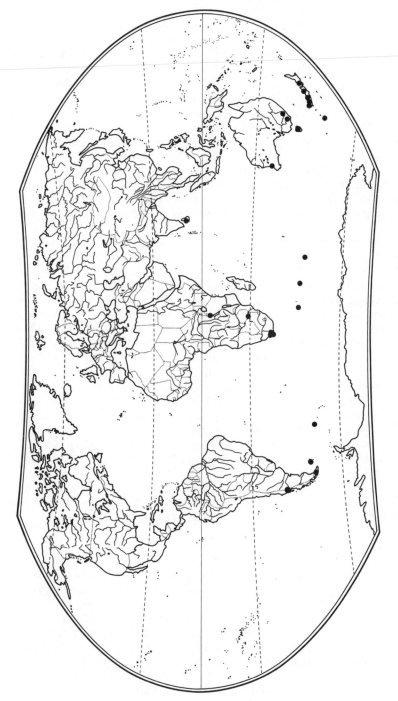

Figure 5. The distribution of *Phreodrilidae*, the only aquatic family endemic in the southern temperate zone.

have recently been introduced by man over the ocean to North
America (*Paranais frici, Potamothrix hammoniensis, Isochaetides
newaensis*, etc.) and vice versa (*Limnodrilus cervix, Sparganophilus tamensis*, etc.). Obviously, many species have by the help
of man only recently spread from the northern hemisphere to the
southern temperate zone (in addition to the above mentioned cosmopolitans also, for example, *Vejdovskyella comata, Ophidonais serpentina, Ilyodrilus templetoni*, etc.).

The majority of the species strictly specific to a region are
distributed in a very limited area, many of them in one water body.
Thus, 59 species are endemic for Lake Baikal with its outflow, 17
for the ancient lakes of Macedonia, 7 for the Caspian Sea, etc.
Only one locality (in giant Lakes - one part of the lake) is known
for about 300 inland water oligochaetes of all families (42% of
the total number), i.e. they are endemic in the narrowest sense.

Are the endemic species relicts of the earlier fauna or young
species arisen on spot? The problem has widely been discussed in
connection with Lake Baikal. At present it is supposed that only
a limited number of ancestral species could adapt themselves to
the exceptional conditions of Lake Baikal in olden times. These
species soon branched off into numerous new species (Izossinov,
1962, etc.). The same could be true of the other ancient giant
lakes with peculiar endemic fauna, as Lakes Ohrid, Tahoe, Titicaca,
the Caspian Sea. Only smaller, separate populations of the species with disconnected ranges, such as *Tatriella slovenica* in the
Tatras and Estonia or *Psammoryctides deserticola, Potamothrix caspicus* and other Ponto-Caspian brackish water species in the estuaries of the Black Sea can undoubtedly be regarded as relicts. In
some cases the evolution of such relict, isolated populations has
caused the formation of new subspecies and even species (for instance, the subspecies of the Baikal species of *Lamprodilus* in
Karelia and Lake Ohrid, species of *Phreodrilus* in Ceylon and Lake
Tanganyika). Due to considerable isolation especially the phreatic
worms (for example genera of Lumbriculidae) and parasites (Branchiobdellidae) tend to form local, vicarious species.

DISTRIBUTION RANGES OF FAMILIES

Nowadays the Tubificidae are distributed everywhere but they
surely originate from the northern temperate zone. The majority
of the species distributed in only one zoogeographical
region populate the northern temperate zone while 53% of
them - are holarctic. Only single representative have managed,
in the remote past (possibly under the conditions of different
climate and form of continents), to reach the southern temperate
zone. There they gave birth to some new species and genera (*Telmatodrilus pectinatus, T. multiprostatus, Tubifex natalensis,
Rhyacodrilus simplex, Antipodrilus*). Of the 25 genera occurring

in inland waters only 4 have adapted themselves to the torrid zone
(*Branchiura, Bothrioneurum, Siolidrilus, Paranadrilus*) while 3 of
them are monotypic.   Even separate eurytherm species have either
comparatively recently started to colonize the torrid zone (subspecies of *Tubifex tubifex* in South America) or are distributing
there by the help of man at the present time (*Limnodrilus hoffmeisteri*).   Several big genera (*Tubifex, Peloscolex, Rhyacodrilus*, etc.)
have given rise to endemic species for different continents so that
their primary home is difficult to fix.   In the case of some other
genera, the centres of distribution are clearly observable: in the
western part of North America for *Ilyodrilus*, in the eastern part
of North America for *Limnodrilus*, in the Balkan Peninsula for
Psammoryctides (s.s.), in the Ponto-Caspian system for *Potamothrix*
and in South-East Asia for *Branchiura*.

The Naididae also originate from the northern temperate zone
as proven by their thermophobe sexual reproduction and the predominantly northern distribution of several genera (*Stylaria, Chaetogaster, Vejdovskyella, Uncinais*, etc.).   Single antropochorous finds
in the southern temperate zone have not been taken into account
here.   The common asexual reproduction by budding (paratomy) has
enabled the family of Naididae to distribute all over the world.
The number of species distributed in one region is rather evenly
divided between the Holarctic, Sino-Indian and Neotropical regions
(29, 28 and 34%) while very little is left for the others.   The
genera breathing by gills, *Dero, Aulophorus* and *Branchiodrilus* as
well as *Allonais* and *Allodero* are most abundantly represented in
the torrid zone.   The distribution centre of genus *Slavina* lies
in the Neotropical region.   The brackish-water genus *Paranis* seems
to be connected with the ancient mediterranean sea of Tethys.   *Nais*
and *Pristina* are cosmopolitan.

The Lumbriculidae is a thermophobe family of the northern
temperate zone, most of the species being endemic.   Only the antropochorous ubiquist *Lumbriculus variegatus* has also been taken to
the southern temperate zone.   The occurrence of *Stylodrilus heringianus* in the Asswan Reservoir on the Nile also deserves attention.

Of the other families the Branchiobdellidae (crayfish parasites
both in North America and Eurasia), Dorydrilidae (in Europe), Lycodrilidae (in Lake Baikal) and Enchytraeidae (the only genuine limicolous genus *Propappus* in Eurasia) belong to the northern hemisphere
exclusively.

The only water-dwelling genus of Lumbricidae (*Eiseniella*) is
palearctic but distributed by man to both temperate zones.   The
primary home of the Glossoscolecidae is also supposed to be the
northern temperate zone where only few species of the general *Sparganophilus, Criodrilus* and *Biwadrilus* occur nowadays.   In any case,
most species of this partly dry-land megadrile family populate

tropics in the modern time; among them genus *Glyphidrilus* in South
East Asia, *Alma* and *Callidrilus* in Africa, *Drilocrius* and *Glyphi-
drilocrius* in South America are limicolous or, at least, paludi-
colous.

Only 2 megadrile families probably originate from the torrid
zone: Alluroididae and the mostly terrestrial Megascoledae.

The family Phreodrilidae originates from and is distributed
in the southern temperate zone; 2 of its species found in the
torrid zone nowadays are obviously relicts. It is possible that
the small puzzling family of Opistocystidae, at present known in
both Americas, also originates from the southern hemisphere.

Haplotaxidae, possibly the oldest of the present families of
oligochaetes, is truly cosmopolitan, with endemic species on all
continents.

Most of the species of Aeolosomatidae are known in the Holarctic. The reason may well be in the fact that these very small
worms have insufficiently been studied elsewhere. Probably the
family which very easily distributes has been cosmopolitan long
ago like several of its species.

In conclusion we may state that of the 15 families of Oligochaeta populating water bodies 9 obviously originate from the
northern temperate zone, 1 from the southern temperate zone, 2 from
the torrid zone while 2 are cosmopolitan and the origin of 1 family
is obscure. A typical fauna of aquatic oligochaetes rich in forms
can be found in the temperate zones only, while in the southern one
it is to a large extent antropochorous. Few groups originating
from the water bodies of the temperate zones or from dry land
populate the water bodies of the torrid zone.

EVOLUTION OF AQUATIC OLIGOCHAETES

The distribution of separate families reflects the history of
their origin and evolution. The Haplotaxidae, regarded as the
oldest one, is dispersed on all continents and in all zoogeographic
regions. The separate branch of Aeolosomatidae also seems to be
cosmopolitan. Most of the younger families occur in the northern
hemisphere. Their representatives found in the southern temperate
zone are either antropochorous immigrants or very old comers whose
scanty number proves the exceptional character of the natural
immigration. Of the aquatic families only the Phreodrilidae, possibly also the Opistocystidae, originate from the southern temperate zone. In spite of the small number (21) and rather a high
degree of morphological homogenity a noticeable ecological radia-
tion has happened among Phreodrilidae. Like several families of
the northern hemisphere, the Phreodrilidae include at least one

species with gills and several crayfish parasites. It resembles
the convergent evolution of the Australian marsupials and the Pla-
centals of the rest of the world. Of the terrestrial oliogochaete
families, also, obviously one originates from and has evolved on
the former southern continent - Acanthodrilidae.

One cannot observe the origin of aquatic and terrestrial oligochaetes separately as the oligochaetes probably arose in the
intermediate biotope, that of marsh mud and phreatic water where
the majority of Haplotaxidae lives even now. The divergence into
aquatic and terrestrial lines was temporary and the lines have later
repeatedly crossed (for example, the Enchytraeidae have returned to
dry land and partly into water again). However, it is not justified
to consider the Haplotaxidae as the ancestors of the Tubificina.
The Haplotaxidae have the established set of setae (only crotchets,
not more than two per bundle) characteristic of Lumbricina (and
also Lumbriculida and Moniligastrida). The indefinite number of
setae and the diverse, often strikingly similar form to those of
Polychaeta in the case of most species of Tubificina is surely a
primitive character which is most unlikely to have been formed
anew (if the Tubificina were the successors of the Haplotaxidae).

Observing the distribution of Megadrili, Omodeo (1963) finds
that the ranges of these families have differentiated on the supercontinent of Pangaea as early as in the Mesozoic, not later than
100-200 million years ago. Judging by the scheme, he considers
the age of genera to be 30-100 million years (Cretaceous - Paleo-
gene) and that of species millions or tens of millions years. It
is quite justified to assume (and Omodeo does) that the evolution
of Megadrili has been much slower than that of mammals. The above
said should be valid for the Microdrili populating waters as well.

Probably the oligochaetes arose and differentiated before the
formation of Pangaea, in the Perm period when the seas of the
Carboniferous period withdrew, large marshes dried up and the polychaete ancestors of the Oligochaeta populating the marshes partially adapted themselves to the terrestrial conditions. Reptiles
also arose at the same time. The previous big glaciation in the
Perm period may be reflected even nowadays in the thermophobe
character of the sexual reproduction of most families of the oligochaetes. The solid supercontinent of Pangaea existed in the
Triassic. Due to the decelerated evolution the dissociation of
Pangaea in the Jurassic did not lead to the differentiation of
families anew. The ranges of the families formed earlier were
distributed between the new continents and then partially modified
due to the evolution of the climate.

Undoubtedly in some areas there existed and exist periods of
quicker evolution of some groups. *Lamprodrilus* in Lake Baikal and
Trichodrilus in Europe may serve as examples. Among other possible

reasons we could mention quick orographic and hydrographic changes under the conditions of relatively cold climate in the Pleistocene which are still going on. As in the glaciation in the Perm period, that of the Pleistocene may have caused a new, quicker differentiation of the thermophobe aquatic oligochaetes. The small, probably young endemic families of Dorydrilidae and Lycodrilidae as well as a strongly different genus *Kurenkovia* (Lumbriculidae) may be a result of it.

ZOOGEOGRAPHIC DIVISION OF INLAND WATERS

The zoogeographic division of inland waters (fresh waters, brackish water and salt lakes) does not coincide completely with that of dry land. No universal schemes exist. Investigators of different animal groups have determined the regions differently. This has encouraged the author to offer his division which is mainly based on that of Starobogatov (1970) but has taken into account the distribution of the groups and species of the aquatic oligochaetes. The total number of regions is 6 while together with subregions there are 18 divisions.

I. The Holarctic region. The biggest and best studied region. 419 species (60% of the world inland water fauna) have been found here. The distribution of 72% of them is limited to this region. It is the main distribution area of Tubificidae, Lumbriculidae and Branchiobdellidae; about a half of the species of Naididae and many representatives of smaller families occur here. On the basis of the oligochaetes we can distinguish between 6 subregions: the Euro-Siberian, West Balkan, Ponto-Caspian (brackish water), East Siberian, Pacific and Atlantic (the last two ones in North America).

II. The Baikal region. Almost all of the 68 species known in the giant lake (less than 10% of the world fauna) are endemic: if to include the species which drift down the stream into the Angara and Yenisei rivers and partly populate the water reservoirs there – 87%, otherwise 62%. Over one half of the Baikal oligochaete fauna is formed by the Lumbriculidae which quickly diverge here (35 species, almost 1/3 of the family). Some endemic species of Tubificidae, Naididae, Enchytraeidae and Haplotaxidae as well as a separate small family Lycodrilidae also occur here.

III. The Sino-Indian region partly belongs to the temperate zone, partly to the torrid one and is therefore rather heterogeneous. Of the three subregions the Amur-Japanese one could very well belong to the Holarctic as for its fauna, too (several endemic *Peloscolex*, *Rhyacodrilus*, *Lumbriculus*, etc., but also *Branchiura* and *Branchiodrilus*). The Indio-Malayan subregion is characterized by numerous species of Naididae (especially *Dero*) and Glossoscolecidae (*Glyphidrilus*). The relatively poorly studied subregion

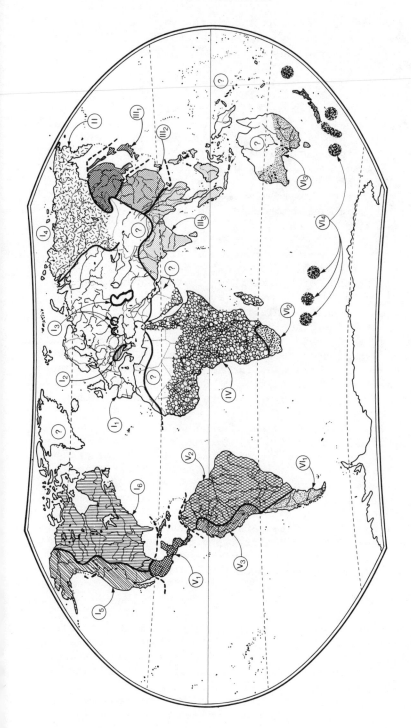

Figure 6. The zoogeographical regions and subregions of the inland waters. I – Holarctic region (I_1 – Euro-Siberian, I_2 – West Balkan, I_3 – Ponto-Caspian, I_4 – East Siberian, I_5 – Pacific, I_6 – Atlantic subregion). II – Balkan region. III – Sino-Indian region (III_1 – Amur-Japanese, III_2 – Chinese, III_3 – Indo-Malayan subregion). IV – Aethiopian region. V – Neotropical region (V_1 – Middle American, V_2 – Brazilian, V_3 – Andian subregion). VI – Antarctic region (VI_1 – Patagonian, VI_2 – Cape, VI_3 – Australian, VI_4 – New Zealand subregion).

of China is intermediate. The Sino-Indian region is the original
home of the peregrine *Branchiura sowerbyi*. The 170 species found
here make up 24% of the world fauna. 51% of the species are
specific for the region.

IV. The Ethiopean region lies fully in the torrid zone and
is thus rather poor in the aquatic oligochaetes (83 species or 12%
of the world fauna; specific for the region are 42% of the species).
Comparatively numerous are Naididae and Glossoscolecidae; the
almost specific group is Alluroididae. As regards the oligochaete
fauna, the African giant lakes do not differ essentially from their
surrounding, smaller water bodies. The permanent stratification of
the lakes makes the deeper layers of water anaerobic. The surrounding phytophilous-paludicolous biocoenoses do not offer much material
for the evolution in the benthal of a giant lake either.

V. 133 species (19% of the world fauna) are known in the
Neotropical region, of them 51% are specific for the region. Of
its three subregions, that of Brazil lying in the torrid zone has
been studied most profoundly. Numerous species of Naididae, also
peregrine and endemic species of Tubificidae (among them local subspecies of *Tubifex tubifex*), Aeolosomatidae, Megascolecidae and
Opistocystidae have been found here. The cool high mountain lakes
of the subregion of the Andes, poorly studied so far, are mostly
populated by the genera distributed in the northern temperate zone
but as endemic species. The subregion of Central America seems
to be a transitional zone from Brazilian to the Atlantic subregion.

VI. The Antarctic region covers the southern part of the
former continent of Gondwana which now belongs to the temperate
zone. The thermophobe family of Phreodrilidae and the occurrence
of several endemic species of Haplotaxidae are characteristic of
it. Still, most numerous here are the antropochorous Naididae and
Tubificidae brought from the northern hemisphere, although there
are some autochtones of these families, too. The division into
the subregions of Patagonia, Cape, Australia and New Zealand is
provisional and based rather on cartography than on differences
in the fauna. So far 85 species (12% of the world fauna) have
been found. Only 41% of them are specific for the region.

MARINE OLIGOCHAETES

The secondary transition to marine life has occurred indepen-
dently in many groups of oligochaetes and in different parts of
the world. No marine family exists but there are several subfamilies (Phallodrilinae, Clitellinae and Paranaidinae) populating
mainly the sea and brackish water, also several genera (*Limnodriloides, Rhizodrilus, Jolydrilus, Grania*) and separate marine
species of the fresh water genera (*Peloscolex, Tubifex, Nais,
Aeolosoma*, etc.). Many amphibious species (from among the

Enchytraeidae, less from among the Glossoscolecidae, etc.) populate
the marine littoral. The gradual increase of the salt tolerance
which enabled the oligochaetes to colonize the sea has happened in
the marine littoral (terrestrial forms) and, in the case of the
fresh water forms, either in the river mouths where the current
takes them or in the inland seas with unstable salinity and salt
lakes, like the Caspian Sea, characterized by low salinity (about
13 °/oo) and relatively high sulphate content. Here we can find
many purely fresh water animals, even 3 species of the otherwise
exceptionally fresh water family the Lumbriculidae (*Stylodrilus
parvus* and 2 endemic species). In the case of the intrusion of
ocean water into such an inland sea, as has happened in the Black
Sea, the endemic brackish water fauna finds shelter in the river
estuaries. The zone of the Black Sea and Caspian Sea (the Ponto-
Caspian subregion of inland waters) is a remnant of the former
Tethys. It is the distribution centre of the brackish water sub-
family Paranaidinae. Nowadays several brackish water forms from
the fresh water genera *Potamothrix* and *Psammoryctides* are evolving
here.

It is impossible to fix the time and place of origin of most
marine forms; one can only assume surely that it has not happened
simultaneously. The opposite process also occurs: in marine groups
(e.g. in genus *Phallodrilus*) single fresh water species have arisen
anew. In the sea, in its turn, an extensive ecological differentiation of the oligochaetes has taken place. Most of them prefer
biotopes with lower salinity: the littoral and shallow brackish
water seas as the Baltic (the majority of Enchytraeidae, also
Clitellio arenarius, *Paranais litoralis*, etc.). Rather many of
them can be found in shelf seas with the ocean salinity (35 °/oo),
at the depths of tens and hundreds of metres (species of *Grania*,
Limnodriloides, *Phallodrilus*, etc.). Quite recently first oligochaetes were found even in the abyssal of the ocean, at the depth
of several kilometres (*Peloscolex aculeatus*, *Adelodrilus voraginus*,
Bathydrilus asymmetricus, etc.).

It should be comparatively easy for the marine and littoral
oligochaetes to expand their ranges to different parts of the
world: the ocean is not so serious a barrier for them as for the
fresh water oligochaetes. Nevertheless, so far only single cosmopolitans, mainly from among the inhabitants of the littoral (*Enchytraeus albidus*, *Paranais litoralis*, *Pontoscolex corethrurus*) are
known. Some well known forms reveal a clearly regional distribution. Thus the distribution of *Tubifex costatus* is limited to the
European coast while *Clitellio arenarius* is amphiatlantic. The
species of *Grania,* at least partially, seem to vicariate in different parts of the world ocean.

Most of the information on the marine oligochaetes has become available only during the last 15 years. For a generalized summary it is too scanty for the present.

FORMATION OF THE OLIGOCHAETE FAUNA OF THE WATER BODIES OF THE NORTH-WESTERN PART OF THE SOVIET UNION

This is an attempt to outline the changes of distribution ranges with an evolution of natural conditions on a concrete territory. The area under observation covers the basins of the Baltic and White Seas in the western part of the Russian plain, in the belt of coniferous and mixed forests.

During the last glaciation the area was covered by the Scandinavian glacier. The refuges of the hydrofauna from where the repopulation of the area proceeded were located to the south and east (in Central and North-East Europe). Under the glacier the earth crust subsided, rivers flowed towards the glacier where they, together with melting waters, formed periglacial lakes, the outflows of which proceeded along the edge of the ice. Obviously the rivers and periglacial lakes in the vicinity of the glaciers were populated by various oligochaetes, although due to the insufficiency of feed their abundance would be low. In any case, they constantly entered downstream periglacial lakes. 18,000 years ago the glaciers, melting, began to withdraw and during about 8,000 years slowly withdrew from the observed area, leaving a complicated set of periglacial lakes behind it. The lakes of glacial origin left behind surely got the nucleus of their oligochaete fauna immediately from the periglacial lakes.

10,000 years ago the glacier withdrew to the Baltic hollow. For the first 2,000 years the Baltic Sea was a fresh water periglacial lake. It became fresh even later, when the sea temporarily detached itself from the ocean again. Therefore it was a favourable route for the distribution for the fresh water fauna.

In Estonia *Rhyacodrilus coccineus* (common in the rivers and brooks flowing directly into sea) and a part of fresh water fauna of the islands have distributed via the Baltic Sea during its freshwater stages. It was more difficult for the oligochaetes to distribute to the Scandinavian Peninsula as the glacier withdrew uphill and no periglacial lakes were formed. Therefore the fauna of the large Scandinavian lakes is poorer than that of the Karelian lakes (Ladoga, Onega, etc.). Already before the withdrawal of the glacier the Kola Peninsula was detached by the salty White Sea. Here lies the reason why some fresh water species (among others *Lamprodrilus isoporus*) which occur in Karelia and even on the islands of the White Sea are absent on the peninsula.

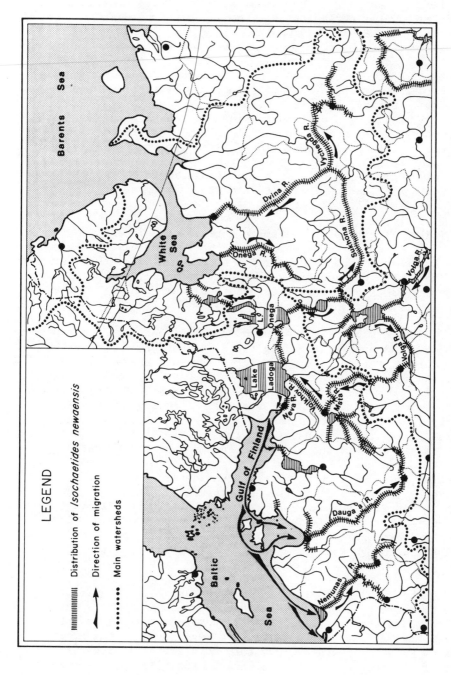

Figure 7.   The distribution and directions of migration of *Isochaetides newaensis* in the north-western part of the Soviet Union.

The water bodies of the Early Holocene were all oligotrophic,
their fauna being poor in specimens and rich in species. At
present such water bodies can even be found in the northern part
of the area, in Karelia and on the Kola Peninsula. Even in the
profundal many, among them exigent, species, can found: characteristic are *Stylodrilus heringianus*, *Peloscolex ferox*, *Tubifex
tubifex* while *Rhyacodrilus coccineus*, *Lamprodrilus isoporus*, *Lumbriculus variegatus*, *Alexandrovia onegensis*, etc. occur less
frequently. During the Holocene, noticeably quicker in the south,
the eutrophication and dystrophication of the lakes has happened
and the oxyphilous animals have disappeared from the profundal.
Only *Tubifex tubifex* and *Potamothrix hammoniensis* populate the
profundal of big mesotrophic and eutrophic lakes to the south of
the Finnish Gulf while only *P. hammoniensis* is to be found in
small eutrophic lakes. The latter tolerates long stagnation
periods best of all.

With warming up of the climate during the Holocene the local
fauna has been enriched by some late-comers (probably *Bothrioneurum
vejdovskyanum* and species of *Dero*). At the same time the thermophobe and oxyphilous species have either gradually become extinct
or their areas have decreased or become dismembered (*Tatriella
slovenica*, *Lamprodrilus isoporus*, etc.).

Nowadays human activities have rapidly accelerated eutrophication, simultaneously favouring the distribution of species by
the construction of canals and water reservoirs as well as by
navigation. The newcomers from the Ponto-Caspian subregion are
colonizing new water bodies at present: *Isochaetides newaensis*,
I. michaelseni, *Potamothrix moldaviensis*, *P. bavaricus*, *P. heuscheri*,
P. vejdovskyi, *Psammoryctides moravicus*, *Paranais frici*. For
example, *I. newaensis* probably got from the Volga to the basin
of the Baltic Sea via the navigable canals over the Valdai Heights
in the 18th century. Nowadays it is common in big navigable rivers
and mouths of small rivers of the north-western part of the Soviet
Union while it is absent in the upper reaches. In about 1970 *I.
newaensis* and *I. michaelseni* appeared in Lake Peipsi-Pihkva (Minina,
1975) and *P. moldaviensis* in Lake Vortsjarv for the first time
where they were absent earlier. By now they have been found
repeatedly although each of them in one spot only for the time
being.

The thermophile tropic species (some species of *Aeolosoma*,
Branchiura sowerbyi, perhaps *Aulophorus furcatus* and *Nais christi-
nae* also) form the last wave of antropochorous immigrants into
northern Europe. Earlier some of them were found in the green-
houses of botanical gardens, now they populate cooling water
bodies of electric stations. As regards the oligochaetes such
biotopes have been insufficiently studied in the area under
observation.

REFERENCES

Benham, W.B., 1903, A Note on the Oligochaeta of the New Zealand
    Lakes. Trans. Proc. N.Z. Inst., 36:192-198.
Bretscher, K., 1903, Tiergeographisches uber die Oligochaten.
    Biol. Centralblatt, 23:618-625.
Brinkhurst, R.O. and B.G.M. Jamieson, 1971, Aquatic Oligochaeta of
    the World. Edinburgh, 860 p.
Cekanovskaja, O.V., 1962, The Aquatic Oligochaeta of the USSR.
    (In Russian). Moskva-Leningrad, 411 p.
Gates, G.E., 1958, Contribution to a Revision of the Earthworm
    Family Lumbricidae. II. Indian Species. Breviora Mus. Comp.
    Zool., 91:1-16.
Izossimov, V.V., 1962, Oligochaetes of the Family Lumbriculidae
    (in Russian). Trudy Limnol. Inst., 1:3-126.
Jarvekulg, A., 1957, Uber das Vorkommen der Krebskiemenegel
    (Gattung Branchiobdella Odier) als Parasiten des Edelkrebses
    (Astacus astacus (L.), in Estland. (In Estonian). Eesti NSV
    Tead. Akad. Loodusuurijate Seltsi Aastaraamat, 50:249-256.
Michaelsen, W., 1903, Die geographische Verbreitung der Oligo-
    chaeten. Berlin, 187 S.
Minina, N.G., 1975, On the Horizontal Distribution of the Zoo-
    benthos in the Lake Pskovsko-Chudskoje.(In Russian). Trudy
    Pskov. otd. GosNIORH, 1:75-83.
Omodeo, P., 1963, Distribution of the Terricolous Oligochaetes
    on the Two Shores of the Atlantic. North Atlantic Biota and
    Their History. Oxford-London-N.Y.-Paris, 127-151.
Starobogatov, J.I., 1970, Fauna of Mollusca and the Zoogeogra-
    phical Division of the Continental Water Bodies. (In Russian).
    Leningrad, 372 p.
Stephenson, J., 1930, The Oligochaeta. Oxford, 978 p.
Zacharias, O., 1889, Ueber Anpassungserscheinungen im Hinblick
    auf passive Migration. Biol. Centralblatt, 9:107-113.

THE AQUATIC OLIGOCHAETA OF ARGENTINA:

CURRENT STATUS OF KNOWLEDGE

Danilo Héctor Di Persia

Centro de Ecología Aplicada del Litoral
Consejo Nacional de Investigaciones
Ciéntificas Y Técnicas (Conicet)
República Argentina

ABSTRACT

The distribution and ecology of the aquatic Oligochaeta of Argentina are reviewed, including the families of Naididae, Tubificidae, Opistocystidae, Phreodrilidae, Haplotaxidae, Alluroididae and Aeolosomatidae. Habitat preferences are described for pleuston and benthos.

INTRODUCTION

The aquatic fauna of Oligochaeta has been relatively little studied in the Argentine Republic, as well as in the rest of South America, where the current state of its knowledge, even when considerably improved during the last years, is still far from being satisfactory. Several studies carried on by foreign scientists during the end of the last century and first half of this one comprised small collections, mostly geographically restricted. The interest in this group has increased during the last decade, but still a large part of the Argentine territory has not been studied at all in this aspect of its fauna, and that is shown by the fact that there are no records of aquatic oligochaetes from eleven of the twenty three Argentine provinces.

As my activities take place basically at the NE of Argentina,
my knowledge of that region is deeper than on the rest of the
country, particularly concerning the Naididae, Aeolosomatidae.
Opistocystidae and Tubificidae, families which have received most
attention. The species of Haplotaxidae collected from that region
is also the only record of that family. From the NW of the country
(particularly the provinces of Tucumán, Salta and Jujuy) all that
is known, specially about Tubificidae, is the result of studies
carried on by Dr. K. Gavrilov and collaborators. The few known
species of Alluroididae and Phreodrilidae have only been reported
once.

SYSTEMATICS

Family Aeolosomatidae

Revised by Bunke (1967) and considered by Van der Land (In
Brinkhurst and Jamieson, 1971) it is known from Argentina with a
single genus, represented by the following species (Figure 1).

Genus *Aeolosoma* Ehrenberg, 1831

Aeolosoma flavum Stolc, 1903

It has been collected from several habitats of the Argentine
fluviatic littoral, as follows: Entre Ríos Province, Paraná
Department, pools of Bajada Grande, on the algal-crust developed
upon gasteropods of the genus *Ampullaria* Lamarck (Di Persia and
Radici de Cura, 1973). Santa Fe Province, La Capital Department,
Don Felipe oxbow Pond and La Guardia and Redonda Ponds, being
part of the pleustonic fauna on carpets of *Azolla caroliniana*
and *Salvinia herzogii*, and on the algal-crust upon Ampullariidae
(Di Persia, 1974a, b; Di Persia and Radici de Cura, 1973).
Corrientes Province, San Cosme Department, town of San Cosme,
Cabrera and Sotelo Ponds. In the former one, numerous populations
of it were living together with other species of Aeolosomatidae
and diverse invertebrates upon colonies of the poriferan *Corvomeyenia australis*, developed on submerged parts of the gramineous
Glyceria multiflora. In Sotelo Pond the species was found among
roots of *Salvinia herzogii* (Di Persia, 1976b).

Distribution: it has been recorded from Germany and Czechoslovakia, while in South America it is known only from the above
mentioned Argentine localities.

Aeolosoma beddardi Michaelsen, 1900

This species has been found in different aquatic habitats of
the NW of Corrientes Province: San Cosme Department, town of San
Cosme, on submerged stems of *Glyceria multiflora,* species which

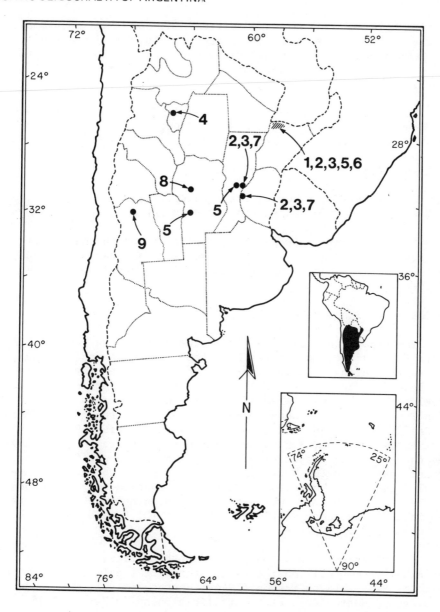

Figure 1. Geographic distribution of Aeolosomatidae
1. *Aeolosomatidae*; 2. *A. evelinae*; 3. *A. flavum*;
4. *A. hemprichi*; 5. *A. marcusi*; 6. *A. sawayai*;
7. *A. travancorense*; 8. *A. quaternarium*;
9. *Aeolosoma* sp.

forms a marginal belt around Cabrera Pond; San Cosme Department,
nearby Santa Ana, in ponds of Parque Leconte, among roots of
Eichhornia crassipes; Capital Department, a few kilometers away
from Corrientes City, Brava Pond, on *E. crassipes* (Di Persia,
1976b).

Distribution: it was found in England, China and, in South
America, in Uruguay (Cordero, 1951; Dioni, 1960), Brazil (Marcus,
1944), and Surinam (Van der Land, 1971).

Aeolosoma hemprichi Ehrenberg, 1828

Cordero (1951) reported its finding at Tucumán Province, and
it has not been recorded subsequently in the country.

Distribution: cosmopolitan species, in South America it was
registered from Brazil, states of Sao Pablo and Paraná (Marcus,
1944) and Río Grande do Sul (Cordero, 1951), and from Uruguay
(Cordero, 1951).

Aeolosoma evilinae Marcus, 1944

The records of this species in the country are several. At
Santa Fe Province, La Capital Department, Redonda Pond, among
roots of floating aquatic plants (Dioni, 1967), and La Guardia
and Redonda Ponds and Don Felipe oxbow Pond, as part of pleuston
developed on *Salvinia herzogii* and *Azolla caroliniana*. Entre
Ríos Province, Paraná Department, from pools in the neighbourhoods
of Bajada Grande, on *Ludwigia peploides* (Di Persia, 1974a).
Corrientes Province, San Cosme Department, Cabrera Pond, living
together with other species of this genus on the poriferan *Corvomeyenia australis* (Di Persia, 1976b). It was also found on the
algal-crust upon the valves of the three species of gasteropods
of the genus *Ampullaria* which thrive in high numbers in different
habitats of the above mentioned provinces (Di Persia and Radici
de Cura, 1973).

Distribution: known only from South America, it was
described from Brazil on the basis of material collected from the
state of Sao Paulo by Marcus (1944); subsequently it was found
in Uruguay, Montevideo, at an artificial pond in Parque Rodó
(Dioni, 1960).

Aeolosoma travancorense Aiyer, 1926

This species was recorded at Santa Fe Province, La Capital
Department, among roots of floating aquatic plants in Redonda
Pond (Dioni, 1967). I have collected it at the same province,
from El Alemán, Redonda and La Guardia Ponds, Don Felipe oxbow
Pond, and temporary and permanent pools by National Road 168,

way to the subfluvial Tunnel Hernandarias, being part of the
pleustonic fauna developed on *Eichhornia crassipes, A. caroliniana*
and *S. herzogii*, and also at Entre Rios Province, Parana Department, from Laguna Grande and pools at Bajada Grande, on *Ludwigia
peploides,* on submerged parts of stems of *Solanum malacoxylon*,
and in soft muddy bottoms (Di Persia, 1974a). It was found with
other species of *Aeolosoma* among the epibiontic algae growing
upon the molluscs Ampullariidae (Di Persia and Radici de Cura,
1973), in different water bodies connected to the alluvial valley
of Parana River, along the tract between the cities of Corrientes
and Santa Fe.

Distribution: it is known from Asia, Europe and South America.
The records in this continent comprise Brazil, states of Sao Paulo
and Paraná (Marcus, 1944); Uruguay, Montevideo City, from the artificial pond in Parque Rodó (Dioni, 1960), and Surinam, where it is
the most common species of the genus, according to Van der Land
(1971).

Aeolosoma sawayai Marcus, 1944

I have found this species only at the NW of Corrientes
Province, at San Cosme Department, town of San Cosme, on submerged
stems of *Glyceria multiflora*, together with colonies of poriferans,
which thrive in high numbers in Cabrera Pond, and at Itatí Department, in pools nearby the road to the town of Itatí, where it was
collected from stems and the back of leaves of *Nimphoides indica*
(Di Persia, 1976b).

Distribution: South American species, it is known from Brazil,
states of Paraná and Sao Paulo (Marcus, 1944; Du Bois-Reymond
Marcus, 1944); Uruguay, Canelones Department, in small pools by
the road between bathing resorts (Dioni, 1960); Surinam, where it
is very common (Van der Land, 1971).

Aeolosoma marcusi Van der Land, 1971

This is the first record of the species from Argentina, where
it was collected at Santa Fe Province, La Capital Department, a
few kilometers from Recreo Town, in pools with vegetation at the
left side of Salado River, close to National Road 166, 7 November
1976, col. B.R. de Ferrádas, G. Martines and J.C. Poledri. 2.
Corrientes Province, basin of Riachuelo Stream, from several
lentic water bodies at the NW of the province, 17 January 1976,
col. D.H. Di Persia and N.T. Roberto. 3. Córdoba Province,
Río Cuarto Department, small pond with vegetation about 30 km
away from Río Cuarto City, January 1979, col. C.M. Gualdoni.

These specimens, visible to the naked eye, reached 3.5-7 mm
as single individuals, and may attain 9-10 mm in chains. I have
not found more than two zooids in collected samples and in material kept in laboratory since 1976 up to date, even when the
literature mentions up to five zooids. Width varies from 200 μ
to 450 μ on the same individual, being   fully stretched or
contracted. The epidermal glands, of different size and irregular
shape, are regularly scattered on the body, accumulating at the
prostomium and the caudal end. Colour shows several shades of
green and bluish green. Colourless glands are also observed,
mostly on the prostomium. Prostomium rounded, slightly wider
than following segments, with cilia. Two lateral ciliated fields
are present, with latero-dorsal continuations and large sensory
pits. Mouth placed posteriorly, with shape of pentagon or half
moon, followed by a globulose pharynx which continues in a narrow
intestine, widening abruptly from the posterior part of IV to XI,
then narrowing backwards. This widened zone of the specimens
shows longitudinal irregularities which give it a striated aspect.
The caudal end stretches and contracts constantly during displacement, and shows two dorsally placed ciliated lobules with adhesive
functions, which allow the worms to remain adhered to the glass
walls, when the container is shaken. Fission-zone usually in
segment XIV or XV (XIII to XVIII in the literature). Setal
bundles of only hairs, smooth, of variable length, reaching about
500 μ at most, 8-10 (occasionally 11) per bundle. Very evident
dissipiments, specially in second zooid, where segments are
shorter, there is no intestinal dilatation and dissipiments
appear as clearly distinguishable walls. Nephridia usually present in all segments, beginning in II/III.

Distribution: Brazil, Sao Paulo State, Sao Paulo City and
neighbourhoods, among roots of *E. crassipes* (Marcus, 1944), and
Uruguay, from an artificial Pond at Parque Rodó, Montevideo
(Dioni, 1960).

Aeolosoma quaternarium ? Ehrenberg, 1831

When referring to the Argentine microscopic fauna, Frenzel
(1891) mentioned the presence of this species at Córdoba Province,
even with doubts. It has been recorded posteriorly in the country.

Distribution: as pointed out by Van der Land (In Brinkhurst
and Jamieson, 1971) it is a species known with certainty only
from some European countries.

Aeolosoma sp.

Several specimens clearly attributable to this genus were
collected at Mendoza Province, Capital Department, Parque General
San Martín of Mendoza City, from an artificial pond, among

filamentous algae. This material could not be further identified
for its deterioration due to fixation. It is the first record of
the genus from the W of the country.

Family Naididae

It comprises four subfamilies (Sperber, 1948; Brinkhurst and
Jamieson, 1971) of wide distribution all over the world, three of
which are present in Argentina with the ten following genera:
*Chaetogaster, Amphichaeta, Homochaeta, Nais, Slavina, Stylaria,
Allonais, Dero, Stephensoniana* and *Pristina* (Di Persia, 1978).
Their distribution in Argentine Provinces is shown in Figure 2.

Subfamily Chaetogastrinae Sperber, 1948

Genus *Chaetogaster* von Baer, 1827

Chaetogaster cristallinus Vejdovsky, 1883

Santa Fe Province, La Capital Department, at Don Felipe oxbow
Pond, among roots of *S. herzogii* at La Guardia and Redonda Ponds,
on floating vegetation (*S. herzogii, A. caroliniana* and *E. crassipes*) and at Saladillo 3° Stream, on carpets of *Hydromistria stolonifera, Azolla caroliniana* and *S. herzogii*. Corrientes Province,
Itati Department, pools close to Itatí, on *Nimphoides indica*;
Capital Department, Brava Pond, among roots of *Eichhornia crassipes* and *S. herzogii*; San Cosme Department, Cabrera Pond, on submerged stems of *Glyceria multiflora*, living together with several
species of Aeolosomatidae and Naididae, on freshwater poriferans.
It was also found in several temporary and permanent pools and
ponds within the basin of Riachuelo Stream, at NW of that Province
(Di Persia, 1975b).

Distribution: it was recorded from Europe, Asia, North America
and Africa. In South America it is known only from the above
mentioned localities.

Chaetogaster limnaei von Baer, 1827

Registered at several lentic and lotic water bodies of the
Paraná River's alluvial valley, along the tract between the
cities of Corrientes and Santa Fe, being part of the algal-crust
upon molluscs of the genus *Ampullaria* (Di Persia and Radici de
Cura, 1973).

Distribution: present in Europe, Asia, North America, Australia and, in South America, in Brazil, Sao Paulo State, on Planor-
bidae of the genus *Australorbis* (Du Bois-Reymond Marcus, 1947),
and Uruguay, on *Ancylus* (Cordero, 1951).

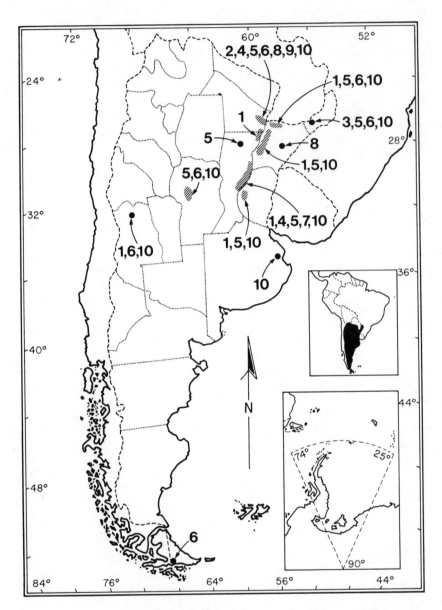

Figure 2.   Geographic distribution of genera of Naididae
            1. *Chaetogaster*; 2. *Amphichaeta*; 3. *Homochaeta*
            4. *Stylaria* 5. *Dero* 6. *Nais* 7. *Allonais*
            8. *Slavina* 9. *Stephensoniana* 10. *Pristina*.

Chaetogaster diaphanus (Gruithuisen, 1828)

This species was recorded at Mendoza Province, Maipú Department, nearby Maipú Town, from the margins of Mendoza River, among filamentous algae, together with other Naididae (*Nais elinguis* and *Pristina proboscidae*) (Di Persia, in press c).

Distribution: it is found in several countries in Europe, Asia, and North America; this record from Mendoza Province is the first one in the country; previously known from Uruguay (Cordero, 1951).

Genus *Amphichaeta* Tauber, 1879

Amphichaeta leydigii Tauber, 1879

It was collected at Chaco Province, from Negro River, short before its flowing into Barrangueras River, where it was part of the bottom fauna (Di Persia, 1977c).

Distribution: it is known from several European countries (Denmark, Poland, Germany, Czechoslovakia, Switzerland, Russia). Its presence in Argentina is the first record of the genus in the continent.

Subfamily Naidinae Lastockin, 1924

Genus *Homochaeta* Bretscher, 1896

Homochaeta lactea (Černosvitov, 1937)

Described by Černosvitov (1937b) from a single specimen collected from Misiones Province, near to the town of Loreto, where it was found in a small creek which dries up in summer. It has not been recorded again, and is considered as *incerta sedis* (In Brinkhurst and Jamieson, 1971).

Distribution: known only from the typic locality.

Genus *Nais* Müller, 1773

Nais communis Piguet, 1906

It was mentioned from the Patagonia by Michaelsen and Boldt (1932) without giving locality, and it has not been recorded again.

Distribution: cosmopolitan species, it has been found in South America at Peru, Huaron Lake (Piguet, 1928) and Brazil, states of Sao Paulo and Pará (Marcus, 1943, 1944; Du Bois-Reymond Marcus, 1949).

Nais variabilis Piguet, 1906

Recorded previously at Misiones Province, Candelaria Department, Loreto Town, from a small creek, where one sexually mature
specimen was collected (Černosvitov, 1937b). I have subsequently
found it at Córdoba Province, Punilla Department, San Roque Reservoir, from littoral benthic fauna; at Entre Ríos Province, Concordia Department, Auí Grande Stream (Di Persia, 1975c), and at Chaco
Province, on the bottom of Negro River (Varela et al, in press).

Distribution: cosmopolitan, in South America it was recorded
from Peru, Huaron Lake (Piguet, 1928).

Nais pseudobtusa Piguet, 1906

Corrientes Province, San Cosme Department, town of San Cosme,
Cabrera Pond, living with other species of Aeolosomatidae and
Naididae upon colonies of *Corvomeyenia australis* (Di Persia,
1975c).

Distribution: known from several countries of Europe, North
America, Africa and Asia. There has not been any other record
from South America so far.

Nais elinguis Müller, 1773

Michaelsen (1903) reported about it from the actual Territorio Nacional de Tierra del Fuego, Antártida e Islas del Atlántico Sur. I have found it at Mendoza Province, Maipú Department,
by Mendoza River, next to the town of Maipú; Capital Department,
Parque General San Martín, in samples collected from an artificial
pond built with recreational purposes; Guaymallén Department,
close to Aldea La Primavera, and Las Heras Department, from ditches
at El Challao. In all these collections, the specimens were among
filamentous algae and in benthos with plant debris. It was by far
the best represented species at Mendoza Province, being always in
high numbers (Di Persia, in press c).

Distribution: cosmopolitan, in South America it was registered
in Titicaca Lake, Peru (Černosvitov, 1939).

Genus *Slavina* Vejdovsky, 1883

Slavina evelinae (Marcus, 1942)

Even when Du Bois-Reymond Marcus (1947) mentioned Argentina
within this species distribution, according to Cernosvitov, *in
litt.*, there were no certain records of it. I have found it at
Chaco Province, La Cava Pond, with fauna associated with the roots
of *Pistia stratiotes*; Barranqueras Pond, in carpet of *E. crassipes*,

and Tragadero River, little before its mouth, within the bottom
fauna. Corrientes Province, heads of Mirinay channel, in benthic
fauna of "embalsados" with abundant debris of *Typha angustifolia*,
T. latifolia and *Fuirena robusta* (Di Persia and Varela, in press).
"Embalsados" are formations which constitute sorts of floating
islands, being sometimes displaced by currents of the waters or
the action of the winds, formed by an organic soil of variable
thickness, with very dense and varied vegetation, dominated by
amphibian species, including even shrubs when reaching a large
size. On that soil it is usual to find small shallow pools where
a many varied biota may develop. Burkart (1957), when considering
the embalsados, stated that they are true thick mattresses of
rhizomes, roots and debris, strongly bound together, which float
and may drift.

Distribution: South American species, it is known from Brazil,
states of Sao Paulo and Ceará (Marcus, 1942), Pernambuco State,
among *Eichhornia* (Marcus, 1944) and Pará State (Du Bois-Reymond
Marcus, 1947, 1949), and from Surinam, close to Paramaribo, in
sites of standing water with dense vegetation (*Typha, Cyperus
gigantea* and *Eichhornia*) (Harman, 1974).

Slavina isochaeta Černosvitov, 1939

This species, recorded now for the first time from Argentina,
was collected at Chaco Province, Libertad Department, the benthos of
Negro River nearby the bridge of National Road 16, 19 August 1976,
col. N.T. Roberto. That place is within a section where the river
does not receive any sort of pollutant effluents derived from human
activities. It shows a narrow course, flanked by low ravines,
with marginal patches of *Paspalum repens, Hydromistria stolonifera*
and *S. herzogii* and small isolated groups of *Victoria cruziana*.

The specimens are large, up to 6.5 mm of length, and appear
covered by cutaneous secretions with adhering foreign matter.
Papillae of about 20 μ high and 10 μ of maximum width are also
observed. The prostomium is rounded. Dorsal setae beginning at
IV, usually 3 hairs (occasionally 2), and 2-3 needles per bundle.
Hairs smooth, stout and stiff, of 3.7-5 μ width next to their
bases, with length of 275-420 μ, not varying much within the same
bundle. Černosvitov (1939) states that the setae vary between
180 and 350 μ. Needles tapering towards the distal end, of 77-95
μ length, this being a wider range than the one stated in the
original description (70-80 μ). Ventral setae, 4-5 per bundle;
in II-V usually 4 per bundle, 120-137μ (in the specimens from
Peru 105-115 μ), with distal tooth about one third longer than
the proximal, and thinner; slightly proximal nodulus. The width
varies between 3.7 and 5 μ. On the following segments, usually
5 setae per bundle (sometimes 4), of 87-100 μ length, in a range
slightly wider than the one stated by Cernosvitov, 92-98 μ.

Width similar to the anterior setae. Distal tooth thinner than
proximal, and a little longer. In the present material, measurements of the setae as well as body length of preserved worms
(2.5-4 mm in the specimens from Peru, up to 6.5 mm in the Argentine
ones) are larger than what is stated by Černosvitov.

Distribution: known before now only from Titicaca Lake, Peru,
from where it was considered as endemic species by Brinkhurst (In
Brinkhurst and Jamieson, 1971).

Genus *Stylaria* Lamarck, 1816

Stylaria lacustris (Linnaeus, 1767)

It was found at Chaco Province, San Fernando Department, in
lentic water bodies nearby Negro River, where the specimens were
upon *Victoria cruziana*, and at Santa Fe Province, La Capital
Department, Saladillo 3° Stream, among *Hydromistria stolonifera*
and *Salvinia herzogii* (Di Persia, 1975b).

Distribution: known from Europe, W of Asia and North America.
This is the only record from South America.

Genus *Dero* Oken, 1815

Subgenus *Dero* Oken, 1815

Dero (Dero) digitata (Muller, 1773)

Collected at Corrientes Province, Itatí Department, from
sideroad trenches on the way to Itatí Town, on the back of leaves
of *Nimphoides indica,* inside of mucous tubes with adhering foreign
matter, clustered together in parallel way, living with other tube-
building species of this genus (Di Persia, 1977a). Misiones
Province, Guayavera Creek, tributary of Yabebíri Stream, described
as *Dero quadribranchiata* by Černosvitov (1937b), species considered
as synonym by Sperber (1948) and Brinkhurst and Jamieson (1971).

Distribution: cosmopolitan. Found at South America in Paraguay (Michaelsen, 1905b), Uruguay (Cordero, 1931, 1951), Brazil,
states of Pará and Amazonas (Du Bois-Reymond Marcus, 1947, 1949),
and Surinam, where it would be abundant (Harman, 1974).

Dero (Dero) cooperi Stephenson, 1932

Santa Fe Province, La Capital Department, collected from pond
by National Road 168 between Santa Fe and Paraná Cities, on floating hydrophytes, together with other species of Naididae (Di Persia
and R. de Ferradás, in press).

Distribution: species known before only from Africa (Abyssinia) (Stephenson, 1932). This is the only record from South America.

Dero (Dero) nivea Aiyer, 1929

The specimens were found in the bottom fauna, at Entre Rios Province, Concordia Department, upper section of Ayuí Grande Stream (Di Persia, 1977a); Chaco Province, Negro River (Varela et al, in press), and Santa Fe Province, La Capital Department, Saladillo 3° Stream, where they were also on carpets of *A. caroliniana*, *S. herzogii* and *Hydromistria stolonifera* (Di Persia, 1977a).

Distribution: present in Europe, Asia, North America, Africa, Australia and South America, where it was reported in several localities of the N of Surinam (Harman, 1974).

Dero (Dero) obtusa d'Udekem, 1855

It was mentioned for Argentina by Cordero (1951) without stating locality, and later on found at two provinces. Córdoba, Departments of Punilla and Calamuchita, in San Roque, Los Molinos and Embalse del Río Tercero man-made lakes, as part of the benthic fauna in shallow areas of the three reservoirs. It was particularly numerous in the zone under the influence of San Antonio River and Los Chorrillos Stream, tributaries of San Roque Lake, where abundant plant debris was present (Bonetto and Di Persia, in press), and in samples collected nearby Los Reartes River and Del Medio River, in Los Molinos Lake, where it was the dominant species of oligochaetes. At Embalse del Río Tercero Lake it was present in low numbers (Bonetto et al, 1976). Corrientes Province, Itati Department, in sideroad pools on the way to Itatí, living together with other tube-building species of oligochaetes, on the back of leaves of *Nimphoides indica* (Di Persia, 1976a).

Distribution: known from Europe, North America, Asia, Africa and South America, where it was found in Brazil, Sao Paulo State (Marcus, 1943) and in different localities of Surinam (Harman, 1974).

Dero (Dero) multibranchiata Stieren, 1892

Recorded at Corrientes Province, Itatí Department, from pools close to Itati Town, on *N. indica*. Santa Fe Province, La Capital Department, from La Guardia Pond, on carpets of *A. caroliniana* and *S. herzogii*; from several pools by National Road 168 between Santa Fe and Paraná, on *E. crassipes*, *S. herzogii* and *A. caroliniana*, and from Saladillo 3° Stream, among *A. caroliniana*, *S. herzogii* and *Hydromistria stolonifera* (Di Persia, 1977a). Chaco Province, in the bottom fauna of Negro River (Varela et al, in press).

Distribution: it is known from Neotropical America, where it
was previously recorded from Trinidad (Stieren, 1892) and Brazil,
state of Pernambuco, Pará and Amazonas (Marcus, 1944; Du Bois-
Reymond Marcus, 1947, 1949).

Dero (Dero) plumosa Naidu, 1962

Up to now it has been found only at Corrientes Province,
Itati Department, nearby Itatí Town, in ponds with *N. indica*, on
the back of their leaves, together with other species of oligochaetes (Di Persia, 1976a).

Distribution: known from Asia, described on material from the
S of India (Naidu, 1962b). This is the only record from South
America.

Dero (Dero) evelinae Marcus, 1943

Cordero (1951) mentioned it for Argentina, and I have found it
at the NE of the country: Santa Fe Province, La Capital Department,
La Guardia Pond and Don Felipe oxbow Pond, and several pools by
National Road 168, in pleuston developed on carpets of *S. herzogii*
and *A. caroliniana*; Entre Ríos Province, Paraná Department, temporary pools at Bajada Grande (Di Persia, 1974a, 1974b).

Distribution: described from the Brazilian states of Sao Paulo
and Río Grande do Sul (Marcus, 1943), posteriorly was recorded from
the states of Paraná and Pará (Du Bois-Reymond Marcus, 1949). It
has also been mentioned for Uruguay (Cordero, 1951).

Dero (Dero) botrytis Marcus, 1943

To the present, it had been found at Corrientes Province,
San Cosme Department, in vegetated pools at Parque Leconte, as
part of the pleustonic fauna developed among roots of *S. herzogii*
(Di Persia, 1976a).

New Record: Santa Fe Province, La Capital Department, Saladillo 3º Stream and nearby pools, 26 April 1976, col. B.R. de
Ferradás, G. Martinez and D.H. Di Persia, on *S. herzogii*, *H. stolonifera* and *A. caroliniana,* present in low numbers, over 500 km
southern of the former record.

Distribution: described by Marcus (1943) upon material from
Brazil, states of Sao Paulo and Rio Grande do Sul, and later on
mentioned for Pernambuco State (Marcus, 1944), and for Uruguay
(Cordero, 1951).

Subgenus *Aulophorus* Schmarda, 1861

Dero (Aulophorus) furcatus (Muller, 1773)

Recorded at Misiones Province from a creek next to Loreto (Cernosvitov, 1937b), and at Entre Ríos Province (Cernosvitov, 1942). I found it at Santa Fe Province, La Capital Department, where it was collected from Don Felipe oxbow Pond and from the ponds Redonda, La Guardia and El Alemán, in bottom and among roots and stems of *E. crassipes* and *Reussia rotundifolia* (Di Persia, 1974a, 1974b).

Distribution: known from Europe, Africa, Asia, North America and Neotropical America, where it is widespread, with records from Trinidad (Stieren, 1892); Uruguay (Cordero, 1931); Bonaire (Michaelsen, 1933); Brazil, states of Sao Paulo (Marcus, 1943) and Pernambuco (Marcus, 1944, and Surinam (Harman, 1974).

Dero (Aulophorus) hymanae Naidu, 1962

In the country, the only record of this species is from Ayuí Grande Stream, Concordia Department, Entre Ríos Province, where it was found in the benthic fauna (Di Persia, 1975a).

Distribution: Asia, S of India (Naidu, 1962). This is the only record from South America.

Dero (Aulophorus) carteri Stephenson, 1931

Recorded at Corrientes Province, San Cosme Department, from several ponds of Parque Leconte, nearby Santa Ana Town, and from Sotelo Pond, San Cosme Town, where the specimens were collected from carpets of *S. herzogii* (Di Persia, 1975a), and Capital Department, Brava Pond, where *S. herzogii* was forming a marginal belt (Poi de Neiff, 1977).

Distribution: species known only from South America: Paraguay (Stephenson, 1931); Brazil, Sao Paulo State (Marcus, 1943), states of Pernambuco and Piauí (Marcus, 1944), and Surinam (Harman, 1974).

Dero (Aulophorus) indicus Naidu, 1962

Corrientes Province, San Cosme Department, Cabrera Pond, found together with several Aeolosomatidae and Naididae on poriferans developed upon submerged parts of *Glyceria multiflora* (Di Persia, 1975a).

Distribution: known from Asia, S of India, where it was described by Naidu (1962c). This is the only record from South America.

Dero (Aulophorus) schmardai (Michaelsen, 1905)

The only record of this species in Argentina was made at
Santa Fe Province, La Capital Department, in a pond close to the
subfluvial Tunnel Hernandarias, nearby National Road 168, as part
of the pleuston. That pond was covered by floating hydrophytes,
with dominance of *A. caroliniana, S. herzogii* and *E. crassipes*
(Di Persia and R. de Ferradas, in press).

Distribution: there is certain confusion about this species,
considered as *species inquirenda* by Brinkhurst and Jamieson (1971),
and recorded under different names from Paraguay (Michaelsen,
1905a; Piguet, 1928), Peru (Piguet, 1928) and Brazil (DuBois-
Reymond Marcus, 1944).

Dero (Aulophorus) superterrenus Michaelsen, 1912

This species is here reported in its first record from this
country, where it was collected in several places of the fluvial
littoral. 1. Entre Ríos Province, Diamante Department, town of
Villa Libertador General San Martín, in the water-basins of terrestrial bromeliads of the genus *Bromelia*, 13 November 1977, col.
D.H. Di Persia and S.M. Storani. 2. Entre Ríos Province, Paraná
Department, at the locality of Paracao, in bromeliads of the same
genus, 18 March 1978, col. D.H. Di Persia. 3. Corrientes Province,
Capital Department, 10 May 1978, col. D.H. Di Persia and L. Benetti.
4. Chaco Province, General Donovan Department, 15 January 1977,
col. N.T. Roberto and D.H. Di Persia. 5. Santa Fe Province,
Vera Department, along roads nearby the Reserve Provincial Laguna
La Loca, 1 February 1979, col. G. Martinez and D.H. Di Persia.
In these cases, the specimens were collected from water-basins on
leaves of *Aechmea distichantha*.

Preserved specimens of 10-15 mm of length, formed by 53-98
segments, with prostomium rounded or somewhat conical, at times.
Dorsal setae beginning in IV, 1 (less frequently 2) hair, smooth,
of 350-440 μ length, and 1-2 bifid needles, of 52.5-65 μ, per
bundle. Teeth of the needles small. Nodulus very distinct, on
the distal fourth of setae. Ventral setae bifid, 5-7 per bundle
in II-V, 70-87.5 μ in length; in the following segments usually
6, scattering to 4 per bundle posteriorly, of 80-87.5 μ. Marcus
(1943) stated the number of setae to be 8-10 per bundle in II-V,
scattering backwards. In my specimens, the number of setae agrees
with that reported by Michaelsen (1912) from Costa Rica (5-7)
although his worms showed a wider range of length than mine (up
to 95 μ in VI). Ventral setae all of similar shape, with proximal
tooth a little thicker than distal and some longer in the first
segments, becoming then of about the same length in the posterior
segments. Nodulus distal in the anterior segments, approximately
median or slightly distal backwards, in the middle and posterior

segments. Marcus (1943) reported the presence of penial setae in worms with clitellum, not present in the Argentine specimens; these setae would be longer (120 μ), straight and considerably thicker, with distal nodulus and a single prong.

Distribution: in Neotropical America it has been recorded at Costa Rica, on epiphytic bromeliads of the genus *Vriesea* (Michaelsen, 1912; Picado, 1913); Brazil, states of Sao Paulo, Paraná and Pernambuco, on epiphytic as well as terrestrial bromeliads (Marcus, 1943, 1944). The species is also known from Malaysia, Selangor, where it was collected from a hollow on a tree (Stephenson, 1913a).

Genus *Allonais* Sperber, 1948

Allonais lairdi Naidu, 1965

It was collected at Entre Ríos Province, Concordia Department, from the upper section of Ayuí Grande Stream, in bottom fauna (Di Persia, 1975b).

Distribution: species known from Asia (S of India) (Naidu, 1965), and South America, with this only record.

Allonais paraguayensis (Michaelsen, 1905)

The only record of this species was made at Santa Fe Province, La Capital Department, from a pond by National Road 168, next to the Subfluvial Tunnel Hernandarias, where it was part of the pleustonic fauna developed on *S. herzogii, A caroliniana* and *E. crassipes*. (Di Persia and R. de Ferradas, in press).

Distribution: its presence is known in Asia, Africa, North America and South America. In this continent, the records were made at Paraguay (Michaelsen, 1905a; Stephenson, 1931); Brazil, states of Sao Paulo (Marcus, 1943), Alagôas, Pernambuco and Ceará (Marcus, 1944), Pará and Amazonas (Du Bois-Reymond Marcus, 1947, 1949); Uruguay (Cordero, 1951), and several localities in Surinam, where it would be abundant (Harman, 1974).

Genus *Stephensoniana* Černosvitov, 1938

Stephensoniana trivandrana (Aiyer, 1926)

The species were found at Chaco Province, as part of the bottom fauna of the Negro River (Di Persia, 1977b).

Distribution: it was previously known from India, Palestina (actual Israel), Republic of South Africa, and in South America from Surinam (Harman, 1974).

Subfamily Pristininae Lastockin, 1924

Genus *Pristina* Ehrenberg, 1828

Pristina osborni (Walton, 1906)

Córdoba Province, Punilla Department, where it was collected
from San Roque man-made Lake (Di Persia, 1973), in low numbers,
in shallow sampling sites with plant debris (Bonetto et al, 1976;
Bonetto and Di Persia, in press). It was recently recorded at
Mendoza Province, Capital Department, Parque General San Martin,
as part of the bottom fauna on an artificial pond (Di Persia,
in press c).

Distribution: it comprises Asia, North America, Africa and,
in South America, Brazil, states of Sao Paulo (Marcus, 1943) and
Minas Gerais (Righi, 1973) where it was recorded as *Pristina
minuta*, found in moist soils close to Capivara River.

Pristina proboscidea Beddard, 1896

Found at Misiones Province, Candelaria Department, creek
close to Loreto Town (Černosvitov, 1937b); Santa Fe Province, La
Capital Department, Redonda and La Guardia Ponds and Don Felipe
oxbow Pond, in the bottom fauna as well as on floating aquatic
plants (*S. herzogii* and *A. caroliniana*) (Di Persia, 1974a);
Corrientes Province, Capital Department, Brava Pond, on *E. crassipes*; Entre Ríos Province, Concordia Department, heads of Ayuí
Grande Stream, of dense vegetation (Di Persia, in press b);
Mendoza Province, Maipu Department, nearby Maipu Town, from the
margins of Mendoza River, among filamentous algae (Di Persia,
in press c).

Distribution: known from Asia, Africa and Australia, it was
recorded from several South American countries: Chile, Valparaiso,
(Beddard, 1896); Paraguay (Michaelsen, 1905c); Brazil, states of
Pará, Ceará and Amazonas (Marcus, 1943; Du Bois-Reymond Marcus,
1947), and Surinam (Harman, 1974).

Pristina americana Černosvitov, 1937

In Argentina, it has been recorded only from Misiones Province, close to Loreto, in the Guayavera Creek, tributary of
Yabebirí Stream, in Pastora Creek and in a small pool of stagnant
water which dries up in summer (Cernosvitov, 1937b).

Distribution: it may be present in Africa, Chad and Central
African Republic (Grimm, 1974). In South America, it is known
from Brazil, states of Sao Paulo (Marcus, 1943), Pará and Alagôas
(Marcus, 1944; Du Bois-Reymond Marcus, 1947, 1949), and Surinam
(Harman, 1974).

Pristina notopora Černosvitov, 1937

The species was described from mature specimens collected from Misiones Province, Candelaria Department, close to Loreto, in Pastora Creek, which becomes dried in summer (Černosvitov, 1937b). It has not been collected again.

Distribution: there is a recent record of it from Africa, Central African Republic (Grimm, 1974).

Pristina menoni (Aiyer, 1929)

It was found at Entre Ríos Province, Concordia Department, heads of Ayuí Grande Stream, among abundant aquatic plants, floating and rooted. Corrientes Province, Capital Department, Brava Pond, on roots of *E. crassipes*. Santa Fe Province, La Capital Department, pools with dense vegetation by National Road 168, close to Santa Fe City (Di Persia, in press b).

Distribution: known from Europe, Asia and Africa; in South America it has also been recorded from Surinam (Harman, 1974).

Pristina breviseta Bourne, 1891

I have collected this species at Santa Fe Province, La Capital and Las Colonias Departments, from pools at both margins of Salado River, a few kilometers from the town of Recreo, on carpets of *S. herzogii*, and at Chaco Province, in the bottom fauna of Negro River (Di Persia, in press b).

Distribution: species present at the S and E of Asia, North America and, in South America, at Brazil, states of Sao Paulo (Marcus, 1943) and Pará (Du Bois-Reymond Marcus, 1949).

Pristina jenkinae (Stephenson, 1931)

The first record of this species for the country was recently made by myself (Di Persia, in press b) at Chaco Province, San Fernando Department, from a lentic water body nearby the mouth of Negro River, in the neighbourhoods of Barranqueras Town, in water accumulated upon leaves of *Victoria cruziana*.

Distribution: it is found in Asia, Africa and, possibly Europe, while in South America it had been previously recorded from Brazil, states of Sao Paulo and Pernambuco (Marcus, 1943, 1944).

pristina macrochaeta Stephenson, 1931.

    Entre Ríos Province, Concordia Department, Ayuí Grande Stream,
in marginal places with *E. crassipes, A. caroliniana* and *S. herzogii*; Corrientes Province, pools close to Goya City, among filamentous algae, and Itatí Department, in pools nearby Itati, among
leaves of *Nimphoides indica*, living together with other naidids;
Santa Fe Province, La Capital Department, Don Felipe oxbow Pond
and Saladillo 3º Stream, on aquatic floating plants with dominance
of *S. herzogii, A. caroliniana* and *H. stolonifera* (Di Persia, in
press b).

    Distribution: Asia and, in South America, Paraguay (Stephenson,
1931; Carter and Beadle, 1931); Brazil, states of Sao Paulo, Ceará
and Pernambuco (Marcus, 1943, 1944); Uruguay (Cordero, 1951) and
Surinam (Harman, 1974).

Pristina evelinae Marcus, 1943

    It was first recorded in Argentina at Santa Fe Province, La
Capital Department, La Guardia and Don Felipe oxbow Pond, among
E. crassipes (Di Persia, 1973), and posteriorly from several temporary and permanent pools next to National Road 168, between the
cities of Santa Fe and Paraná, being part of the pleuston (Di Persia, 1974a). At recent date, it was found at Chaco Province, San
Fernando Department, from a pond placed nearby the town of Barranqueras, close to the nouth of Negro River, on *Victoria cruziana*
(Di Persia, in press b).

    Distribution: described from Brazil, where it is known from
the states of Sao Paulo (Marcus, 1943) and Pará (Du Bois-Reymond
Marcus, 1949). It is known from Asia and possibly also from Europe
(Brinkhurst and Jamieson, 1971). Harman (1974) has considered it
as a junior synonym of *P. aequiseta*.

Pristina leidyi Smith, 1896

    In a recent paper, Harman and McMahan (1975) abolished the
subspecies concept as in *Pristina longiseta*, establishing *P. leidyi*
Smith, 1896, for those species reported for North and South America and Hawaii. In Argentina, it has been mentioned for the
northern region by Michaelsen (1921), and for Misiones Province,
Candelaria Department, from a swamp in the vicinities of Loreto,
by Cernosvitov (1942), when describing *P. longiseta bidentata*, I
have found it posteriorly at Corrientes Province, San Cosme
Department, Cabrera Pond, together with poriferans developed upon
submerged parts of *G. multiflora*; Chaco Province, San Fernando
Department, Barranqueras Pond, on *E. crassipes*; Santa Fe Province,

La Capital Department, Saladillo 3° Stream, on floating hydrophytes
(Di Persia, in press b); probably present also at Buenos Aires
Province, Partido Chascomus, Yalca Pond (Vucetich, 1973).

Distribution: in accordance with the recent re-evaluation
made by Harman and McMahan (1975), it comprises North America,
Hawaii and South America, where it has been recorded from Chile
(Michaelsen, 1903a); Paraguay (Michaelsen, 1905a); Peru (Piguet,
1928); Uruguay (Cordero, 1931a, 1931b, 1951); Brazil, states of
Sao Paulo (Marcus, 1943), Alagoas, Pernambuco (Marcus, 1944) and
Amazonas (Du Bois-Reymond Marcus, 1947), and from Surinam (Harman,
1974).

Family Tubificidae

Of the seven subfamilies composing it (Brinkhurst and
Jamieson, 1971), four are present in Argentina. The species
present in the country and their distribution are shown in
Figure 3.

Subfamily Tubificinae Eisen, 1879

Genus *Tubifex* Lamarck, 1816

Tubifex tubifex (Muller, 1774)

Mentioned from Argentina by Brinkhurst (1963). I found it at
Córdoba Province, in the reservoirs San Roque, Los Molinos and
Embalse del Río Tercero, from shallow areas and in low numbers
(Bonetto et al, 1976; Bonetto and Di Persia, in press).

Distribution: being a cosmopolitan species, in South America
it has been recorded from Paraguay (Stephenson, 1931), and from
Brazil, states of Sao Paulo (Marcus, 1942), Río Grande do Sul
(Marcus, 1944), Pará and Amazonas (Du Bois-Reymond Marcus, 1947,
1949).

Genus *Limnodrilus* Claparede, 1862

Limnodrilus hoffmeisteri Claparede, 1862

Its records from Argentina are several: Misiones Province
(Cernosvitov, 1939). Boundary between the province of Salta and
Tucumán, Tala River, from benthos: Tucumán Province, near Tafi
del Valle (Gavrilov and Paz de Tomsic, 1950). Córdoba Province,
benthos in the man-made lakes San Roque, Los Molinos and Embalse
del Rio Tercero, eutrophic reservoirs where the species is strongly
dominant, being the only oligochaete found at depth (Bonetto et
al, 1976; Bonetto and Di Persia, in press). Chaco Province,

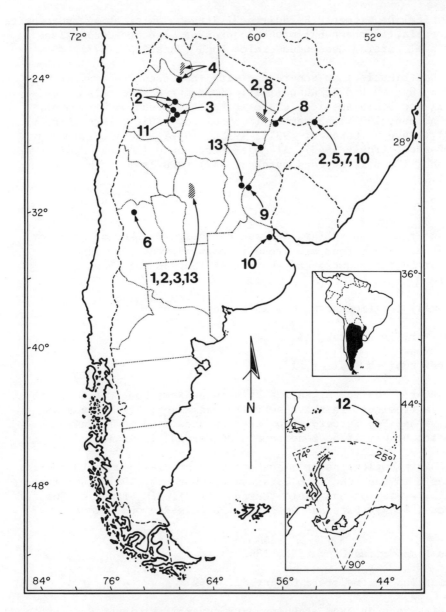

Figure 3.   Geographic distribution of Tubificidae
        1. *Tubifex tubifex*; 2. *Limnodrilus hoffmeisteri*;
        3. *L. udekemianus*; 4. *L. claparedeianus*; 5. *L. neotro-*
        picus; 6. *Peloscolex pigueti*; 7. *Aulodrilus limnobius*
        8. *A. pigueti*; 9. *Paranadrilus descolei*; 10. *Bothrio-*
        neurum americanum; 11. *B. pyrrhum*; 12. *Rhyacodrilus*
        palustris; 13. *Branchiura sowerbyi*.

lower section of Negro River, where it gets to be the only present
oligochaete of the benthic fauna in places undergoing a strong
deficit of dissolved oxygen due to pollution by industrial effluents
(Varela et al, in press).

Distribution: cosmopolitan, it has been recorded from Peru
(Černosvitov, 1939, 1945) and Brazil, states of Sao Paulo, Rio
Grande do Sul (Marcus, 1942, 1944) and Pará (Du Bois-Reymond
Marcus, 1947).

Limnodrilus udekemianus Claparéde, 1862

Found at Tucumán Province (Gavrilov and Paz, 1949), and at
Córdoba Province, eutrophic man-made lakes San Roque, Los Molinos
and Embalse del Rio Tercero, where the worms were part of the
bottom fauna, usually in marginal sampling stations, of moderate
depth, in low numbers with respect to other oligochaetes (Bonetto
et al, 1976; Bonetto and Di Persia, in press).

Distribution: cosmopolitan, it has also been recorded from
Brazil, Sao Paulo State (Marcus, 1942).

Limnodrilus claparedeianus Ratzel, 1868

This species was recorded from two provinces of the NW of the
country: Jujuy Province, in several localities along National Road
34, between the towns of Libertador General San Martin and Pampa
Blanca, and Salta Province, 3 km away from the town of Guemes,
along the same road (Gavrilov, 1970, 1977).

Distribution: cosmopolitan species, in South America was
collected from Uruguay, Montevideo, Arroyo Miguelete (Cordero,
1931a, 1931b).

Limnodrilus neotropicus Černosvitov, 1939

It is known only from Misiones Province in the NE of the
country (Černosvitov, 1939) and has not been recorded again.

Distribution: South American species, it is also present in
Peru, Titicaca Lake (Černosvitov, 1939) and Paraguay (Cernosvitov,
1939).

Genus *Peloscolex* Leidy, 1851

Peloscolex pigueti Michaelsen, 1933

It was recorded only at Mendoza Province, Capital Department,
from an artificial pond of the Parque General San Martín, at Mendoza City, among filamentous algae and bottom fauna with abundant
plant debris (Di Persia, in press c).

Distribution: known before only from Peru, Junín Department, Naticocha Lake (Piguet, 1928), this is the first record of the species from Argentina (Di Persia, in press c). Considering that both localities are placed at de Cordillera de los Andes, separated by great distance, it is possible that the species distribution were wider, related specially with ecosystems of altitude.

Subfamily Aulodrilinae Brinkhurst, 1971

Genus *Aulodrilus* Bretscher, 1899

Aulodrilus limnobius Bretscher, 1899

Recorded from Misiones Province, Loreto Town, Pastora Creek (Cernosvitov, 1937b).

Distribution: cosmopolitan species, in South America it has also been found in Brazil, states of Sao Paulo, Pará and Paraná (Marcus, 1944; Du Bois-Reymond Marcus, 1947).

Aulodrilus pigueti Kowalewski, 1914

Chaco Province, Negro River, as part of the benthic fauna in places variably affected by pollution - particularly organic - of industrial effluents (Varela et al, in press). Corrientes Province, Capital Department, benthos of Brava Pond and its embalsados (Varela, in press). At Negro River, when the habitat does not bear severe conditions of pollution, this species and *Limnodrilus hoffmeisteri* reach similar proportions, and the collections show the presence of gill-bearing oligochaetes (Naididae and Opistocystidae). But when the dissolved oxygen deficit is severe, *Aulodrilus pigueti*, the Naididae and the Opistocystidae disappear (Varela et al, in press).

Distribution: cosmopolitan, recorded previously in South America from Brazil, Pará State (Du Bois-Reymond Marcus, 1947).

Subfamily Rhyacodrilinae Hrabe, 1963

Genus *Paranadrilus* Gavrilov, 1955

Paranadrilus descolei Gavrilov, 1955

Recorded at Entre Rios Province, Paraná Department, from benthos of Las Tunas Stream, a few kilometers away from Paraná City (Gavrilov, 1955a, 1955b, 1958, 1977).

Distribution: to the present, it is only known from the type locality.

Genus *Bothrioneurum* Stolc, 1888

Bothrioneurum americanum Beddard, 1894

Described from material collected from Buenos Aires Province, in the neighbourhoods of Buenos Aires (Beddard, 1894, 1895, 1896); posteriorly mentioned for Misiones Province by Cernosvitov (1938, 1939). It has not been recorded again in this country.

Distribution: known only from South America, it has been collected from Paraguay (Černosvitov, 1938) and Peru-Bolivia, Titicaca Lake (Černosvitov, 1939).

Bothrioneurum pyrrhum Marcus, 1942

It was found at Tucumán Province, nearby Bella Vista (Gavrilov, 1948, 1977).

Distribution: species strictly South American, it was described on material collected from Brazil, state of Sao Paulo (Marcus, 1942).

Genus *Rhyacodrilus* Bretscher, 1901

Rhyacodrilus palustris (Ditlevsen, 1904)

Considered *species inquirenda* by Brinkhurst (In Brinkhurst and Jamieson, 1971), it was recorded from the actual Territorio Nacional de Tierra del Fuego, Antártida e Islas del Atlantico Sur, being part of the fauna of oligochaetes in the South Georgia Islands (Michaelsen, 1905).

Distribution: present in Europe and possibly in Asia.

Subfamily Branchiurinae Hrabe, 1966

Genus *Branchiura* Beddard, 1892

Branchiura sowerbyi Beddard, 1892

This species was recorded by myself at Córdoba Province, Punilla Department, as part of the benthos of San Roque man-made Lake, in shallow places (Bonetto et al, 1976; Bonetto and Di Persia, in press).

New Records: 1. Corrientes Province, close to Goya City, in benthos of pools along Provincial Road 27, 5 March 1975, col. D.H. Di Persia and J.C. Poledri. 2. Santa Fe Province, La Capital Department, Flores oxbow Pond, 21 August 1977 and 2 February 1978, col. D.H. Di Persia. It was part of the benthic fauna and among roots of *E. crassipes*, which covered up completely the surface of the water bodies.

Distribution: it is a species of world wide range, introduced in several countries by human action.

Family Opistocystidae

This family - which comprised *Opistocysta*, described by Cernosvitov (1936) from material collected at the Argentina Province of Misiones (Harman, 1969; Brinkhurst and Jamieson, 1971; Di Persia, in press a), as single genus until recently - has recently been re-evaluated by Harman and Loden (1978), who erected two new genera, *Trieminentia* and *Crustipellis*, both monospecifics. Two of the three species regarded as valid have been found in Argentina, at the NE region (Figure 4).

Genus *Opistocysta* Černosvitov, 1936

Opistocysta funiculus Cordero, 1948

Known from Misiones Province according to Cernosvitov (1936), and found lately by myself at the NW of Corrientes Province, nearby the town of San Luis del Palmar, where it was part of the benthic fauna of a small tributary of the Riachuelo Stream (Di Persia, in press a).

Distribution: in South America, Brazil, states of Sao Paulo, Pernambuco (Marcus, 1944), Amazonas and Pará (Du Bois-Reymond Marcus, 1947, 1949). It might possibly be present also in Africa (Sudan) according to Brinkhurst (1966).

Genus *Trieminentia* Harman and Loden, 1978

Trieminentia corderoi (Harman, 1969)

It was found in Santa Fe Province, La Capital Department, Timbo Island, within the Paraná River's alluvial valley, in a temporary pool, close to National Road 168, between Santa Fe and Parana Cities. The worms were part of the benthos, with abundant detritus of *S. herzogii*, *A. caroliniana*, *H. stolonifera* and *Ricciocarpus natans* (Di Persia, in press a).

New Records: 1. Chaco Province, Negro River close to Puerto Tirol, 20 August 1976 and 23 January 1977, col. M.E. Varela, D.H. Di Persia and N.T. Roberto. In that place, the river, of little volume and digressive course, is variably affected by organic pollution due to the effluents from slaughterhouses and related industries (Varela et al, in press). 2. Corrientes Province, San Cosme Department, San Cosme Town, benthos of Totoras Pond, 21 April 1977, col. N.T. Roberto and M.E. Varela. 3. Corrientes Province, Esteros del Ibará, channel of Galarza Pond, 11 August 1976, col. J.J. Neiff and R.T. Roberto. 4. Corrientes Province,

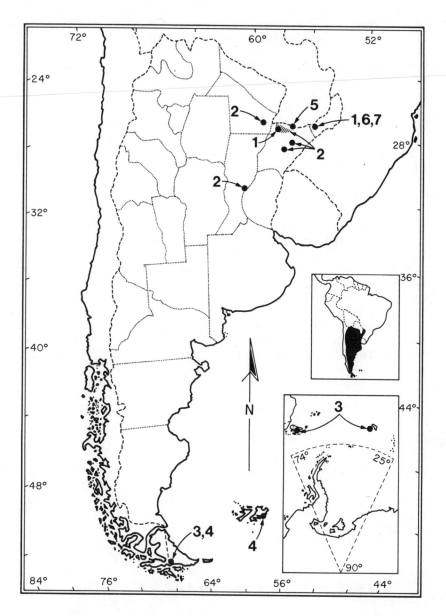

Figure 4.   Geographic distribution of Opistocystidae, Phreodrilidae,
Haplotaxidae and Alluroididae.
Opistocystidae: 1. *Opistocysta funiculus*; 2. *Trieminentia
corderoi*.   Phreodrilidae: 3. *Phreodrilus crozetensis*
4. *P. niger*. Haplotaxidae: 5. *Haplotaxis gordioides*
6. *Haplotaxis* sp.   Alluroididae: 7. *Brinkhurstia
americanus*.

Esteros del Iberá, Iberá Pond, 11 August 1976, col. J.J. Neiff and
N.T. Roberto. 5. Corrientes Province, nearby the town of Itá
Ibaté, Sirena Pond, 21 March 1977, col. J.J. Neiff. 6. Corrientes
Province, González Pond, 23 February 1977, col. N.T. Roberto, C.
Bonetto and L. Benetti. In all of these cases the worms were part
of the benthic fauna, in sediments rich in plant debris.

Distribution: species typical of Neotropical America, it is
known from Uruguay (Cordero, 1948; Harman, 1969; Brinkhurst and
Jamieson, 1971; Harman and Loden, 1978), and from Costa Rica,
Heredia Province (Harman and Loden, 1978).

Family Phreodrilidae

When making a taxonomical revision of this family, known from
the Southern Hemisphere and Ceylon, Brinkhurst (1965) considered
that it should be restricted to a single genus Phreodrilus, and
created two new subgenera, Antarctodrilus and Insulodrilus, which
were added to the recognized ones Phreodrilus, Astacopsidrilus,
Phreodriloides, Gondwanaedrilus and Schizodrilus. In that paper,
Brinkhurst regarded as synonyms of Phreodrilus niger the following
species: Hesperodrilus albus and H. pellucidus, described by Beddard
(1894) for the southern extreme of South America, and Phreodrilus
africanus, which Goddard and Malan (1913b) described for the S. of
Africa. Two species of Phreodrilidae are known from Argentina and
their distribution in the country is shown in Figure 4.

Genus Phreodrilus Beddard, 1891

Subgenus Phreodriloides Benham, 1907

Phreodrilus (Phreodriloides) crozetensis Michaelsen, 1905

In Argentina it was collected at the actual Territorio Nacional de Tierra del Fuego, Antártida e Islas del Atlántico Sur, in
South Georgia Island (Michaelsen, 1905a, 1905b, 1921) and in Tierra
del Fuego, in neighbourhoods of Ushuaia (Michaelsen, 1921).

Distribution: Crozet Island (Michaelsen, 1905a, 1905b; Brinkhurst, 1965) and possibly S. of Chile, Londonberry Island (Cernosvitov, 1934).

Subgenus Antarctodrilus Brinkhurst, 1965

Phreodrilus (Antarctodrilus) niger (Beddard, 1894)

As the former species, it was found at the Territorio Nacional
de Tierra del Fuego, Antártida e Islas del Atlántico Sur, where it
was collected nearby Usuaia (Tierra del Fuego) and Port Stanley,
Malvinas Islands (Beddard, 1894, 1895, 1896).

Distribution: Republic of South Africa (Goddard and Malan, 1931b; Brinkhurst, 1965; Brinkhurst and Jamieson, 1971).

Family Haplotaxidae

The Haplotaxidae were reviewed by Brinkhurst (1966) and are not known much from Argentina, their distribution being shown in Figure 4.

Genus *Haplotaxis* Hoffmeister, 1843

Haplotaxis gordioides (Hartman, 1821)

This species was recorded at the upper stretch of Paraná River, Saltos de Apipé, nearby the town of Ituzaingó, where it was collected during a pronounced lowering of the waters, being part of the benthos, 29 November 1971, col. J.C. Poledri and U. Molet. It has been collected again from that area (Varela, personal communication) but I have not obtained that material for its study.

Distribution: it is a species of wide holarctic distribution. The Argentine material needs more investigation.

Haplotaxis sp.

Cernosvitov (1939) reported the finding of immature specimens at the neighbourhoods of Loreto Town, Misiones Province, which he attributed to *Phreoryctes* sp. One of the specimens is clearly referable to *Haplotaxis*, in accordance with Brinkhurst (1966).

Family Alluroididae

Established by Michaelsen (1900) and reviewed by Brinkhurst (1964) and Jamieson (1968), this family comprises two subfamilies, Syngenodrilinae -restricted to Kenya- and Alluroidinae, of wide distribution in the Ethiopian Region, registered from the Neotropical Region (NE of Argentina) by Černosvitov (1936) as *Alluroides americanus no. nud.* The species was described by Brinkhurst (1964) as *Alluroides americanus* and posteriorly was made the type of a new monotypic genus, *Brinkhurstia*, by Jamieson (1968). It is possible that the latter family were more numerous and of wider range in aquatic habitats, in Africa as well as in South America, as pointed out by Jamieson (In Brinkhurst and Jamieson, 1971), being necessary further studies.

Subfamily Alluroidinae

Genus *Brinkhurstia* Jamieson, 1968

Brinkhurstia americanus (Brinkhurst) Jamieson, 1968

Misiones Province, town of Loreto, nearby Santa Ana, in the SW of the province, from Pastora Creek, collected by Černosvitov at several times during 1931 and 1932 (Černosvitov, 1936; Brinkhurst, 1964). This is the only member of the family known up to date in South America.

Distribution: known only from the above mentioned localities, it has not yet been recorded again.

CONCLUSIONS

To the present, the seven families of aquatic oligochaetes here considered include in Argentina thirty-nine species of Naididae, thirteen Tubificidae, nine Aeolosomatidae, two Opistocystidae, two Phreodrilidae, one Haplotaxidae and one Alluroididae, which makes sixty-seven species in the whole.

It is easy to assume that this number of species may be incomplete, considering the unequal knowledge on the different families and country's territory, it is therefore impossible to speculate about zoogeographical aspects.

However, it may be pointed out that, together with several species of cosmopolitan or wide distribution, there are others known also from Africa -as the naidids *Dero (Dero) cooperi, Pristina notopora,* possibly *P. americana* - or from Asia, as the naidids *D. (D.) plumosa, D. (Aulophorus) hymanae, Allonais lairdi* and *Pristina macrochaeta,* showing a distribution on the Southern Hemisphere, as happens with the phreodrilids. Others such as haplotaxids and alluroidids require investigation.

Fifteen species are strictly South American, among them the aeolosomatids *Aeolosoma evelinae, A. sawayai* and *A. marcusi* (which is recorded for the first time from the country); the naidids *Homochaeta lactea, Slavina evelinae, S. isochaeta* (species previously considered endemic of Titicaca Lake and now recorded from a locality other than the typic one for the first time), *D. (D.) evelinae, D. (D.) botrytis,* and *D. (A.) schmardae;* the tubificids *Limnodrilus neotropicus, Peloscolex pigueti, Paranadrilus descolei, Bothrioneurum americanum* and *B. pyrrhum,* and the alluroidid *Brinkhurstia americanus.* There are also present three species recorded from the Neotropical Region, the naidids *D. (D.) multibranchiata* and *D. (A.) superterrenus* (this one being recorded for the first time from Argentina), and the opistocystid *T. corderoi.*

The knowledge of the great rivers of the Cuenca del Plata (Parana, Paraguay, Uruguay and their tributaries) is still fragmentary, and further studies will probably reveal the presence of

more species, widening the distribution of some of them. This may
happen with those species dwelling on dense carpets of floating
plants developed on water bodies close to the rivers, which become
widely spread over extensive areas when periodic floods carry
away groups of hydrophytes.

On the other hand, there are very few records of aquatic oligochaetes south of parallel 36°, and it is necessary to intensify the
studies on worms from the diverse lentic and lotic habitats of the
different regions of the country, from most of which there is no
information on the subject.

ACKNOWLEDGEMENTS

I feel especially grateful to Dr. Walter Harman, Department of
Zoology, Louisiana State University, U.S.A., for his kindness in
presenting this paper at the First International Symposium on
Aquatic Oligochaete Biology, in spite of all the inconveniences
that it might have meant for him.

REFERENCES

Beddard, F.E., 1894, Preliminary notice of South American Tubifi-
    cidae collected by Dr. Michaelsen, including the description
    of a branchiate form. Ann. Mag. Nat. Hist., 13:205-210.
Beddard, F.E., 1895, "A monograph of the order Oligochaeta."
    Claredon Press, Oxford.
Beddard, F.E., 1896, Naididen, Tubificiden und Terricolen. I.
    Limicole Oligochaeten. Erg. Hamb. Magalh. Sammelr., 1:5-18.
Bonetto, A.A., M.C. Corigliano, D.H. Di Persia, and R. Maglianesi,
    1976, Características limnológicas de algunos lagos eutró-
    ficos de embalse de la región central de Argentina. Ecosur,
    3:47-120.
Bonetto, A.A. and D.H. Di Persia, The San Roque reservoir and
    other man-made lakes in the central region of Argentina, In
    "Lake and Reservoir Ecosystems." F.B. Taub (ed.) Elservier
    Scientific Publishing Company, Amsterdam, (in press).
Brinkhurst, R.O., 1963, Taxonomical studies on the Tubificidae
    (Annelida, Oligochaeta). Int. Revue ges. Hydrobiol., 48:1-89
Brinkhurst, R.O., 1964, A taxonomic revision of the Alluroididae
    (Oligochaeta). Proc. zool. Soc. Lond., 142:527-536.
Brinkhurst, R.O., 1965, A taxonomic revision of the Phreodrilidae
    (Oligochaeta). J. Zool., 147:363-386.
Brinkhurst, R.O., 1966, A taxonomic revision of the family Haplo-
    taxidae (Oligochaeta). J. Zool. Lond., 150:29-51.
Brinkhurst, R.O. and B.G.M. Jamieson, "Aquatic Oligochaeta of
    the World." Oliver and Boyd, Edinburgh.

Carter, M.A. and L.C. Beadle, 1931, The fauna of the swamps of the Paraguayan Chaco in relation to its environments III. J. Linn. Soc. Lond., 37:379-386.

Černosvitov, L., 1934, Oligochètes. Rés. Voy. Belgica (Zool.), 1935:1-11.

Černosvitov, L., 1936, Oligochaeten aus Südamerica. Systematische Stellung der Pristina flagellum Leidy. Zool. Anz., 113:75-84.

Černosvitov, L., 1936, Notes sur la distribution mondiale de quelques oligochètes. Mem. Soc. Zool. Tchécosl., 3:16-19.

Černosvitov, L., 1937b, Notes sur les Oligochaeta (Naididées et Enchytraeidées) de l'Argentine. Anal. Mus. Nac. Cienc. Nat. B. Rivadavia, 39:135-157.

Černosvitov, L., 1938, Oligochaeta. Mission Scientifique de l'Omo, 4:255-318.

Černosvitov, L., 1939, Oligochaeta. The Percy Sladén Trust expedition to Lake Titicaca in 1937. Trans. Linn. Soc. Lond., (3) 1:81-116.

Černosvitov, L., 1942, Oligochaeta from various parts of the world, Proc. Zool. Soc. Lond., (B) 111:197-236.

Cordero, E.H., 1931a, Die Oligochaten der Republik Uruguay I. Zool. Anz., 92 (11-12):333-336.

Cordero, E.H., 1931b, Notas sobre los Oligoquetos del Uruguay. Anal. Mus. Nac. Histo. Nat. B. Rivadavia, 36:343-357.

Cordero, E.H., 1948, Zur Kenntnis der Gattung Opisthocysta Cern. (Archiologochaeta). Comun. Zool. Mus. Hist. Nat. Montev., 2 (50):1-8.

Cordero, E.H., 1951, Sobre algunos oligoquetos limicolas de Sud America. Inst. Inv. Cienc. Biol. Montev., 1:231-240.

Dioni, W., 1960, Notas hidrobiológicas I. El genero Aeolosoma en el Uruguay. Actas Trab. Primer Congr. Sudam. Zool., 2:107-119.

Dioni, W., 1967, Investigación preliminar de la estructura básica de las associaciones de la micro y mesofauna de las ráices de las plantas flotantes. Acta Zool. Lilloana, 23:111-137.

Di Persia, D.H., 1973, Notas sobre oligoquetos dulceacuícolas argentinos. Physis, 32 (85):279-285.

Di Persia, D.H., 1974a, Sobre algunos oligoquetos dulceacuicolas del área comprendida entre Santa Fe y Paraná. Rev. Asoc. Cienc. Nat. Litoral, 5:33-44.

Di Persia, D.H., 1974b, Géneros, subgéneros y especies de oligoquetos dulceacuícolas de las familias Naididae y Aeolosomatidae registrados en cuencas leníticas próximas a las ciudades de Santa Fe y Paraná. Rev. Asoc. Cienc. Nat. Litoral, 5:89-98.

Di Persia, D.H., 1975a, Oligoquetos del subgénero Aulophorus Schmarda nuevos para la fauna acuática argentina (Dero, Naididae). Comun. Cient. Cecoal, 2:1-7.

Di Persia, D.H., 1975b, Sobre algunas especies y un genero de oligoquetos acúaticos nuevos para Argentina. Comun. Cient. Cecoal, 4:1-8.

Di Persia, D.H., 1975c, Acerca de dos especies del género *Nais* de
    Argentina (Oligochaeta, Naididae). Neotropica, 21:131-134.
Di Persia, D.H., 1976a, Nuevas citas del género *Dero* s.s. (Naididae,
    Tubificoidea) para la oliogoquetofauna acuática argentina.
    Physis, 35 (90):1-7.
Di Persia, D.H., 1976b, El género *Aeolosoma* Ehrenberg, 1828 en
    algunos ambientes leníticos del noroeste de la provincia de
    Corrientes, Argentina. Physis, 35 (90):9-15.
Di Persia, D.H., 1977a, Adición al conocimiento del subgénero
    Dero (Oligochaeta, Naididae). Neotropica, 23 (69):11-16.
Di Persia, D.H., 1977b, Presencia en Argentina de *Stephensoniana
    trivandrana* (Oligochaeta, Naididae). Comun. Cient. Cecoal,
    5:1-4.
Di Persia, D.H., 1977c, *Amphicaeta leydigii* Tauber en la fauna
    bentónica del río Negro, provincia de Chaco (Oligochaeta,
    Naididae). Neotrópica, 23 (70):117-122.
Di Persia, D.H., 1978, Clave para la determinación de géneros y
    subgéneros de Naididae del nordeste argentino. Rev. Asoc.
    Cienc. Nat. Litoral, 9:1-12.
Di Persia, D.H., Especies argentinas del género *Opistocysta* (Oligo-
    chaeta, Opistocystidae). Acta Zool. Lilloana (in press a).
Di Persia, D.H., El género *Pristina* (Naididae, Oligochaeta) en la
    República Argentina. Physis (in press b).
Di Persia, D.H., Sobre algunos oligoquetos acúaticos del norte de
    Mendoza. Physis (in press c).
Di Persia, D.H. and M. Radici de Cura, 1973, Algunas consideracio-
    nes acerca de los organismos epibiontes desarrollados sobre
    Ampullaridae. Physis, 32 (85):309-319.
Di Persia, D.H. and B. R. de Ferradás, Notas sistemáticas y bioeco-
    lógicas sobre la oligoquetofauna vinculada a la vegetación
    acuática en un ambiente lenítico del Paraná medio (in press).
Di Persia, D.H. and M.E. Varela, *Slavina evelinae* (Oligochaeta,
    Naididae) en las provincias argentinas de Corrientes y Chaco.
    Physis (in press).
Du Bois-Reymond Marcus, E., 1944, Notes on freshwater Oligochaeta
    from Brazil. Comun. Zool. Mus. Hist. Nat. Montev., 1(20):1-8
Du Bois-Reymond Marcus, E., 1947, Naidids and tubificids from
    Brazil. Comun. Zool. Mus. Hist. Nat. Montev., 2(44):1-18.
Du Bois-Reymond Marcus, E., 1949, Further notes on naidids and
    tubificids from Brazil. Comun. Zool. Mus. Hist. Nat. Montev.,
    3(51):1-11.
Frenzel, J., 1891, Untersuchungen über die mikroskopische fauna
    Argentiniens. Arch. Mikr. Anat., 38:1-24.
Gavrilov, K., 1948, Sobre la reproducción uni y biparental de los
    Oligoquetos. Acta Zool. Lilloana, 5:221-311.
Gavrilov, K., 1955a, Ein neuer spermathekenloser Vertreter der
    Tubificiden. Zool. Anz., 155 (11-12):294-302.
Gavrilov, K., 1955b, Uber die uniparentale Vermehrung von *Parana-
    drilus*. Zool. Anz., 155 (11-12):302-306.

Gavrilov, K., 1958, Notas adicionales sobre *Paranadrilus*. Acta
    Zool. Lilloana, 16:149-236.
Gavrilov, K., 1970, Nota faunística sobre *Limnodrilus claparedea-*
    nus Ratzel (Tubificidae, Oligochaeta). Acta Zool. Lilloana,
    26:145-156.
Gavrilov, K., 1977, Oligochaeta, In "Biota acuatica de Sudamerica
    Austral," S.H. Hurlbert (ed.), San Diego State University,
    San Diego, California.
Gavrilov, K. and N.G. Paz, 1949, *Limnodrilus inversus* n. sp. y su
    reproducción uniparental. Acta Zool. Lilloana, 8:537-565.
Gavrilov, K. and N.G. Paz de Tomsic, 1950, Nota adicional sobre
    la reproducción de *Limnodrilus*. Acta Zool. Lilloana, 9:533-568.
Goddard, E.J. and D.E. Malan, 1913b, Contribution to a knowledge of
    South African Oligochaeta. Part II. Description of a new spe-
    cies of *Phreodrilus*. Trans. R. Soc. S. Afr., 3:242-248.
Grimm, R., 1974, Einige Oligochaeten aus Nigeria, dem Tschad und
    der Zentralafrikanischen Republik. Mitt. Hamburg. Zool. Mus.
    Inst., 71:95-114.
Harman, W., 1974, The Naididae (Oligochaeta) of Surinam. Zool.
    Bijdr. Leiden, 133:1-36.
Harman, W., 1969, Revision of the family Opistocystidae (Oligo-
    chaeta). Trans. Amer. Micros. Soc., 88:472-478.
Harman, W. and M.L. McMahan, 1975, A reevaluation of *Pristina*
    longiseta (Oligochaeta, Naididae) in North America. Proc.
    Biol. Soc. Wash., 88 (17):167-178.
Harman, W. and M.S. Loden, 1978, A reevaluation of the Opistocys-
    tidae (Oligochaeta) with descriptions of two new genera.
    Proc. Biol. Soc. Wash., 91 (2):453-462.
Jamieson, B.G.M., 1968, A taxonometric investigation of the Allu-
    roididae (Oligochaeta). J. Zool. Lond., 155:55-86.
Marcus, E., 1942, Sobre algumas Tubificidae do Brasil. Bol. Fac.
    Fil. Cienc. Let. Univ. S. Paulo, Zool., 25 (6):153-254.
Marcus, E., 1943, Sobre Naididae do Brasil. Bol. Fac. Fil. Cienc.
    Let. Univ. S. Paulo, Zool., 32 (7):3-247.
Marcus, E., 1944, Sobre Oligochaeta limnicos do Brasil. Bol. Fac.
    Fil. Cienc. Let. Univ. S. Paulo, Zool., 43 (8):5-135.
Michaelsen, W., 1900, Oligochaeta. Das Tierreich, 10:1-575.
Michaelsen, W., 1903, Hamburgische Elb-Untersuchung IV. Oligo-
    chaeten. Jb. Hamb. wiss. Anst., 19 (2):169-210.
Michaelsen, W., 1905a, Die Oligochaeten der deutschen Südpolar-
    Expedition 1901-1903 nebst Erörterung der Hypothese über
    einen früheren grossen, die Sudspitzen der Kontinente verbin-
    denen antarktischen Kontinent. Deutsche Südpol-Exped. 1901-
    1903, 9 Zool., 1:5-7.
Michaelsen, W., 1905b, Die Oligochaeten der Schwedischen Südpolar-
    Expedition, 1901-1903. Wiss Ergebn, schwed. Südpolar-Exp.,
    5 (3):1-12.
Michaelsen, W., 1905c, Zur Kenntnis der Naididen, In Untersuchun-
    gen uber die Süsswasser-Mikrofauna Paraguays, E.v. Daday (ed.)
    Zoologica, 18 (44):350-361.

Michaelsen, W., 1912, Über zentralamerikanische Oligochäeten. Arch. Naturgesch., 78A, 9:112.

Michaelsen, W., 1921, Neue und wenig bekannte Oligochäeten aus skandinavischen Sammlungen. Ark. Zool. Stockholm., 13 (19): 1-25.

Michaelsen, W., 1933, Süss- und Brackwasser-Oligochaeten von Bonaire, Curaçao and Aruba. Zool. Jb. Syst., 64:327.

Michaelsen, W. and W. Boldt, 1932, Oligochaeten der deutschen limnologischen Sunda-Expedition, In Tropische Binnengewasser II, A. Thienemann, (ed.). Arch. Hydrobiol. Supl., 9:587-622.

Naidu, K.V., 1962b, Studies on the freshwater Oligochaeta of South India. I. Aeolosomatidae and Naididae, Part III. J. Bombay Nat. Hist. Soc., 59:520-546.

Naidu, K.V., 1962c, Studies on the freshwater Oligochaeta of South India. I. Aeolosomatidae and Naididae, Part IV. J. Bombay Nat. Histo. Soc., 59:897-921.

Naidu, K.V., 1965, Some freshwater Oligochaeta of Singapure. Bull. Nat. Mus. St. Singapure, 33:13.

Picado, C., 1913, Les Broméliacées épiphytes considerées comme milieu biologique. Bull. Scient. France et Belgique, 47:215-360.

Piguet, E., 1928, Sur quelques Oligochètes de l'Amerique du Sud et de l'Europe. Bull. Soc. neuchât. Sci. nat. (N.S.), 52:78-101.

Poi de Neiff, A., 1977, Estructura de la fauna asociada a tres hidrófitos flotantes en ambientes leníticos del nordeste argentino. Comun. Cient. Cecoal, 6:1-16.

Righi, G., 1973, On *Pristina minuta* (Oligochaeta, Naididae) from Brazilian soil and its epizoic *Rhabdostyla pristinis* sp. n. (Ciliata, Epistylidae). Zool. Anz., 191 (5-6):295-299.

Sperber, C., 1948, A taxonomical study of the Naididae. Zool. Bidr. Uppasala, 28:1-206.

Stephenson, J., 1931, The Oligochaeta from Brazil and Paraguay. J. Linn. Soc. Lond., 37:291-326.

Stephenson, J., 1931a, Oligochaeta from the Malay Peninsula. J. fed. Malay St. Mus. Kuala Lumpur, 16:261.

Stephenson, J., 1932, Report on the Oligochaeta: Mr. Omer-Cooper's investigations of the Abyssinian freshwater. Proc. Zool. Soc. Lond., 1932:227-256.

Stieren, A., 1892, Über einige *Dero* aus Trinidad. Protok. Obshch. Estest. Yurev., 10:103-123.

Van der Land, J., 1971, Family Aeolosomatidae, In: "Aquatic Oligochaeta of the World," R.O. Brinkhurst and B.G.M. Jamieson, ed., Oliver and Boyd, Edinburgh.

Varela, M.E., Primer hallazgo de *Aulodrilus pigueti* (Tubificidae, Oligochaeta) en la República Argentina (in press).

Varela, M.E., D.H. Di Persia and A.A. Bonetto, La fauna macrobentónica y su relación con la contaminación orgánica en el río Negro, provincia de Chaco. Estudio preliminar. Ecosur (in press).

THE AQUATIC OLIGOCHAETA OF THE ST. LAWRENCE

GREAT LAKES REGION

D.R. Spencer

Ecological Sciences Division
NUS Corporation
1910 Cochran Road
Pittsburgh, Pennsylvania 15220

ABSTRACT

An extensive amount of research has been conducted on the
freshwater Oligochaeta of the St. Lawrence Great Lakes, making
this region perhaps the best known in the continental United
States. Prior to the advent of useable systematic keys, little
attempt was made to identify Great Lakes oligochaete species.
Taxonomic problems that have materialized in the literature are
presented. The zoogeography of the Great Lakes fauna is discussed along with the question of endemic species. A state-of-
the-art review on the distribution of the aquatic Oligochaeta
occurring in the 5 major Great Lakes and inland waters of the
region is presented. Future areas of research should center
around resolving the various systematic problems within the
Naididae and Tubificidae, and a revision of the Enchytraeidae
in North America is proposed. More intense surveys of the
Great Lakes fauna, especially on inland waters, is required to
provide further baseline information on the fauna before environmental degradation occurs, and continued monitoring of
previously studied areas is indicated to determine changes in
species associations.

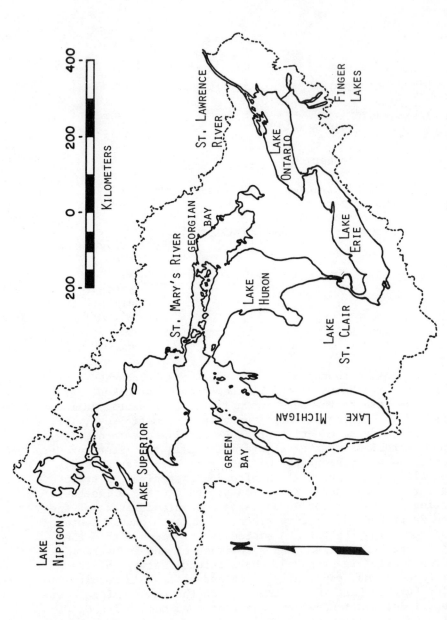

Figure 1. The St. Lawrence Great Lakes Region (broken line represents the limits of the watershed).

INTRODUCTION

The Great Lakes region of eastern North America encompasses
766,640 square kilometers and contains some of the largest and
deepest freshwater lakes in the world (Figure 1). The oligochaetes of the region are perhaps the best known in North America,
despite the difficulty of obtaining collections from these freshwater seas. The author has reviewed a large number of recently
published papers, which contain information on the aquatic Oligochaeta of the region. The majority of these reports have centered
on surveys conducted exclusively on the five major Great Lakes.
Interestingly, very few of these studies were on Oligochaeta but
were concerned with the entire benthic community. The present
paper provides a "state-of-the-art" report on the Oligochaeta of
the Great Lakes region as to systematics, zoogeography and the
question of endemics, distribution, and areas for future research.

SYSTEMATICS

Historically, the systematics of the aquatic Oligochaeta of
the Great Lakes region were quite disorganized before the mid-
1960's. Few attempts were made to provide taxonomic keys for
species in the Great Lakes region. Moore (1906) presented a key
based on cursory collections obtained from a variety of locations
within the region, as did Walton (1906) for the Naididae of Cedar
Point, Ohio. The above studies initiated an attempt to classify
the worms of the Great Lakes but, because of incomplete taxonomic
knowledge of the group at that time, these keys are of little
value to us today. A period of sixty years would elapse before a
firm foundation for Great Lakes oligochaete systematics could be
developed.

The work of Dr. Ralph O. Brinkhurst, then of the University
of Liverpool, provided an impetus for present day Great Lakes
oligochaete systematics. His studies on the North American
Oligochaeta (Brinkhurst, 1964, 1965; Brinkhurst and Cook, 1966)
provided students of the group with descriptions of 75% of all
known species ever recorded from the Great Lakes. Simplified
taxonomic keys based mainly on easily recognized internal and
external characteristics were provided which stimulated further
investigation within the region.

During the late 1960's to the present, a number of papers which
treat various aspects of Great Lakes oligochaete systematics have
been published (Hiltunen, 1967, 1969a; Howmiller and Beeton, 1970;
Howmiller, 1974b; Stimpson, et al., 1975; Loden, 1978a; Spencer,
1978a). In addition, some taxonomic keys have been produced which
treat certain genera (Cook and Hiltunen, 1975; Spencer, 1977;
Spencer 1978b) and useful regional keys to the families Naididae
and/or Tubificidae (Brinkhurst, et al., 1968; Brinkhurst, 1968;
Hiltunen, 1973; Howmiller and Loden, 1976; Brinkhurst, 1976).

Seven families of aquatic Oligochaeta containing over 90 taxa are known from the Great Lakes region (See Table 1). The Aeolosomatidae have been occasionally recorded, but have been excluded from this report conforming to the recent decision by Brinkhurst and Jamieson (1971) to eliminate the family from the Oligochaeta. The taxonomy of the group is in fairly good order and conforms to Brinkhurst and Jamieson (1971). However, several systematic problems have been discovered, mainly through the work of J.K. Hiltunen on the Naididae and Tubificidae, which will probably require additional investigation.

Within the family Haplotaxidae, *Halplotaxis gordioides gordioides* remains an enigma. The species has never been collected in a fully developed sexual condition in North America (Cook, 1975b) thus making its identification in the Great Lakes region, or other areas of North America doubtful.

Some Great Lakes Naididae have presented systematic problems. Hiltunen (1973) observes that according to Sperber and her predecessors, *Chaetogaster diaphanus* has been separated from *C. crystallinus* by the presence of an antero-dorsal, shallow invagination on the prostomium occurring in the latter. However, he has seen specimens which otherwise fit the description of *C. diaphanus* but have the antero-dorsal incision. Because of the similarity of characters, Hiltunen maintains that the separation of the two species must be based on other anatomical characters. Hare and Carter (1977) also believe *C. diaphanus* and *C. crystallinus* not to be separable using the characters given by Brinkhurst and Jamieson (1971). They also mentioned that Kasprzak (1972) had synonymized these two species as *C. diaphanus*. Howmiller and Loden (1976) support the idea that *Amphichaeta* is inadequately studied in the Great Lakes region. Hiltunen (1973) stated that probably all specimens of *Amphichaeta* in the Great Lakes fit the description of *A. leydigi*. Later, however, Hiltunen, in Howmiller and Loden (1976), believed that more than one species of worm resembling *A. leydigi* may have been present in collections from Wisconsin. This belief was recently reconfirmed by Hiltunen that more than one species is, indeed, present in the region (personal communication). Hiltunen (1967) questioned the accurate interpretation of all North American species of *Nais* recorded in the literature. This problem had also concerned other investigators. Howmiller and Loden (1976) believed in *N. variabilis* to be a variable species, possibly indistinguishable from *N. communis* or *N. elinguis*. Hiltunen (1973) considered *N. pardalis* as a morphological variant of *N. bretscheri*. Hare and Carter (1977) found specimens of *N. communis* and *N. variabilis* not to be consistently separable and referred to this material as *N. communis-variabilis*. Brinkhurst (1976) stated that *N. communis* and *N. variabilis* are difficult to distinguish in preserved material, the descriptions in the literature are confusing, and that more detailed systematic

Table 1.   The aquatic Oligochaeta of the St. Lawrence Great Lakes and their occurrence within the region.

Taxa	Inland Waters	Lake Ontario	Lake Erie	Lake St. Clair including St. Clair River	Lake Huron	Lake Michigan	Lake Superior including St. Mary's River
Lumbriculidae							
Lumbriculus variegatus variegatus (Müller)	x	x	x		x	x	x
L. variegatus inconstans (Smith)	x		x	x	x		x
Stylodrilus heringianus Claparède		x	x	x	x	x	x
Eclipidrilus lacustris (Verrill)	x (Cayuga Lake)						x
Haplotaxidae							
Haplotaxis gordioides gordioides (Hartmann)	x						
Naididae							
Chaetogaster diastrophus (Gruithuisen)	x		x	x	x	x	x
C. langi Bretscher		x	x		x		x
C. diaphanus (Gruithuisen)	x	x	x	x	x	x	x
C. setosus Svetlov		x	x				
C. limnaei von Baer	x	x	x	x		x	x
Amphichaeta sp.						x	x
Paranais litoralis (Müller)		x			x		x
P. frici Hrabě	x		x			x	
Bratislavia unidentata (Harman)			x				
Specaria josinae (Vejdovsky)	x		x	x		x	x
Uncinais uncinata (Ørsted)	x		x	x		x	x
Ophidonais serpentina (Müller)	x		x			x	x

Table 1. (Continued)

Taxa	Lake Superior including St. Mary's River	Lake Michigan	Lake Huron	Lake St. Clair including St. Clair River	Lake Erie	Lake Ontario	Inland Waters
Nais communis Piguet	×		×		×	×	×
N. variabilis Piguet	×		×			×	×
N. simplex Piguet	×		×		×	×	×
N. bretscheri Michaelsen	×		×			×	×
N. pardalis Piguet	×	×	×			×	
N. elinguis Müller	×	×	×	×	×	×	×
N. alpina Sperber	×					×	
N. barbata Müller	×		×			×	
N. pseudobtusa Piguet	×		×			×	×
N. behningi Michaelsen							×
Slavina appendiculata d'Udekem	×	×	×	×	×	×	×
Vejdovskyella comata (Vejdovsky)	×		×		×	×	×
V. intermedia (Bretscher)		×		×	×	×	×
Arcteonais lomondi (Martin)	×	×	×	×	×	×	×
Stylaria lacustris (Linnaeus)	×	×	×	×	×	×	×
S. fossularis Leidy				×	×	×	×
Piguetiella michiganensis Hiltunen	×	×			×	×	×
Haemonias waldvogeli Bretscher							×
Dero digitata (Müller)	×	×	×	×	×		×
D. furcata (Müller)							×

Table 1. (Continued)

Taxa	Inland Waters	Lake Ontario	Lake Erie	Lake St. Clair including St. Clair River	Lake Huron	Lake Michigan	Lake Superior including St. Mary's River
D. vaga (Leidy)	x		x				
Allonais pectinata (Stephenson)				x			
Pristina idrensis Sperber					x		x
P. osborni (Walton)	x		x		x		x
P. acuminata Liang			x				
P. forelia (Piguet)	x	x		x	x		x
P. plumaseta Turner	x						
P. aequiseta Bourne	x	x					x
P. leidyi (Smith)	x	x	x	x			x
Tubificidae							
Tubifex tubifex (Müller)	x	x	x		x	x	x
T. ignotus (Stolc)	x	x				x	x
T. newaensis (Michaelsen)			x		x		
T. kessleri americanus Brinkhurst and Cook	x	x			x	x	x
Limnodrilus hoffmeisteri Claparède	x	x	x	x	x	x	x
L. udekemianus Claparède	x	x	x	x	x	x	x
L. claparedeianus Ratzel	x	x	x	x	x	x	x
L. profundicola (Verrill)	x	x	x		x	x	x
L. cervix Brinkhurst	x	x	x	x	x	x	x
L. silvani Eisen		x					

Table 1. (Continued)

Taxa	Lake Superior including St. Mary's River	Lake Michigan	Lake Huron	Lake St. Clair including St. Clair River	Lake Erie	Lake Ontario	Inland Waters
L. maumeensis Brinkhurst and Cook	x	x	x	x	x	x	
L. angustipenis Brinkhurst and Cook	x (St. Mary's River)	x	x		x	x	
Psammoryctides californianus Brinkhurst	x (St. Mary's River)						x Cayuga Lake, Black River, MI
Potamothrix moldaviensis Vejdovsky and Mrazek	x (St. Mary's River)	x	x	x	x	x	x
P. hammoniensis (Michaelsen)		x				x	
P. bavaricus (Öschmann)		x				x	x
P. bedoti (Piguet)		x	x	x	x	x	x
P. vejdovskyi (Hrabě)	x	x	x		x	x	x
Ilyodrilus templetoni (Southern)	x	x	x	x	x	x	x
Peloscolex variegatus Leidy	x		x				x
P. curvisetosus (Brinkhurst and Cook)	x (St. Mary's River)				x	x	
P. multisetosus multisetosus Brinkhurst and Cook	x	x	x	x	x	x	x
P. multisetosus longidentus Brinkhurst and Cook		x				x	x
P. ferox (Eisen)	x	x	x	x	x	x	x
P. superiorensis Brinkhurst and Cook	x	x	x			x	
P. freyi Brinkhurst	x (St. Mary's River)	x	x	x	x	x	
Aulodrilus limnobius Bretscher		x	x	x	x	x	x
A. pluriseta (Piguet)	x	x	x		x	x	x
A. pigueti Kowalewski	x	x	x	x	x	x	x
A. americanus Brinkhurst and Cook	x	x	x	x	x	x	x

Table 1. (Continued)

Taxa	Inland Waters	Lake Ontario	Lake Erie	Lake St. Clair including St. Clair River	Lake Huron	Lake Michigan	Lake Superior including St. Mary's River
Bothrioneurum vejdovskyanum Stolc	×	×	×				
Rhyacodrilus coccineus (Vejdovsky)		×			×	×	
R. sodalis (Eisen)		×			×		×
R. punctatus Hrabe							×
R. montana (Brinkhurst)	×	×			×	×	×
Branchiura sowerbyi Beddard	×		×	×			
Phallodrilus hallae Cook and Hiltunen					×		×
Enchytraeidae							
Mesenchytraeus sp.					×		×
Achaeta sp.		×					
Lumbricillus sp.		×				×	
Glossoscolecidae							
Sparganophilus tamesis Benham	×	×	×	×	×	×	
Lumbricidae							
Aporrectodea trapezoides (Duges)		×			×		
A. tuberculata (Eisen)			×		×		
Eiseniella tretraedra (Savigny)			×		×		×
Lumbricus terrestris Linnaeus		×					

analysis is required. With respect to the genus *Stylaria*, some
workers in the Great Lakes region prefer to separate the species,
S. lacustris and *S. fossularis* (Hiltunen, 1973; Howmiller and
Loden, 1976; Howmiller, 1974b), but Brinkhurst, in Brinkhurst and
Jamieson (1971), considers the genus monotypic, recognizing only
S. lacustris. Brinkhurst (1976) reemphasized that *S. fossularis*
should be regarded as a form of *S. lacustris* because both have
been found together in many habitats in Europe and North America.
Brinkhurst and Jamieson (1971) conclude that *Vejdovskyella intermedia* is a synonym of *V. comata* because of overlapping taxonomic
characters. Others (Hiltunen, 1973; Howmiller and Loden, 1976)
believe that the two should be maintained as distinct species.
A number of Great Lakes investigators have recorded *V. intermedia*
as the species found (Hiltunen, 1967, 1969a,c; Stimpson, et al.,
1975; Spencer, 1978b). Brinkhurst (1976) stated that the differences between *V. comata* and *V. intermedia* have become less obvious
with the descriptions of various subspecies or variants of an
intermediate nature. Recently, Brinkhurst (1978) points to a
"complex" referred to as *V. comata* which includes *V. intermedia,
V. macrochaeta,* and possibly *V. simplex* that form a series of
entities in Europe. Brinkhurst mentioned that in North America
the *comata* and *intermedia* forms are said to be separable by
J.K. Hiltunen, and presents the differences Hiltunen uses to separate the two entities. He concludes that a revision of North
American and European material would probably lead to a separation
of the various forms at the species level when more differences
are recognized.

Two changes in nomenclature in the Naididae effecting Great
Lakes oligochaetologists have recently been published. Harman
and McMahan (1975) refer all forms of *Pristina longiseta* reported
from North America, South America, and Hawaii to *P. leidyi,* and
Harman and Loden (1978) transferred *Pristina unidentata* to the
recently described genus, *Bratislavia*.

The family Tubificidae has several problem areas, some of
which are quite complex. The definition of species within *Limnodrilus* has created probably the most controversy between Great
Lakes oligochaetologists. Brinkhurst, in Brinkhurst and Jamieson
(1971), believes the genus to be one of the most clearly defined
but the species are not always easy to separate owing to the
simplicity of the setae. They further state that there are often
distinct forms within any one locality which may be held to represent valid species but comparisons with material from other sites
soon blurs the picture. Hiltunen (1967) considers *Limnodrilus* to
be the largest and perhaps more complex genus of tubificids in
the Great Lakes region. *L. cervix* has been noted as being found
in nature as two forms ("variants") similar to each other except
for the heads of the penis sheath (Hiltunen, 1967, 1969a, 1973).
Hiltunen (1969a) notes that the taxonomic relationship between

the 2 forms of *L. cervix* remain unexplained. He theorized that
one form could result from self-fertilization which may account
for the morphological differences observed, but the existence
elsewhere of monotypic populations of "typical" *L. cervix* (Kennedy,
1965) casts doubt on this thesis. Hiltunen (1969a) believes it
unlikely that any type of self-fertilization would occur in one
location and not another area. Hiltunen (1969a) also recognized
two forms of *L. claparedeianus* in western Lake Erie. Intermediate
forms between *L. claparedeianus* and *L. cervix* have been recorded
(Brinkhurst and Cook, 1966; Brinkhurst, 1976; Howmiller and Beeton,
1970; Howmiller and Loden, 1976). *L. hoffmeisteri* has been recognized by Hiltunen (1967) as appearing in the Great Lakes as three
different types, all differing in the form of the penis tube.
Brinkhurst (1965) stated that "this, the commonest tubificid any-
where, shows considerable variation in the apparent form of the
penis sheath, which is often due to the angle at which it is
viewed.....many intermediate forms are known, and it is not possible to recognize distinct varieties of the species." Hiltunen
(1969a) reported from western Lake Erie a problematical taxon
termed *L. spiralis* which is presently recognized as a form of *L.
hoffmeisteri* by Brinkhurst and Jamieson (1971). A number of other
Great Lakes investigators (Kinney, 1972; Howmiller, 1974b; Stimpson
et al., 1975; Howmiller and Loden, 1976; Pliodzinskas, 1978) have
followed Hiltunen's observations on *L. spiralis* and separated it
from *L. hoffmeisteri*. *L. spiralis* has been suggested as an interspecific hybrid of *L. claparedeianus* and *L. hoffmeisteri* (Cook
and Johnson, 1974). Recently, Brinkhurst (1978) reasserted the
synonymy of the two entities. Previously, the genus *Peloscolex*
was characterized by possessing a body wall covered with papillae.
However, Brinkhurst (1963) defined the genus on the basis of the
male efferent duct, the most reliable single character, and used
the papillate body wall as a character of secondary importance.
Two species of *Peloscolex (P. freyi* and *P. superiorensis)* that
have been described from the Great Lakes region by Brinkhurst
(1965) and Brinkhurst and Cook (1966), lack a body wall with
papillae. The transfer of *Psammoryctides curvisetosus* to *Peloscolex* by Loden (1978a) brings the number of non-papillate *Peloscolex* species occurring in the Great Lakes region to three.
Hiltunen (1973) called for a complete revision of the genus on
the basis of this difference in external characteristics. This
revision is currently being undertaken by Dr. C. Holmquist,
Swedish Museum of Natural History (See Holmquist, 1978). A
controversy has existed in the genus *Potamothrix*, not only in
North America but also in Europe, ever since Michaelsen (1926)
questioned the specific validity of *P. bedoti*. Michaelsen
suggested that *P. bedoti* was only a variant form of *P. bavaricus*
which was supported by other workers (Brinkhurst, 1963, 1965;
Brinkhurst and Jamieson, 1971). In the Great Lakes region, investigators were uncertain how to approach the problem. Howmiller
and Beeton (1970) noted an intermediate form between *P. bavaricus*

and *P. hammoniensis*. They stated that the spermathecal saetae
were more anterior than segment ten, suggesting that the specimens
were *P. bedoti*. Howmiller and Loden (1976) had difficulty separating *P. hammoniensis, P. bavaricus* and *P. bedoti* and believed them
to be morphologically similar. Hiltunen (1973) suggested that the
taxa needed further research to solidify its identity. Hiltunen
(1967, 1969c) recorded, as a convenience, the taxon of *P. bavaricus* in his collections from Lake Michigan and Lake Ontario. The
specimens, along with other similar material, were kindly loaned
to the author by Hiltunen during a reappraisal of the "*P. bavaricus - bedoti* complex" in North America. The *P. bavaricus* recorded
by Hiltunen (1967, 1969c) were reidentified as *P. bedoti,* and his
additional material from Lakes Erie and Huron were also attributed
to this species. The work of Hrabě (1967), Timm (1970) and more
recently, Spencer (1978b) has indicated that *P. bedoti* and *P.
bavaricus* should be considered separate species. Brinkhurst and
Cook (1966) recognized *Rhyacodrilus sodalis* from Lake Ontario as
appearing as three distinct variants, differing in somatic setal
morphology. Hiltunen (personal communication) believes that this
extreme variation in setal morphology in the Lake Ontario material
suggests a degree of doubt in placing the taxon in *R. sodalis*.

The Enchytraeidae has had no modern revision in North America
and have received little attention elsewhere. Within the Great
Lakes region, Hiltunen (1967) did not discuss the group because
the taxonomy was inadequate nor did Spencer (1978b). Howmiller
(1974a) recorded *Lumbricillus* and specimens of the *Henlea - Enchytraeus* group from Lake Michigan; Cook (1975a) and Loveridge and
Cook (1976) reported *Mesenchytraeus* from Lake Superior, Georgian
Bay, and the North Channel of Lake Huron; and Judd and Bocsor
(1975) collected *Achaeta* from Lake Ontario. These generic records
of Enchytraeidae from the Great Lakes, should be considered as
tentative until a North American revision is undertaken.

ZOOGEOGRAPHY

The aquatic oligochaete fauna of the St. Lawrence Great Lakes
region closely resembles that occurring in the Palaearctic region.
For example: of the 82 taxa comprising the Lumbriculidae, Naididae
and Tubificidae, 77% occur within the Palaearctic while 40% or
less are common to other zoogeographic regions (Table 2). Of the
taxa occurring within the above 3 families, twenty-three (28%)
are Holarctic, 21 (26%) are cosmopolitan and 14 (17%) are exclusively Nearctic species (Table 3). The Great Lakes Lumbriculidae,
Naididae and Tubificidae can be divided into the following zoogeographical classifications determined by their present worldwide
distribution:

Table 2.   Percentage of Great Lakes taxa within the Lumbri-
           culidae, Naididae, and Tubificidae occurring in
           other zoogeographical regions of the world.

Zoogeographical Regions	Percentage
Palaearctic	77
Neotropical	26
Oriental	30
Ethiopian	40
Australian	22

Table 3.   Zoogeographical affinity of the Great Lakes
           oligochaete taxa within the families Lumbriculidae,
           Naididae, and Tubificidae.

Zoogeographical Regions	Number of Taxa	Percentage
Holarctic	23	28
Cosmopolitan	21	26
Other Nearctic Areas	14	17
Holarctic, Ethiopian	8	10
Holarctic, Australian	3	4
Holarctic, Oriental, Ethiopian	3	4
Holarctic, Neotropical	2	2
Holarctic, Oriental	2	2
Nearctic, Neotropical	2	2
Great Lakes Endemics ?	2	2
Nearctic, Neotropical, Oriental, Ethiopian	1	1
Holarctic, Ethiopian, Australian	1	1

Taxa which are Holarctic: *S. heringianus, E. lacustris, C. setosus, Amphichaeta* sp., *S. josinae, U. uncinata, N. bretscheri, N. alpina, N. behningi, V. intermedia, A. lomondi, S. lacustris, P. acuminata, T. newaensis, T. kessleri americanus, L. cervix, P. moldaviensis, P. hammoniensis, P. bedoti, P. vejdovskyi, P. ferox, R. sodalis* and *R. punctatus.*

Taxa which are cosmopolitan: *C. diastropus*, *C. langi*, *C. limnaei*, *N. communis*, *N. variabilis*, *N. elinguis*, *S. appendiculata*, *H. waldvogeli*, *D. digitata*, *D. furcata*, *A. pectinata*, *P. aequiseta*, *T. tubifex*, *L. hoffmeisteri*, *L. udekemianus*, *L. claparedeianus*, *L. profundicola*, *A. limnobius*, *A. pluriseta*, *A. pigueti* and *B. sowerbyi*.

Taxa which are exclusively Nearctic: *L. variegatus inconstans*, *P. michiganensis*, *D. vaga*, *P. plumaseta*, *L. maumeensis*, *L. angustipenis*, *P. californianus*, *P. variegatus*, *P. curvisetosus*, *P. multisetosus multisetosus*, *P. multisetosus longidentus*, *P. freyi*, *A. americanus* and *R. montana*.

Eight species have been recorded as being Holarctic, but distribution also includes the Ethiopian region: *P. litoralis*, *P. frici*, *N. simplex*, *N. pseudobtusa*, *V. comata*, *T. ignotus*, *I. templetoni* and *B. vejdovskyanum*.

Seven other zoogeographical groupings could be determined from world distributional records. Three species are recorded as Holarctic, but have also been collected in the Australian region: *P. idrensis*, *P. bavaricus* and *R. coccineus*. Three species were Holarctic, with distribution extending to the Oriental and Ethiopian regions: *C. diaphanus*, *S. fossularis* and *P. foreli*. Two species were noted as Holarctic and Neotropical: *O. serpentina* and *N. pardalis*. Species found to be Holarctic and Oriental were *N. barbata* and *L. silvani*. *B. unidentata* and *P. leidyi* were taxa distributed in the Nearctic and Neotropical regions. *P. osborni* appears to be found in the Nearctic, Neotropical, Oriental and Ethiopian regions; and *L. variegatus variegatus* is widely distributed in the Holarctic, Ethiopian and Australian regions.

The distribution of many oligochaetes, particularly in the Neotropical and Australian regions is incomplete. Thus, the apparent affinity of the Great Lakes fauna to the Palaearctic should be approached with caution until further data are available on a worldwide basis.

The Great Lakes fauna appears to consist mainly of Holarctic, cosmopolitan and exclusively Nearctic taxa, the question now is how did they arrive here? Speculation on the dispersal of aquatic oligochaetes into the Great Lakes has been addressed by Cook and Johnson (1974) and Cook and Hiltunen (1975).

Cook and Johnson (1974) believe that Segerstrale's theory of an all freshwater route into North America from the Old World is supported, in part, by the similarity of oligochaete taxa in both areas. They theorized that oligochaetes, along with other benthic components, arrived in North America during or before the penultimate (Illinoian) glacial period and achieved their present distribution through proglacial lakes and their fluctuating drainage patterns.

Cook and Johnson (1974) cite Brinkhurst et al. (1968) as suggesting the introduction by man of some oligochaetes into the Great Lakes region such as the tubificids *P. ferox*, some or all *Potamothrix*, and the lumbriculid *S. heringianus*. Recently, however, Brinkhurst (1976) notes *S. heringianus* as occurring in a variety of localities in North America other than the Great Lakes and state that these records dispose of the introduction theory for (at least) this species.

The lumbriculid, *E. lacustris*, was thought by Cook and Johnson (1974) to have originated from Mississippi drainage populations. This theory could have merit in a comparison of the Great Lakes and Mississippi oligochaete fauna (see below).

Cook and Hiltunen (1975) speculated that the tubificid, *P. hallae* may have gained entrance to the Great Lakes region 1) via Hudson Bay like the other glaciomarine relict species (Ricker, 1959); however the latter were widely distributed in the proglacial lakes along the entire margin of the Wisconsin ice sheet, so that present distributions coincide with the southern limit of the last glaciation; or 2) through a late Wisconsin marine transgression of the St. Lawrence lowlands to the level of Montreal and Lake Champlain, and possibly Lake Ontario. With both hypotheses, however, the assumption must be made that environmentally induced pressures have restricted *P. hallae* to the upper Great Lakes.

Of the aquatic Oligochaeta occurring in the St. Lawrence Great Lakes, the majority occur in other Nearctic areas. For example: all of the Lumbriculidae, 39 out of 41 taxa of Naididae, and 31 out of 37 Tubificidae have been recorded from areas other than the Great Lakes region. *C. setosus* and *N. alpina* (unconfirmed) appear to be the only Naididae restricted to the Great Lakes region. Six species of Tubificidae are not known from other Nearctic areas: *T. ignotus*, *T. newaensis*, *P. bedoti*, *P. superiorensis*, *R. punctatus* and *P. hallae*. Brinkhurst (1976), however, recently cited an unconfirmed report of *P. superiorensis* from the Tobique River, New Brunswick.

The drainage basin of the Great Lakes region is surrounded by 3 major watersheds: Hudson Bay, Atlantic, and Mississippi. Detailed data on the aquatic oligochaete fauna in these drainages is scarce. Recently, Brinkhurst (1976, 1978) introduced new distribution records for Canadian waters, including those bordering the Great Lakes region. However, species composition and distribution of the fauna in the watersheds within the United States in the vicinity of the Great Lakes basin is lacking. Foremost of

the United States watersheds is that of the extensive Ohio - upper
Mississippi drainage basin which borders the Great Lakes basin to
the south and west. The author has been studying the oligochaete
fauna of the Ohio River - upper Mississippi areas, and together
with the work of others (Howmiller, 1974b; Maciorowski et al.,
1977; Loden, 1978b; M.S. Loden, personal communication), compiled
a preliminary species list of the area (Table 4). Three taxa *P.
michiganensis, L. maumeensis,* and *P. multisetosus longidentus* are
reported here from areas outside the Great Lakes region for the
first time.

        The fauna of the Great Lakes and Ohio - upper Mississippi
drainage basins appear to be quite similar (Table 5). Cook and
Johnson (1974) believe one species of Lumbriculidae to have originated from Mississippi drainage populations and that the rich pele-
cypod fauna of inshore Lake Erie was derived from similar areas,
possibly through the Maumee River during former connections. It
is entirely possible that many oligochaete taxa were dispersed
between the two areas by this or similar routes. Interestingly,
many common Great Lakes species, like *S. heringianus,* have not
been recorded from the Ohio - upper Mississippi watershed. The
water quality in the watershed may limit some of these species,
but others may be collected from some of the more pristine waters,
especially in the Appalachian region.

        Cook and Johnson (1974) found that the Great Lakes differ
from each other quantitatively, not qualitatively, that is in
abundance, not kind of animals inhabiting the benthic environment.
This appears to be true for the oligochaetes of the region. Most
of the taxa are widespread or common throughout the Great Lakes
drainage (Table 6). With a few exceptions, little north-south
differences in distribution between the upper and lower Great
Lakes is evident.

        A zoogeographic discussion of a region would not be complete
without addressing the question of endemism. With the advent of
modern studies on North American oligochaete systematics, many new
taxa were described from the Great Lakes region, and additional
species described later (Table 7). As in most initial investigations of a region many of these species appeared to be restricted
to the Great Lakes basin. However, as surveys in other Nearctic
areas progressed, the majority of species originally described
from the Great Lakes region were found elsewhere. Presently, it
appears that only two species within the Lumbriculidae, Naididae,
and Tubificidae are restricted to the St. Lawrence Great Lakes
region. However, if an unconfirmed record of *P. superiorensis,*
mentioned by Brinkhurst (1976), is validated, the distribution of
only one species (*Phallodrilus hallae*) would be limited to the
Great Lakes basin.

Table 4.   A preliminary species list of Lumbriculidae, Naididae, and Tubificidae recorded from the Ohio and Upper Mississippi River drainage.

Lumbriculidae
  Lumbriculus variegatus variegatus (Müller)
  Eclipidrilus sp.
Naididae
  Chaetogaster diastrophus (Gruithuisen)
  C. diaphanus (Gruithuisen)
  C. limnaei von Baer
  Amphichaeta sp.
  Paranais frici Hrabě
  Bratislavia unidentata (Harman)
  Ophidonais serpentina (Müller)
  Nais communis Piguet
  N. variabilis Piguet
  N. simplex Piguet
  N. bretscheri Michaelsen
  N. elinguis Muller
  N. behningi Michaelsen
  Slavina appendiculata d'Udekem
  Vejdovskyella intermedia (Bretscher)
  Arcteonais lomondi (Martin)
  Stylaria lacustris (Linnaeus)
  Stylaria fossularis Leidy
  Piguetiella michiganensis Hiltunen
  Haemonais waldvogeli Bretscher
  Dero digitata (Müller)
  Dero furcata (Müller)
  Dero vaga (Leidy)
  Allonais pectinata (Stephenson)
  Stephensoniana trivandrana (Aiyer)
  Pristina idrensis Sperber

Naididae
  Pristina acuminata Liang
  Pristina synclites Stephenson
  Pristina foreli (Piguet)
  Pristina aequiseta Bourne
  Pristina leidyi Smith
Tubificidae
  Tubifex tubifex Muller
  Limnodrilus hoffmeisteri Claparède
  Limnodrilus udekemianus Claparède
  Limnodrilus cervix Brinkhurst
  Limnodrilus maumeensis Brinkhurst and Cook
  Psammoryctides californianus Brinkhurst
  Potamothrix moldaviensis Vejd. and Mrazek
  P. hammoniensis (Michaelsen)
  P. bavaricus (Öschmann)
  P. vejdovskyi (Hrabě)
  Ilyodrilus templetoni (Southern)
  Peloscolex variegatus Leidy
  P. curvisetosus (Brinkhurst and Cook)
  P. multisetosus multisetosus Br. and Cook
  P. multisetosus longidentus Br. and Cook
  P. freyi Brinkhurst
  Aulodrilus limnobius Bretscher
  A. pluriseta (Piguet)
  A. pigueti Kowalewski
  A. americanus Brinkhurst and Cook
  Bothrioneurum vejdovskyanum Stolc
  Rhyacodrilus c.f. *coccineus* (Vejdovsky)
  Branchiura sowerbyi Beddard

Table 5.   Number and percentage of Ohio River-upper
           Mississippi aquatic oligochaete taxa within the
           Lumbriculidae, Naididae, and Tubificidae which
           are common to the St. Lawrence Great Lakes.

Family	Number of Taxa	Percentage
Lumbriculidae	2	100
Naididae	29	94
Tubificidae	24	100

Table 6.   Distribution index of taxa within the Lumbriculidae,
           Naididae, and Tubificidae in six (6) areas[1] of the
           St. Lawrence Great Lakes region.

Index	Number of Taxa	Percentage
Widespread[2]	37	45
Common[3]	27	33
Rare[4]	18	22

[1]Lake Superior, Lake Michigan, Lake Huron, Lake Erie, Lake
 Ontario, and inland waters
[2]Occurs in 5-6 areas
[3]Occurs in 3-4 areas
[4]Occurs in 1-2 areas

HISTORICAL INVESTIGATIONS

     Early studies on the aquatic oligochaetes within the Great
Lakes basin were limited before the mid-1960's. Many studies
recorded the group as simply Oligochaeta (Adamstone et al., 1923;
Adamstone, 1924; Thomas, 1966), Tubificidae (Eggleton, 1936, 1937;
Henson, 1962), or *Tubifex* and *Limnodrilus* (Wright and Tidd, 1933;
Shelford and Boesel, 1942; Brown, 1953; Wright, 1955).

     Initial attempts at describing oligochaete species in the
Great Lakes occurred during the latter portion of the nineteenth
century. Smith (1874) and Smith and Verrill (1871) collected
several oligochaete species from the depths of Lake Superior.
They described 6 new species, including 4 from two obscure genera
(*Saenuris* and *Chirodrillus*), which Michaelsen (1900), as cited in
Cook and Johnson (1974), considered dubious taxa of Enchytraeidae.
Nicholson (1873) also recorded *Saenuris* from Lake Ontario, described *S. canadensis*, and found, in his estimation, a second
species of *Saenuris* and *Lumbriculus*. Nicholson, even at this
early date, recognized the importance of Annelida in the Great

Table 7. Aquatic Oligochaeta within the Lumbriculidae, Naididae, and Tubificidae originally described from the Great Lakes region.

Taxa	Type-Locality	Areas Collected Other Than The Great Lakes Drainage
Lumbriculidae		
Eclipidrilus lacustris (Verr.)	St. Ignace Island (Lake Superior)	Great Britain, Quebec, Canada
Naididae		
Piguetiella michiganensis Hilt.	Lake Michigan	Ohio–upper Mississippi drainage
Pristina osborni (Walton)	Lake Erie	India, Brazil, Africa, many Nearctic areas
Tubificidae		
Tubifex kessleri americanus Br. and Cook	Lake Superior and Lake Michigan	4 Canadian provinces, California, Soviet Union
Limnodrilus profundicola (Verr.)	Lake Superior	California, 7 Canadian provinces, Palaearctic, Neotropical
Limnodrilus maumeensis Br. and Cook	Lake St. Clair	Ohio River, Louisiana
Limnodrilus angustipenis Brinkhurst and Cook	Saginaw Bay (Lake Huron)	2 Canadian provinces, Louisiana
Peloscolex curvisetosus (Brinkhurst and Cook)	Eastern Lake Erie	Pennsylvania, Ohio River, North Carolina, Alabama
Peloscolex multisetosus longidentus Brinkhurst and Cook	Lake St. Clair, Western Lake Erie	Ohio River drainage, Louisiana
Peloscolex superiorensis Brinkhurst and Cook	St. Mary's River (Lake Superior)	*
Aulordilus americanus Brinkhurst and Cook	North Channel, St. Mary's River (Lake Superior)	6 Canadian provinces, Ohio River Tributary
Phallodrilus hallae Cook and Hiltunen	Lake Superior	*

* Possible endemics

Lakes fauna. He noted that "of all forms obtained in the dredgings none are of greater interest than the Annelides." He further added "Some of the Oligochaetes Annelides, also, have an extremely wide range, extending to nearly the greatest depths examined, and being rarely absent when the bottom is of a muddy nature." He believed the species of oligochaetes found to be of the family Naididae. H.B. Ward (in Reighard, 1894) sent a number of oligochaete speci- mens from Lake St. Clair to Dr. G. Eisen, then of the California Academy of Sciences. Eisen gave a preliminary report on the results stating that there were 7 genera of Oligochaeta and specu- lated that all species collected were new to science.

During the first quarter of the present century, six papers were published on Great Lakes oligochaetes. Moore (1906) recorded 13 species from collections taken in Lake St. Clair, the upper end of Lake Michigan, the small lakes south of the Straits of Mackinac, and Lake Erie. A description of one new species was presented in that paper. Walton (1906) surveyed the Naididae of Sandusky Bay on Lake Erie. One new species of *Chaetogaster* was described, as were 4 new species of *Nais*, 1 of *Pristina*, and 1 of *Naidium*. The first published report on the oligochaetes of Douglas Lake, Michi- gan was by Smith and Green (1916). Many of the species were ter- restrial, but a number of aquatic oligochaete taxa were noted. Smith (1919) recorded *Lumbriculus variegatus inconstans* from Lake Superior and placed *Lumbricus lacustris* in the genus *Mesoporodrilus* (=*Eclipidrilus*). Hayden (1922) collected 19 naidid taxa from Douglas Lake, Michigan, describing 2 new species and one new variety. Baker (1924) requested Frank Smith to identify some Oligochaetes from the Lake Winnebago region of Wisconsin. Six taxa were determined from these collections.

The period 1930 to the mid-1960's provided 8 more papers on Great Lakes Oligochaeta. Eggleton (1931) also requested Frank Smith to identify material from 2 Michigan lakes. Two species were found in these samples, *Limnodrilus hoffmeisteri* and *L. claparedeianus*. Krecker and Lancaster (1933) found *Limnodrilus, Nais* and *Lumbricillus* occurring at inshore areas of Lake Erie, and Moore (1930) recorded 7 oligochaete taxa from Douglas Lake, Michigan. Kenk (1949) investigated the invertebrates of ponds near Ann Arbor, Michigan and reported 10 oligochaete taxa from these areas. Langlois (1954) summarized the hydrobiological work in the Ohio region of Lake Erie. He noted that only 16 taxa of Lake Erie Oligochaeta had been reported by investigators. The bottom fauna of Lake Huron was investigated by Teter (1960). The identification of his Oligochaeta was made by E.W. Surber, who recorded *Naidium* sp. and *Limnodrilus claparedeianus*. Surber noted *L. claparedeianus* as the most numerous and that *Naidium* sp., the less abundant worm, was probably a new species. He stated that "the fact that it (*Naidium* sp.) has missed the attention of the taxonomist until now, even though it exists by the millions in

Lake Huron and probably the other Great Lakes, emphasizes how
little work has been done in these important bodies of water."
Merna (1960) recorded 3 oligochaete taxa from Lake Michigan.
Sparganophilus, Peloscolex sp. and *Limnodrilus udekemianus*
(most abundant, occurred in 98% of his samples. Hung (1962)
collected some oligochaetes from the Detroit River, but recorded
only *Limnodrilus claparedeianus* and Lumbriculidae from his material.

RECENT INVESTIGATIONS

The revision of the North American fauna made available the
systematic tools necessary for distributional and ecological
studies within the Great Lakes. These works together with a growing movement toward specific level identification associated with
benthological studies and an increased awareness of the deterio-
rating water quality within the Great Lakes system stimulated this
type of research in the region.

Lake Superior Including the St. Mary's River

The approximate locations of recent investigations of the
Lake Superior Oligochaeta are presented in Figure 2.

The first detailed, documented study on the open lake oligochaetes of Lake Superior was part of the work by Cook (1975a).
A major contribution of this investigation to Great Lake oligochaete research was the discovery of *Phallodrilus hallae*, a new
tubificid believed of marine ancestry. The description, distribution and ecology of *P. hallae* has been presented elsewhere (Cook
and Hiltunen, 1975). In addition to *P. hallae*, fourteen other
taxa were reported by Cook from nine lakewide sampling zones in
Lake Superior. In the Duluth basin, *S. heringianus* was the only
oligochaete collected. *S. heringianus* was the most abundant
oligochaete in the Apostle Islands region, but the fauna also
included *Mesenchytraeus, P. variegatus,* other Enchytraeidae, and
P. hallae being collected. Fauna similar to the southwest shore
area was reported at the Keweenaw to Grand Island region. The
eastern quarter region had *Mesenchytraeus* as the dominant taxon
with *S. heringianus*, other Enchytraeidae, *P. hallae* and *P. variegatus* being commonly collected. In Whitefish Bay, *S. heringianus*
and *Mesenchytraeus* were co-dominant oligochaetes. Although collected in lower proportions than in other locations, *Mesenchytraeus*
dominated the fauna with *P. hallae*, other Enchytraeidae and *S.
heringianus* at the Marathon region. The Northern Bays fauna consisted of *Mesenchytraeus* as the most abundant species, with *S.
heringianus, P. hallae,* other Enchytraeidae, *R. sodalis* and *B.
comata* being collected. The ninth zone, designated as the open
lake region, probably represents the primitive, truly profundal
benthic community of the upper Great Lakes according to Cook (1975a).

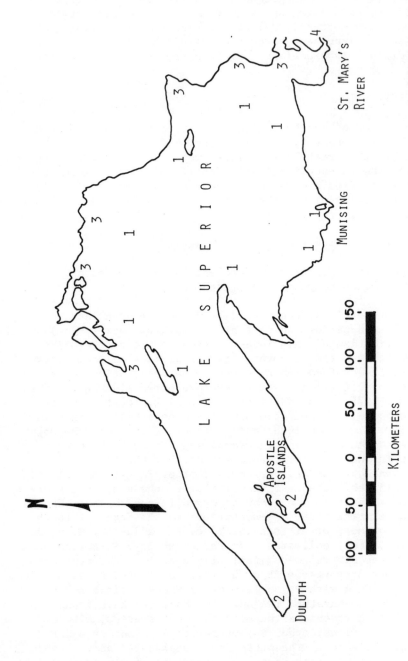

Figure 2.   Approximate locations of recent studies on aquatic Oligochaeta in Lake Superior.
Whole lake surveys (not numerically designated) – Cook (1975a), 1. Adams and
Kregear (1969), 2. Hiltunen (1969b), 3. Barton and Hynes (1978), 4. Veal, 1968.

Mesenchytraeus was the most commonly collected oligochaete, with
P. hallae, other Enchytraeidae and *S. heringianus* frequently found.
Rare taxa collected in the open lake region were *P. variegatus,*
R. sodalis and *R. punctatus.* The discovery of *R. punctatus* from
Lake Superior represented an important find. The species was
known previously only from Ohrid Lake in the Balkans, a site
renowned for its endemic species (Brinkhurst, 1978), and a spring
in Yugoslavia (Hrabě, 1973).

The oligochaete fauna of the eastern basin of Lake Superior
was investigated by Adams and Kregear (1969) in conjunction with
other macroinvertebrate and sedimentary studies. Three environments were recognized from field and laboratory descriptions and
dominant physical and sedimentary processes. The "boundary"
environment was defined as a zone from 2 to 10 km wide along the
southern periphery of the eastern basin and portions of the eastern and western shorelines of the Keweenaw Peninsula, occurred to
a depth of 60-90 meters and consisted of a sand or bedrock substrate. *S. heringianus* was collected sparsely in the sand substrate at only one location, a depth of 63 meters. They also noted
the occurrence of some unidentified Tubificidae and *Nais* and stated
that the Oligochaeta were not as abundant in the boundary environment as in pelagic areas. In the bedrock areas, oligochaetes were
quite abundant with the naidid, *S. lacustris* being of primary
importance. The authors observed that bedrock locations without
algal growth were devoid of Naididae; therefore, the assumption
was made that the presence of Naididae on bedrock is dependent on
the occurrence of algal growth. The shoal environment was determined to be the most restricted of the sedimentary environments
recognized in the eastern basin. The shoal was defined as occurring on the top of shallow aereally restricted positive topographic
features which project to within 30 meters of the water surface
and may be isolated from the land by deeper water. The only oligochaete found was *Chaetogaster* sp. They believed the shoal
environment to be unstable and, hence, unsuitable for most freshwater invertebrates. The pelagic environment occupied the deep
water portion of eastern Lake Superior lakeward of the boundary
environment and extended to the northern and eastern shore of the
lake. The sediment type in this region was classified as lacustrine mud and appeared more complex than the sediments in the
other two environments previously described. *S. heringianus* was
the dominant oligochaete present in the pelagic environment.

Hiltunen (1969b) observed a number of oligochaetes occurring
in samples from the Apostle Islands region of western Lake
Superior. Twelve oligochaete taxa were recorded with *S. heringianus* being the most numerous and frequently encountered. He
found the Tubificidae, except for *P. variegatus*, not occurring
deeper than 41 meters. Hiltunen believed that the Tubificidae
of the Apostle Islands are qualitatively similar to those found

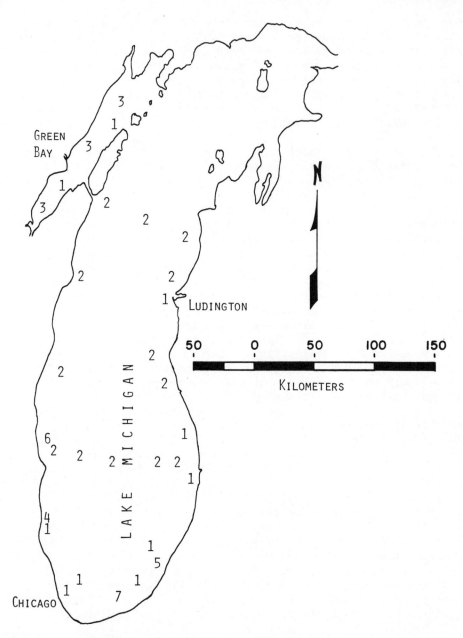

Figure 3.   Approximate locations of recent studies on aquatic
            Oligochaeta in Lake Michigan.   1. Hiltunen (1967),
            2. Howmiller (1974a), 3. Howmiller and Beeton (1970),
            4. Stimpson, et al. (1975), 5. Mozley and Garcia (1972)
            and Mozley and Winnell (1975), 6. Howmiller, in Mozley
            and Howmiller (1977), 7. Rains (1971).

in southern Lake Michigan, but apparently less abundant. He
concluded that the species composition of oligochaetes from the
Apostle Islands region of Lake Superior closely resembled that
reported from the other Great Lakes.

     The harbors and bays, and other inshore areas of Lake
Superior, have received occasional attention in recent years.
Hiltunen (1969b) collected 14 oligochaete taxa from Chequamegon
Bay, and 11 taxa from Duluth and Superior harbors. In the latter
area, Hiltunen stated that the assemblage of oligochaetes (*L. hoffmeisteri* was the dominant species) was similar to that from
Ludington Harbor on Lake Michigan. Adams and Kregear (1969)
reported on a large and diverse benthic fauna, completely atypical
to that in Lake Superior, occurring in the channel southwest of
Grand Island and northwest of Munising. They believed that this
assemblage occurs generally in the developed harbors in eastern
Lake Superior. *P. ferox* was the most abundant species followed
by *Limnodrilus* sp. and then *Tubifex* sp. A recent paper by Barton
and Hynes (1978) reported on the wave-zone macrobenthos of the
St. Lawrence Great Lakes bordering Canada which included Lake
Superior. The Oligochaeta occurring in this habitat were mainly
Naididae, but with the lumbriculid *S. heringianus* and Enchytraei-
dae commonly collected. Naidid oligochaetes commonly found were
C. diaphanus, *N. communis/variabilis*, *N. pseudobtusa*, *N. simplex*,
S. appendiculata, *S. lacustris*, and *U. uncinata*.

     Twenty-eight taxa of Oligochaeta were reported by Veal (1968)
from the St. Mary's River. A wide variety of species like *S.
heringianus*, *A. americanus*, *P. ferox*, *P. freyi* and *P. curvisetosus*
were collected upstream from Sault Ste. Marie. Collections
taken from near highly industrialized and urbanized areas of Sault
Ste. Marie, Ontario, consisted of a poor fauna, with species such
as *T. tubifex*, *L. hoffmeisteri* and *L. cervix* being found. Species
similar to those found above Sault Ste. Marie were not collected
until the river reached Little Lake George and the west channel
of Sugar Island.

Lake Michigan

     The approximate locations of recent investigations on the
Lake Michigan Oligochaeta is presented in Figure 3. Published
investigations have been primarily located in the central and
southern portions of the lake and Green Bay. However, a large
collection of Oligochaeta from a whole lake survey of Lake Michigan has recently been analyzed in the laboratory of Dr. David
White, Great Lakes Research Division, the University of Michigan
at Ann Arbor. Results from this material indicate that, when using
oligochaete assemblages to classify environmental types, some of
the northern portions of the lake appear to be extremely enriched
(Lauritsen, 1979).

The first detailed study of the Lake Michigan oligochaete
fauna was by Hiltunen (1967). Hiltunen found 40 taxa in his
southern Lake Michigan study. Most Naididae were collected at
depths of 6-19 meters, except *V. intermedia* which was taken at
depths down to 37 meters. *P. michiganensis* and *U. uncinata* were
the most commonly collected Naididae. *S. heringianus* occurred
most abundantly on the eastern side of the lake at depths of 37
to 46 meters, and on the western shoreline at 19 meters. The
maximum abundance of *P. variegatus* was at 37 meters on both shorelines. *P. moldaviensis* and *P. vejdovskyi* were widespread and
abundant in southern Lake Michigan, but especially off Waukegan,
Illinois. *Aulodrilus* spp. were not found deeper than 19 meters
and *L. angustipenis*, *L. claparedeianus* and *L. udekemianus* were
not collected deeper than 37 meters. *L. profundicola* was found
to be eurybathic. Hiltunen found *I. templetoni*, *P. freyi*, *P.
multisetosus* and *R. coccineus* not to occur deeper than 40 meters,
and *P. superiorensis* and *T. kessleri americanus* not shallower
than 37 meters. Hiltunen concluded that *S. heringianus*, *P.
variegatus*, *L. profundicola* and possibly *T. kessleri americanus*
are saprophobes; *L. hoffmeisteri*, *P. multisetosus*, *I. templetoni*
and *T. tubifex* are saprophiles; and *L. cervix* and *L. maumeensis*
are saprobionts.

In a short note, Howmiller (1974a) reported the Oligochaeta
in central Lake Michigan. *S. heringianus* comprised about 90%
of the material examined. Howmiller also collected two enchytraeid taxa, which he termed *Lumbricillus* and *Henlea - Enchytraeus* group. The only tubificid species found were *T. tubifex* and
L. hoffmeisteri.

Three harbor areas have been surveyed, one quite extensively.

Howmiller and Beeton (1970) recorded 30 oligochaete taxa,
the majority being Tubificidae, in an extensive survey of Green
Bay. *S. heringianus* occurred only at six stations in the northern
portion of the Bay. *D. digitata*, the most common naidid collected,
was restricted to the lower bay and western shore areas. *A.
lomondi* and *S. appendiculata* were widely distributed, but not in
the extreme lower bay or along the eastern shore. *A. americanus*
was restricted to the western side of the bay, and to sand substrates in the lower bay. *A. pluriseta* was the most widespread
of the genus *Aulodrilus* and associated with sand to mud substrates,
but was absent from the polluted lower bay and the oligotrophic
northern areas. *L. hoffmeisteri* was the most abundant and widespread oligochaete in Green Bay. The authors believed *L. udekemianus* to be common on sandy substrates, and *P. ferox* to prefer
sediments with a large sand component avoiding the more organic
sediments in the path of the Fox River. *P. freyi* was found to
be closely associated with enrichment, as its range extended
farther into the lower bay and reached its greatest relative

abundance near the southern end and near the mouth of the Menominee River. *P. multisetosus* extended farther into the lower
bay, but did not range as far north as *P. ferox* or *P. freyi*. It
was collected in highest density in the vicinity of the Green Bay
harbor entrance light and northward on the highly organic sediments of the eastern portion of the bay. *P. moldaviensis* was distributed similarly to *P. ferox,* being absent from the muddy areas
of the lower bay and eastern half of the middle bay. *P. vejdovskyi* was absent from the lower bay and sporadic in distribution
farther north. *T. kessleri americanus* was collected only at an
oligotrophic northern station. Howmiller and Beeton concluded
that *S. heringianus* and *T. kessleri americanus* were found at the
most oligotrophic northern stations. In the degraded lower bay,
Limnodrilus spp. (especially *L. hoffmeisteri*) and *D. digitata*
were common. As conditions improved in the middle bay, species
like *A. pigueti, A. pluriseta, I. templetoni, P. freyi* and *P.
multisetosus* were found.

     Howmiller, in Mozley and Howmiller (1977), found the oligo-
chaete fauna in Milwaukee Harbor to be similar in species composition to lower Green Bay. Species collected were *L. hoffmeisteri*
as the dominant species, along with *L. cervix, T. tubifex,* and
P. multisetosus.

     Hiltunen (1967) identified 8 oligochaete taxa from Ludington
Harbor. He found that *L. cervix, L. hoffmeisteri, L. maumeensis,*
and their immature stages, comprised a large proportion of the
species in this area and stated that *L. cervix* and *L. maumeensis*
have not been found in the open lake and are restricted to enriched
areas. In this same study, Hiltunen also sampled Green Bay, report-
ing twenty-five oligochaete taxa. Little could be determined from
these collections, but Hiltunen stated that the central portion
of the bay yielded a greater number of species than the southern
portion. Hiltunen concluded that the presence of *S. heringianus*
in the central bay area suggests the environment may be similar
to the open lake, and any adverse effects of the Menominee River
were minimal.

     A few investigations on the inshore distribution and ecology
of Lake Michigan Oligochaeta have been published.

     Rains (1971), as cited in Mozley and Howmiller (1977), found
that oligochaetes comprised 64% of the macroinvertebrates between
5 and 18 meters in Indiana waters of Lake Michigan. *S. heringianus* was rarely collected, and *L. hoffmeisteri, P. moldaviensis,*
and their immatures were dominant species. He found *L. cervix*
and *L. maumeensis* near Burns Ditch at a depth of 5 meters. Ten
meter stations near Gary, Indiana, were the least degraded because
of the frequent collection of *L. profundicola.*

Stimpson et al. (1975) recorded 32 oligochaete taxa from a study of a shallow water area in southwestern Lake Michigan. *P. michiganensis*, *U. uncinata*, *S. heringianus*, *L. hoffmeisteri*, *P. moldaviensis* and *P. vejdovskyi* were widely distributed and comprised the majority of Oligochaeta collected. In comparing depth and type of substrate to oligochaete distribution, they found the Naididae were collected at 3–12 meter depths, with the majority of species taken at 3–6 meters. *S. heringianus* occurred at deeper depths (9–18 meters) than in shallower areas. Slightly higher densities of *S. heringianus* were in areas with a silt substrate. Tubificidae increased markedly with depth and occurred in substrates containing more than 10% silt. The abundance of *L. hoffmeisteri* increased with depth and decreasing sediment particle size. Populations of *P. moldaviensis* also increased with depth, it was the predominant tubificid on sandy substrates. *P. vejdovskyi* had distributional patterns similar to *L. hoffmeisteri*. Densities of *A. americanus*, *A. pluriseta*, *I. templetoni*, *L. cervix*, *L. spiralis* and *P. multisetosus* increased at deep, silty stations. *L. profundicola* occurred only in small numbers and was consistently collected at 3–9 meter depths. They concluded that the 3 and 6 meter, sand bottom communities were characterized by Naididae, principally *P. michiganensis* and *U. uncinata*, and the Tubificidae, primarily *P. moldaviensis*. The 9–18 meter sand bottom community contained *S. heringianus* (usually comprising 50% of the oligochaete population), with *P. moldaviensis*, *L. hoffmeisteri* and *P. vejdovskyi* also present. In contrast to sand substrates, silty sediments at 9–18 meters provided an excellent habitat for *L. hoffmeisteri*, *A. americanus*, *A. pluriseta*, *I. templetoni*, *L. cervix*, *L. spiralis*, *P. multisetosus* and *P. vejdovskyi*.

Mozley and Garcia (1972) reported on the macrobenthos of an area in southeastern Lake Michigan. They collected 10 species of Oligochaeta. *S. heringianus* was the most abundant oligochaete collected and, along with *L. profundicola,* was abundant in deep waters. They also found *L. hoffmeisteri* to be the most abundant tubificid and *L. cervix* was present in some samples, which they believed, implied some environmental deterioration in local areas.

Mozley and Winnell (1975) provided further data on the inshore ecology of Lake Michigan Oligochaeta. *S. heringianus* had no discernable pattern of seasonal fluctuations and its density increased rapidly with depth to about 35 meters. The Naididae were numerous at depths to 16 meters, and the Tubificidae, dominated by *L. hoffmeisteri,* occurred at all depths. They found a total of 21 species of Naididae in the study area and, through analyses of plankton tows, found 8 of these species occurring only above the bottom (additional data on the occurrence of Great Lakes oligochaetes in net tows can be found in Wiley and Mozley (1978) for Lake Michigan and Hiltunen (1969b) for Lake Superior). The authors concluded that the asexual Naididae are especially

well adapted to sudden changes in environment and availability of
food, and species which can grow and mature in a continually cold
environment (S. heringianus) are more dominant in deeper parts of
the inshore areas.

Lake Huron

The approximate locations of recent investigations of the
Lake Huron Oligochaeta are presented in Figure 4.

Lake Huron is technically divided into two sections: the
main basin and Georgian Bay. The open lake (main basin) oligochaete fauna was described by Shrivastava (1974). He determined
that over half of the oligochaete population consisted of S.
heringianus. Other important species collected were L. hoffmeisteri, L. profundicola, P. ferox, T. tubifex, T. kessleri americanus and occasional occurrences of eleven additional taxa.

Two papers have recorded oligochaetes from Georgian Bay.
Brinkhurst et al. (1968) found 10 oligochaete taxa in samples from
Georgian Bay, with S. heringianus being the most widely distributed
species. A more recent report by Loveridge and Cook (1976), was
the first extensive examination of the benthic community of
Georgian Bay and the North Channel. In Georgian Bay, S. heringianus was the dominant oligochaete with Mesenchytreaues and other
Enchytraeidae being of secondary importance. 'In addition, 19
other oligochaete taxa were recorded. The authors found the
distribution of S. heringianus peculiar because it was almost
completely absent from the eastern and southern portions of
Georgian Bay and most dense along the central offshore areas.
Stations with high numbers of S. heringianus and Mesenchytraeus
tended to have low numbers of tubificids, and vice versa. Oligochaetes were fairly restricted to a mean particle size of less
than 5ϕ which, they believed, could account for the absence of
S. heringianus in southern and southeastern Georgian Bay as well
as the relatively low numbers of other oligochaetes in these
areas. In the North Channel, Loveridge and Cook (1976) recorded
22 oligochaete taxa with S. heringianus being dominant. They
cited other literature which indicated that high numbers of amphipods and oligochaetes in the North Channel may be due to the
nutrients discharged through the St. Mary's River and speculated
that this river may also influence macrobenthic populations in
the northwestern end of Georgian Bay as well.

Studies on the oligochaetes occurring in the bays, harbors
and inshore areas of Lake Huron are lacking. Brinkhurst (1967)
reported on the effect of the Saginaw River on the fauna of
Saginaw Bay. Brinkhurst stated that the degree of pollution of
the river and its small size reduced its effect on the bay. He
found D. digitata to be quite common and widely distributed in

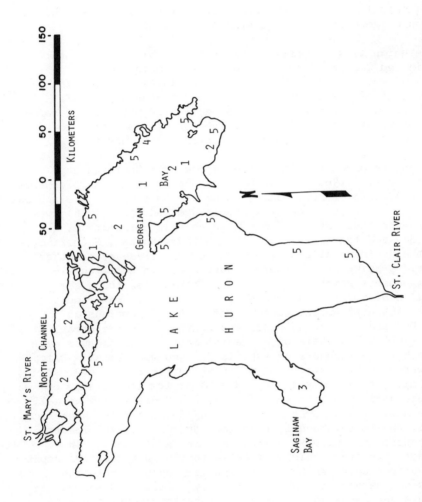

Figure 4. Approximate locations of recent studies on aquatic Oligochaeta in Lake Huron. Main lake survey (not numerically designated) – Shrivastava (1974), 1. Brinkhurst et al. (1968), 2. Loveridge and Cook (1976), 3. Brinkhurst (1967), 4. Hare and Carter (1977), 5. Barton and Hynes (1978).

the inner bay. Statistically, the Tubificidae were distributed
haphazardly, and no correlation existed between species abundance
and sediment particle size. Good correlations, however, were
found between conductivity and the distribution of *L. hoffmeisteri*.
Correlations were positive between *P. ferox* and depth, negative
between the abundance of this species and the amount of organic
matter in the sediments and a positive correlation of *I. templetoni* with depth and sediment organic content. Brinkhurst concluded
that with few exceptions the distribution of tubificid species
cannot be discussed in terms of commonly used ecological factors
and suggested a critical evaluation of the food materials available in the sediments. Brinkhurst further concluded that the
influence of the Saginaw River on the bottom fauna of Saginaw
Bay was less pronounced than the effects of all rivers going
into western Lake Erie. The presence of a polluted inflow into
Saginaw Bay was demonstrated by the presence of *P. litoralis* (a
questionable identification according to Hiltunen, personal communication) and the dominance of *L. hoffmeisteri* in the innermost
part of the bay.

Recently, Hare and Carter (1977) surveyed the Oligochaeta of
Parry Sound, Georgian Bay. They recorded 29 oligochaete taxa
from the Sound. Tubificidae were collected on mud substrates and
S. heringianus and/or Enchytraeidae on firmer substrates, and a
few species of Naididae in the deep region (below 30 meters depth).
At shallower depths, *S. heringianus, P. ferox* and *Nais communis/
variabilis* were found. *R. montanus* and *T. kessleri americanus*
were absent at depths shallower than 15 meters but common at 20-40
meters. *L. profundicola* was not collected shallower than 30 meters,
but was most abundant beyond 60 meters. *L. hoffmeisteri* was common
at all depths especially in shallow areas (60 meters or less).
The four *Aulodrilus* species, *P. freyi, L. udekemianus* and *T. coccineus* were restricted to shallow (less than 14 meters), more productive areas on mud substrates. The Naididae were common at
stations shallower than 15 meters, but the Enchytraeidae were
mainly abundant in deep water. The authors concluded that the
Tubificidae are very abundant in the deeper regions of Parry
Sound. Certain tubificids common in the oligotrophic upper Great
Lakes (*R. montana, T. kessleri americanus, L. profundicola*) are
in the deeper, cooler waters of the Sound. *T. tubifex, L. hoffmeisteri* and *I. templetoni* known from a wide range of habitats
in the Great Lakes, were collected all over the bay. *P. ferox*
appeared to tolerate a wide range of environmental conditions,
being found in degraded sites as well as exposed rocky shores
with *S. heringianus*. This study found *S. heringianus* to be
absent from soft mud substrates at depths greater than 60 meters.
The authors speculated that *S. heringianus* was excluded from most
of the deep water region because the soft sediments have a greater
oxygen demand or possible competition with tubificid species.

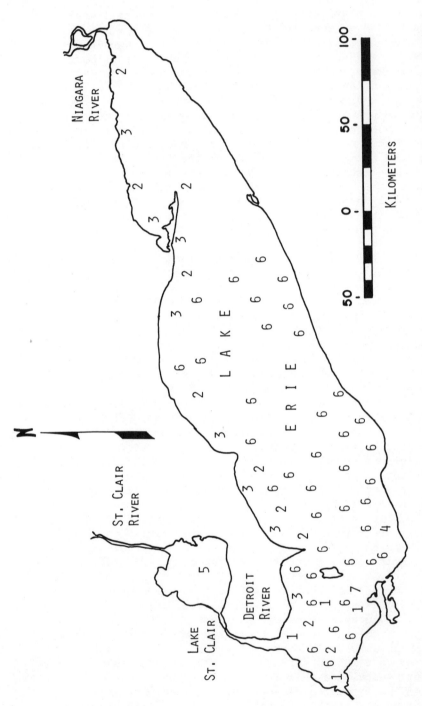

Figure 5. Approximate locations of recent studies on aquatic Oligochaeta in Lake Erie including Lake St. Clair. Whole lake surveys (not numerically designated) - Brinkhurst et al. (1968), 1. Hiltunen (1969a), 2. Veal and Osmond (1968), 3. Barton and Hynes (1978), 4. Spencer (1977), 5. Hiltunen (1971), 6. Pliodzinskas (1978), 7. Smith (1966).

Barton and Hynes (1978) recently reported 30 oligochaete taxa from the wave-zone of Lake Huron. The majority of these taxa were of the Naididae, with *N. communis/variabilis*, *N. paradalis* and *S. lacustris* being most frequently encountered.

Lake Erie including Lake St. Clair

The approximate locations of recent investigations of the Lake Erie and Lake St. Clair Oligochaeta are presented in Figure 5.

The only recent study of Lake St. Clair is that of Hiltunen (1971). He recorded 28 oligochaete taxa from the lake, along with other macrobenthos. The majority of species were Tubificidae, with the Naididae secondary.

Lake Erie is, historically, divided into three basins. From the literature, the distribution of different oligochaete associations appear to correspond to the artificial boundaries of these basins.

In a recent survey of the open water sediments of the western and central basin, Pliodzinskas (1978) recorded 30 oligochaete species with *B. sowerbyi*, *L. cervix*, *L. claparedeianus*, *L. hoffmeisteri*, *L. maumeensis*, *P. ferox*, *P. multisetosus*, and *P. moldaviensis* being commonly found throughout the study. In the central basin, *P. ferox* was the most important species, comprising up to 85% of the oligochaete community, with *P. multisetosus* and *L. hoffmeisteri* being of secondary importance. *Limnodrilus* spp. were dominant in the western basin with species such as *L. hoffmeisteri*, *L. cervix*, *L. maumeensis*, and *B. sowerbyi* being most abundant. *S. heringianus* was noted as occurring only in the far eastern and interior fringe of the central basin.

Hiltunen (1969a), in an investigation of oligochaetes in the western basin of Lake Erie, recorded 23 taxa. The tubificid genus *Limnodrilus* composed 90% of all oligochaetes at 33% of the 40 stations sampled. *L. hoffmeisteri* was the most abundant of the genus and reached its greatest numbers near the mouths of the Detroit, Rasin and Maumee Rivers and declined lakeward. The other species of *Limnodrilus* had similar distributional patterns. He found *P. multisetosus* to be poorly represented in western Lake Erie, except at the mouth of the Detroit River. Hiltunen concluded that the beneficial effect of river outflow on some tubificid species was evident. The relative abundance of *L. hoffmeisteri*, *L. cervix*, *L. maumeensis* and *P. multisetosus* was greatest near the mouth of rivers. He also noted the absence of *S. heringianus* from western Lake Erie.

Veal and Osmond (1968) investigated the bottom fauna of the western basin and the nearshore Canadian waters of the central and eastern basins of Lake Erie. Twenty-one oligochaete taxa were collected, with the Tubificidae constituting the majority of species. In the western basin, Tubificidae comprised 86% of the benthos and were high along the western shore of the basin, especially Maumee Bay and the mouth of the Detroit River. Tubificidae in the central and eastern basins composed 55% and 34%, respectively, of the bottom fauna. Species such as *L. cervix*, *L. hoffmeisteri*, and *L. maumeensis*, increased from the eastern to the western basin. *P. ferox* and *P. moldaviensis* occurred most frequently in the central basin, while *S. heringianus* and *T. tubifex* were more prevalent in the central and eastern basins.

Brinkhurst et al. (1968) found three oligochaete associations occurring in Lake Erie. They summarized the work of Hiltunen (1969a) in the western basin where the species *L. hoffmeisteri*, *L. claparedeianus*, *L. cervix* and *L. maumeensis* were found abundantly, especially near river mouths. Further into the western basin, *Aulodrilus* spp., *Potamothrix* spp., *B. sowerbyi*, *P. ferox* and *P. multisetosus* were found. In the central basin, *P. ferox* was the most frequently collected oligochaete, with *B. vejdovskyanum*, *I. templetoni*, *Potamothrix* spp., and *Aulodrilus* spp. being found. The oligochaetes in the eastern basin of Lake Erie consisted of *Aulodrilus* spp. in shallow waters, along with *S. heringianus*, *T. tubifex*, *P. curvisetosus* and *P. freyi*. *T. tubifex* and *S. heringianus* were noted by Brinkhurst as being typical of the Great Lakes where there is little evidence of eutrophication.

Studies on the oligochaete fauna of the inshore areas of Lake Erie are few. Barton and Hynes (1978) recorded 35 oligochaete taxa from the Canadian wave-zone habitats of Lake Erie. The Naididae especially *N. pardalis*, *N. communis/variabilis*, *N. simplex*, *S. lacustris*, and *U. uncinata* were commonly collected along with the tubificids *L. hoffmeisteri* and *P. moldaviensis*. In a short note, Spencer (1977) provided a species list of 26 oligochaete taxa collected from inshore areas of Lake Erie in the vicinity of Huron and Vermilion, Ohio and recorded *P. unidentata* from the Great Lakes region for the first time. Smith (1966), as cited in Pliodzinskas (1978), examined the oligochaetes of two stations in the Put-in-Bay region of Lake Erie. He collected 15 species at the two sites. *Limnodrilus* spp. were most abundant at his 10 meter station, but more Naididae occurred at his 2 meter location although *Limnodrilus* spp. were still dominant.

Lake Ontario including the upper St. Lawrence River

The approximate locations of recent investigations on the
Lake Ontario and upper St. Lawrence River Oligochaeta are presented in Figure 6. The oligochaetes of Lake Ontario appear to
have been investigated more extensively than the fauna of the
other Great Lakes. Ten papers, where oligochaetes are included,
have been published from the lake since the mid-1960's and four
of these have been whole lake surveys: Brinkhurst et al. (1968),
Hiltunen (1969c), Kinney (1972), and Nalepa and Thomas (1976).

Brinkhurst et al. (1968), found S. heringianus to be the
most common oligochaete species in Lake Ontario, with Tubificidae,
such as L. hoffmeisteri and T. tubifex, widely distributed, especially in grossly polluted bays. P. ferox, Aulodrilus spp., and
Potamothrix spp. had characteristic distribution along the shoreline. They reported I. templetoni, L. cervix, L. claparedeianus,
L. udekemianus and P. multisetosus from polluted stretches of the
lake especially Hamilton Bay, Toronto Bay, below the Niagara River
and the Bay of Quinte.

Hiltunen (1969c) reported 31 oligochaete taxa obtained in a
1964 Lake Ontario survey. S. heringianus was found to be common
throughout the lake, except where the environment was unfavorable,
such as near Toronto, Ontario and Rochester, New York. The
Naididae were collected mainly from shallow waters, except for
V. intermedia which was found at a depth of 48 meters. L. hoffmeisteri was the most common and widespread member of the Tubificidae. P. moldaviensis, P. vejdovskyi, P. ferox, P. multisetosus,
A. americanus, A. pluriseta, L. udekemianus and T. ignotus were
found in shallow waters. L. cervix was found to be saprobiontic,
its distribution thought to be influenced more by water quality
than depth. L. profundicola and T. kessleri americanus were
found only in the two deepest areas. Rhyacodrilus sp. and P.
variegatus were collected in both shallow and deep water, but
could prefer depths between 35 and 195 meters. T. tubifex was
found to be facultative toward water quality, thriving in areas
of urban pollution and in deep waters where the substrate is not
productive. P. vejdovskyi occurred in productive areas, but was
poorly represented off Toronto Harbor. At the mouth of the
Niagara River, Hiltunen reported S. heringianus coexisting with
L. cervix, L. hoffmeisteri, P. multisetosus and T. tubifex. He
speculated that the Niagara River may provide sufficient nutrition
to support saprophiles, but is also free of elements which inhibit
S. heringianus. T. ignotus was also noted as occurring near the
mouth of the Niagara River and was found by Hiltunen (1967) to
be similarly distributed near a river mouth in Lake Michigan.

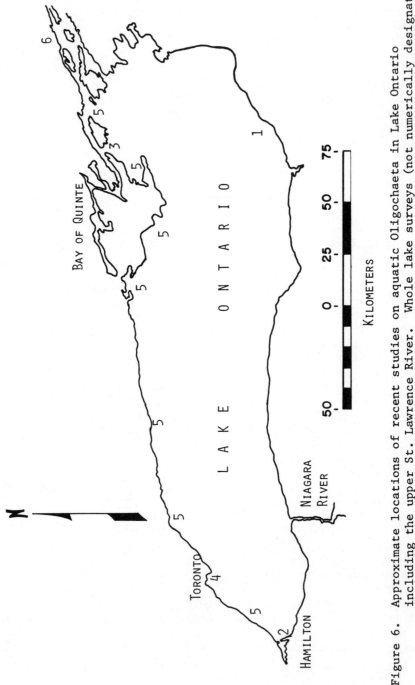

Figure 6.   Approximate locations of recent studies on aquatic Oligochaeta in Lake Ontario
including the upper St. Lawrence River. Whole lake surveys (not numerically designated)
Brinkhurst et al. (1968), Hiltunen (1969c), Kinney (1972), Nalepa and Thomas (1976);
1. Judd and Bocsor (1975) and Bocsor and Judd (1972); 2. Johnson and Matheson (1968);
3. Johnson and Brinkhurst (1971), 4. Brinkhurst (1970), 5. Barton and Hynes (1978),
6. Kinney (1972, in part).

Kinney (1972) found that the oligochaetes of the deeper waters of Lake Ontario were characterized by an oligotrophic fauna similar in composition to the upper Great Lakes, with *S. heringianus* predominating. *L. hoffmeisteri* and *T. tubifex* were common at these depths, along with *L. profundicola*, *L. spiralis* and *T. kessleri americanus*. Samples from his intermediate depth stations yielded *S. heringianus*, Enchytraeidae, *Rhyacodrilus* sp., *P. ferox*, *P. variegatus* and *P. vejdovskyi*. From the Niagara River to Rochester, New York and in the vicinity of the Oswego River discharge the effects of urbanization were found. There was a decrease in the number of *S. heringianus* collected and an increase in *L. hoffmeisteri*, *T. tubifex*, *P. vejdovskyi*, *L. claparedeianus*, *L. udekemianus*, *P. ferox*, *P. multisetosus*, *P. bavaricus* and *P. moldaviensis*. In shallow waters near the Niagara, Genesse and Oswego River discharges, saprophobic forms diminished and were replaced by *L. hoffmeisteri*, *L. spiralis*, *P. moldaviensis*, *P. vejdovskyi*, *T. tubifex*, *L. cervix*, *L. claparedeianus*, *L. udekemianus*, *P. multisetosus*, and *T. ignotus*. At shallow stations in the eastern and northeastern portion of the lake, high densities of saprophobic species were in coexistence with the less sensitive species. Saprophilous and saprobiontic oligochaetes were reported from the Niagara River mouth, lower Genessee River, Rochester Embayment, Little Sodus Bay, Oswego Harbor, Great Sodus Bay, Chaumont and Black River Bays.

In a more recent survey of the bottom fauna of Lake Ontario, Nalepa and Thomas (1976) found *S. heringianus* to be one of the most widely distributed species, accounting for 57% of the oligochaetes collected. This species was not found, however, at the mouth of the Niagara River, the Toronto area, or 2 stations at the far eastern end of the lake. *L. hoffmeisteri* was widely distributed with greatest abundance near the mouth of the Niagara River, near Toronto, and at Rochester, New York. *T. tubifex* was collected from deep water habitats and in polluted areas. Oligochaetes displaying littoral distribution were *Aulodrilus* spp., *P. ferox*, *P. multisetosus*, *P. vejdovskyi* and *P. moldaviensis*. Most of these species were consistently collected along the southern shoreline between the Niagara River and Rochester, New York. The authors statistically related the distribution of some species to various sediment parameters. *S. heringianus* was inversely related to the percent nitrogen and sediment particle size distribution at intermediate depths, *L. hoffmeisteri* was directly related to the percent nitrogen and percent carbon at deep stations and *T. tubifex* was directly related to the percent nitrogen and percent phosphorus at deep stations.

Published information on the inshore oligochaete fauna of Lake Ontario is scanty, but three harbor areas have been studied in detail. Two recent papers (Judd and Bocsor, 1975; Bocsor and Judd, 1972) reported on the inshore benthic fauna of the lake,

including Oligochaete, in relation to the effects of paper plant
pollution and abatement. They found that mats of the algal genus,
Cladophora, were well colonized by the Naididae, which comprised
99% of all oligochaetes collected. The Naididae increased in
population during late summer,. which paralleled the decrease and
deterioration of *Cladophora*. They speculated that this population
increase was due to an increase in reproduction because of an
increased food supply, from the bacteria and detritus associated
with the decomposing *Cladophora*. Barton and Hynes (1978) recorded
31 taxa of Oligochaeta from the Canadian wave-zone areas of Lake
Ontario. Almost 50% of these were of the Naididae, with *C. diaphanus* and *N. pardalis* most frequently encountered.

     Johnson and Matheson (1968) investigated the bottom fauna
of Hamilton Bay and adjacent Lake Ontario. The profundal community of the bay consisted of *L. hoffmeisteri* and *T. tubifex,* both
numerous because of accumulations of organic materials and the
absence of competition. The sediments in this area contained
more than .25% organic nitrogen and .50% phosphorus with loss on
ignition exceeding 10%. Sublittoral sediments had less nitrogen,
phosphorus and a decreased loss on ignition, so oligochaetes were
less numerous. The sediments in the southeastern section of the
bay were rich because of sewage outfalls, however, no biota was
found near steel mills where iron concentrations were high and
dissolved oxygen was low. A large and varied population of
oligochaetes existed in the bay near the canal leading to Lake
Ontario. This was believed to be a result of more favorable
water chemistry and water movement, in combination with high
amounts of organic materials. In the lake, greater oligochaete
numbers were directly related to increasing depth. The authors
speculated that the increase of oligochaetes with depth was
related to an increase of silt and clay, or greater concentration
of nitrogen and organic matter. In the profundal sediments of
the lake, nitrogen levels and loss on ignition was similar to
that in the bay profundal areas. They believed the lower densities of oligochaetes in the lake profundal as compared to the bay
was related to competition with other macroinvertebrates. The
oligochaetes occurring in the lake profundal were mainly *S. heringianus, L. hoffmeisteri* and *T. tubifex.*

     Johnson and Brinkhurst (1971) conducted a study on the bottom
fauna of the Bay of Quinte and adjacent Lake Ontario, in which
three oligochaete associations were described. *P. multisetosus*
and *Limnodrilus* spp. were collected at the inner and middle bay.
A group consisting of *P. ferox, A. pluriseta, P. bavaricus, P.
vejdovskyi* and *P. moldaviensis* occurred in Adolphus Reach and
most of Prince Edward Bay. *S. heringianus* was found only in
samples from deep areas of Lake Ontario. *T. tubifex* occurred
in both the second and third groups. The authors concluded that
the sequence of associations along the Trenton-Lake Ontario

transect resembled that of the west-east transect in Lake Erie. *Limnodrilus* spp., are abundant in western Lake Erie and also in inner Bay of Quinte. This association is replaced toward the central basin of Lake Erie by *P. ferox*, *Potamothrix* spp. and *Aulodrilus* spp., which resembles Adolphus Reach and Prince Edward Bay. The eastern basin of Lake Erie resembles the main basin of Lake Ontario where *S. heringianus* and *T. tubifex* were collected. They also noted that low temperature inhibited diversity in Lake Ontario, although depth and pressure may have also been significant.

Brinkhurst (1970) surveyed Toronto Harbor and found it to be grossly polluted. Species such as *T. tubifex*, *L. hoffmeisteri* and *P. multisetosus* were the most abundant species, with *L. udekemianus*, *L. cervix*, *L. claparedeianus*, *A. pluriseta*, *P. vejdovskyi* and *P. hammoniensis* rarely found. The first 3 species accounted for 93% of the total oligochaetes found. The latter 3 species were restricted to the south or island shores of the study area.

The upper St. Lawrence River was investigated, in part, by Kinney (1972). He reported Enchytraeidae, Lumbriculidae, Naididae, and the tubificids *A. americanus*, *L. hoffmeisteri*, *P. ferox*, *P. moldaviensis*, and *P. vejdovskyi* from this area.

Inland Waters

The approximate locations of recent investigations of the inland water Oligochaeta are presented in Figure 7. Recent studies on the Oligochaeta occurring in the waters of the Great Lakes region, other than the major Great Lakes, are lacking. Brinkhurst et al. (1968), presented a partial listing of aquatic Oligochaeta from Lake Nipigon and the Patricia District Lakes of Ontario. They reported nine oligochaete taxa from these inland waters and stated that 3 of these species (*T. kessleri americanus*, *A. americanus*, and *T. montana*) are probably indigenous northern species, being not found further south than the Great Lakes system. Brinkhurst (1976) recently provided additional records of Oligochaeta from the inland waters of Canada including the St. Lawrence Great Lakes region.

Howmiller (1974b) studied the Oligochaeta of inland waters of Wisconsin, twelve of which are within the Great Lakes drainage. His study recorded 16 oligochaete taxa from various rivers and lakes within the confines of the basin. In general, Howmiller found no consistent correlation of abundance or species composition with lake type from profundal collections. He determined the profundal worm fauna to be species poor with species other than *L. hoffmeisteri*, *I. templetoni* and *T. tubifex* seldom found.

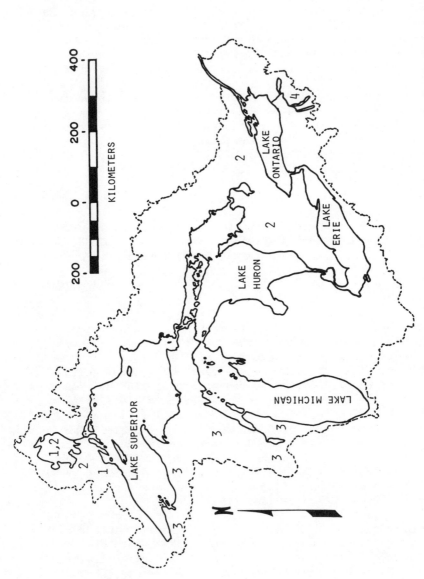

Figure 7. Approximate areas within the Great Lakes region where inland water aquatic Oligochaeta have been recorded in the post-1965 literature. 1. Brinkhurst et al. (1968), 2. Brinkhurst (1976), 3. Howmiller (1974b), 4. Spencer (1978b).

In the first major report on the oligochaete fauna of the
Finger Lakes region of New York State, Spencer (1978b) recorded
40 oligochaete taxa from Cayuga Lake. He found *T. tubifex* and
P. bedoti abundant at profundal depths, along with *T. kessleri
americanus*. *P. vejdovskyi* and *P. moldaviensis* were common littoral inhabitants. The author concluded that the fauna closely
resembles that of the St. Lawrence Great Lakes. He also noted
that *S. heringianus* was absent from Cayuga Lake although present
occasionally in large numbers within the five major Great Lakes,
and *P. bedoti* was found to occur abundantly in Cayuga Lake while
rare in other locations in Europe and North America.

CONCLUSIONS AND THE FUTURE

The aquatic Oligochaeta occurring within the St. Lawrence
Great Lakes region have been well defined in past investigations.
However, there are still many areas requiring additional study,
which will be addressed later in this section. An attempt will
be made now to synthesize the present body of knowledge gained
from Great Lakes oligochaete studies.

Henson (1966) believed the benthic communities of the St.
Lawrence Great Lakes to be composed of a young fauna. He based
this observation on the effects of Pleistocene glaciation, which
occurred as recently as 4,000-8,000 years ago, and the lack of a
rich fauna characteristic of more ancient lakes (few species and
low endemism). The scouring effects of glaciation would have
been devastating to the benthic community, eliminating the fauna
or forcing them south of the glacial margins. The mobility of
the Oligochaeta is restricted, so the pre-glacial forms were
probably eliminated from the region. Cook and Johnson (1974)
believe some oligochaetes were distributed through pro-glacial
lakes and their fluctuating drainage patterns from European
sources. The close resemblance of the Great Lakes species
with the Palaearctic fauna is evident, but other dispersal routes
into the region should also be contemplated, considering the similarity of the fauna in other drainage basins bordering the Great
Lakes region. A more plausible explanation for the present
species distribution in the Great Lakes region would perhaps be
a combination of these dispersal pathways and possibly other
known migration routes. Henson (1966) believed the number of
aquatic species in the Great Lakes to number only 10. Presently,
we have recorded over 80 species from the region, but many are
cosmopolitan forms. Endemism is low with only 1 or possibly 2
species appearing restricted to the region.

The aquatic Oligochaeta differ quantitatively, not qualitatively, in various portions of the Great Lakes region; and appear
to exhibit selective community types, apparently associated with
environmental conditions. There appears to be little difference

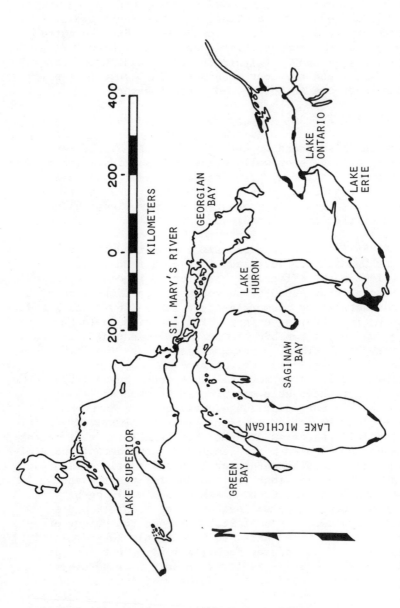

Figure 8. Distribution (in black) of "polluted water" assemblages of tubificids in the
St. Lawrence Great Lakes. There are probably other locations with similar species
associations, but detailed analysis of the oligochaete community in many areas is
not available at the present time. (Modified from Brinkhurst and Cook, 1974).

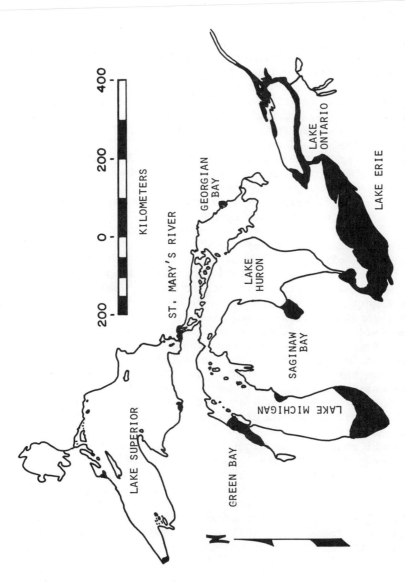

Figure 9. Distribution (in black) of "eutrophic lake" and "polluted water" assemblages of oligochaetes in the St. Lawrence Great Lakes. White areas represent "clean water" oligochaete associations, but may also refer to locations that have not been categorized as to type of oligochaete assemblage. (Modified from Brinkhurst and Cook, 1974).

in the kinds of worms found in the upper and lower lakes, with
the majority being widespread or common throughout the drainage.
The abundance of various species, especially the Tubificidae, does
vary between the 5 major Great Lakes. Brinkhurst et al. (1968)
and Brinkhurst (1969) have noted various species associations
within the region. Their investigations concluded that degraded
or polluted areas have species assemblages dominated by *L. hoffmeisteri* and *T. tubifex*, which include *P. multisetosus, L. cervix,
L. maumeensis, B. sowerbyi* and perhaps *I. templetoni*. The mesotrophic or "eutrophic lake" community consists of *Aulodrilus* spp.,
Potamothrix spp., *P. ferox, P. freyi, L. hoffmeisteri* and *S.
heringianus*. The third association includes "clean water" forms
such as *Rhyacodrilus* sp., *T. kessleri americanus, A. americanus,
L. hoffmeisteri* and species such as *S. heringianus, P. variegatus*.
In addition, *P.hallae* should be included with this listing for
the Lake Superior and northern Lake Huron areas. *T. tubifex* would
also be found in these unproductive situations, occasionally abundant. By applying the above oligochaete classification, it is
possible to delineate certain locations within the region which
are polluted, mesotrophic or oligotrophic (Figures 8 and 9).

An extensive amount of research has been conducted on the
Great Lakes aquatic oligochaete fauna; however, additional knowledge is paramount. The taxonomic and zoogeographic problems,
centering primarily around the Naididae and Tubificidae, warrant
further investigation. The application of promising biochemical
principles presented by Milbrink and Nyman (1973a,b) in conjunction with standard systematic procedures could resolve some of the
taxonomic problems. The Enchytraeidae appear to be a much more
important element in the fauna of the Great Lakes, especially the
upper lakes, than was previously realized. Thus, it is imperative
that the systematics of this family be revised not only in the
Great Lakes region, but the whole of North America. More specific
work should be initiated on the distribution of the Great Lakes
Oligochaeta. Baseline data are non-existent in many areas and
surveys should be made to determine the species present before any
environmental degradation occurs, especially in the upper lakes.
Continual monitoring of the species associations present in areas
already studied could reveal improvement (or deterioration) of
conditions. There is a need for surveys on the species occurring
in the streams, rivers and smaller lakes of the region. More
intense research of these areas will add additional, and perhaps
unusual, data on the distribution of aquatic Oligochaeta in the
St. Lawrence Great Lakes region.

ACKNOWLEDGMENTS

The author is indebted to David G. Cook for his criticism of
this manuscript. He also acknowledges the financial support of
NUS Corporation in various aspects of the paper preparation and

presentation, William R. Cody for photographic expertise, and
Terry R. Rojahn for drafting assistance. This paper is dedicated
to the many scientists, both past and present, who have contributed
to oligochaete research in the St. Lawrence Great Lakes region,
especially Jarl K. Hiltunen of the Great Lakes Fishery Laboratory,
Ann Arbor, Michigan, who not only reviewed this manuscript, but
encouraged the author to study these fascinating invertebrates
some 10 years ago.

REFERENCES

Adams, C.E. and R.D. Kregear, 1969, Sedimentary and faunal envi-
    vironment of eastern Lake Superior. Proc. 12th Conf. Great
    Lakes Res.,1969:1-20.
Adamstone, F.B., 1924, The distribution and economic importance
    of the bottom fauna of Lake Nipigon with an appendix on the
    bottom fauna of Lake Ontario. Univ. Toronto Stud.,Biol. Ser.
    No. 25, Publ. Ont. Fish. Res. Lab., 24:35-100.
Adamstone, F.B. and W.J.K. Harkness, 1923, The bottom organisms
    of Lake Nipigon. Univ. Toronto Stud.; Pub. Ont. Fish. Res.
    Lab., 15:123-170.
Baker, F.C., 1924, The fauna of the Lake Winnebago region. Trans.
    Wis. Acad. Sci., Arts, Letts., 21:109-146.
Barton, D.R. and H.B.N. Hynes, 1978, Wave-zone macrobenthos of
    the exposed Canadian shores of the St. Lawrence Great Lakes.
    J. Great Lakes Res., 4:27-45.
Bocsor, J.G. and J.H. Judd, 1972, Effect of paper plant pollution
    and subsequent abatement on a littoral macroinvertebrate
    community in Lake Ontario: preliminary survey. Proc. 15th
    Conf. Great Lake Res., 1972:21-34.
Brinkhurst, R.O., 1963, Taxonomical Studies on the Tubificidae
    (Annelida, Oligochaeta). Int. Revue ges. Hydrobiol. Syst.
    Beih., 2:7-89.
Brinkhurst, R.O., 1964, Studies on the North American aquatic
    Oligochaeta. I: Naididae and Opistocystidae. Proc. Acad.
    Nat. Sci. Phila., 116:195-230.
Brinkhurst, R.O., 1965, Studies on the North American aquatic
    Oligochaeta. II: Tubificidae. Proc. Acad. Nat. Sci. Phila.,
    117:117-172.
Brinkhurst, R.O., 1967, The distribution of aquatic oligochaetes
    in Saginaw Bay, Lake Huron. Limnol. Oceanogr., 12:137-143.
Brinkhurst, R.O., 1968, Oligochaeta, p. 69-85. In F.K. Parrish
    (ed.) Keys to the Water Quality Indicative Organisms of the·
    Southeastern United States. U.S. Dept. of the Interior,
    Fed. Wat. Poll. Contr. Admin., Cincinnati, Ohio.
Brinkhurst, R.O., 1969, Changes in the benthos of Lakes Erie and
    Ontario. Bull. Buffalo Soc. Nat. Sci., 25:45-71.
Brinkhurst, R.O., 1970, Distribution and abundance of tubificid
    (Oligochaeta) species in Toronto Harbour, Lake Ontario.
    J. Fish. Res. Bd. Can., 27:1961-1969.

Brinkhurst, R.O., 1976, Aquatic Oligochaeta recorded from Canada and the St. Lawrence Great Lakes. Fish. Mar. Serv. Pac. Mar. Sci. Rep., 76-4:1-49.

Brinkhurst, R.O., 1978, Freshwater Oligochaeta in Canada. Can. J. Zool., 56:2166-2175.

Brinkhurst, R.O. and D.G. Cook, 1966, Studies on the North American aquatic Oligochaeta III: Lumbriculidae and additional notes and records of other families. Proc. Acad. Nat. Sci. Phila., 118:1-33.

Brinkhurst, R.O., 1974, Aquatic earthworms (Annelida:Oligochaeta), p. 143-156. In C.W. Hart, Jr. and S.L.H. Fuller (eds.) Pollution Ecology of Freshwater Invertebrates. Acad. Press, New York, London.

Brinkhurst, R.O., A.L. Hamilton and H.B. Herrington, 1968, Components of the bottom fauna of the St. Lawrence Great Lakes. Univ. Toronto Great Lakes Inst. No. PR 33:1-49.

Brinkhurst, R.O. and B.G.M. Jamieson, 1971, Aquatic Oligochaeta of the world. Oliver and Boyd, Edinburgh, 860 p.

Brown, E.H., 1953, Survey of the bottom fauna at the mouths of ten lake Erie south shore rivers. Ohio Dept. Nat. Res. Water Div. L. Erie Poll. Sur., Final Rep., 156-170.

Cook, D.G., 1975a, A preliminary report on the benthic macro-invertebrates of Lake Superior. Fish. Mar. Serv. Tech. Rept., 572:44 p.

Cook, D.G., 1975b, Cave-dwelling aquatic Oligochaeta (Annelida) from the eastern United States. Trans. Am. Microsc. Soc., 94:24-37.

Cook, D.G. and J.K. Hiltunen, 1975, Phallodrilus hallae, a new tubificid oligochaete from the St. Lawrence Great Lakes. Can. J. Zool., 53:934-941.

Cook, D.G. and M.G. Johnson, 1974, Benthic macroinvertebrates of the St. Lawrence Great Lakes. J. Fish. Res. Bd. Can., 31:763-782.

Eggleton, F.E., 1931, A limnological study of the profundal bottom fauna of certain fresh-water lakes. Ecol. Monogr., 1:233-331.

Eggleton, F.E., 1936, The deep water bottom fauna of Lake Michigan. Pap. Mich. Acad. Sci. Arts Lett., 21:599-612.

Eggleton, F.E., 1937, Productivity of the profundal benthic zone in Lake Michigan. Pap. Mich. Acad. Sci. Arts Lett., 22:593-611.

Hare, L. and J.C.H. Carter, 1977, The Oligochaeta, Polychaeta and Nemertea of Parry Sound, Georgian Bay. J. Great Lakes Res., 3:184-190.

Harman, W.J. and M.S. Loden, 1978, Bratislavia unidentata (Oligochaeta: Naididae), a re-description. Southwest. Nat., 23:541-544.

Harman, W.J. and M.L. McMahan, 1975, A reevaluation of Pristina longiseta (Oligochaeta: Naididae) in North America. Proc. Biol. Soc. Wash., 88:167-178.

Hayden, H.E., 1922, Studies on American naid oligochaetes. Trans. Am. Microsc. Soc., 41:167-171.

Henson, E.B., 1962, Notes on the distribution of the benthos in the Straits of Mackinac region. Proc. 5th Conf. Grt. Lakes Res., Univ. Mich. Great Lakes Res. Div. Publ., 9:174 (Abstract).

Henson, E.B., 1966, A review of Great Lakes benthos research. Univ. Mich. Great Lakes Res. Div. Publ., 14:37-54.

Hiltunen, J.K., 1967, Some oligochaetes from Lake Michigan. Trans. Am. Microsc. Soc., 86:433-454.

Hiltunen, J.K., 1969a, Distribution of oligochaetes in western Lake Erie, 1961. Limnol. Oceanogr., 14:260-264.

Hiltunen, J.K., 1969b, Invertebrate macrobenthos of western Lake Superior. Mich. Acad., 1:123-133.

Hiltunen, J.K., 1969c, The benthic macrofauna of Lake Ontario. Great Lakes Fish. Comm. Tech. Rep. No., 14:39-50.

Hiltunen, J.K., 1971, Limnological data from Lake St. Clair, 1963 and 1965. U.S. Dept. Commer., Nat. Oceanic Atoms. Admin., Nat. Mar. Fish. Serv., Data Rep., 54, 45p. on 1 macrofiche.

Hiltunen, J.K., 1973, A laboratory guide: Keys to the tubificid and naidid Oligochaeta of the Great Lakes region. unpublished manuscript. 24 pp.

Holmquist, C., 1978, Revision of the genus *Peloscolex* (Oligo-chaeta, Tubificidae) 1. Morphological and anatomical scrutiny; with discussion on the generic level. Zool. Scripta,7:187-208.

Howmiller, R.P., 1974a, Composition of the oligochaete fauna of central Lake Michigan. Proc. 17th Conf. Great Lakes Res., 1974:589-592.

Howmiller, R.P., 1974b, Studies on aquatic Oligochaeta in inland waters of Wisconsin. Wis. Acad. Sci., Arts, Lett., 62:337-356.

Howmiller, R.P. and A.M. Beeton, 1970, The oligochaete fauna of Green Bay, Lake Michigan. Proc. 13th Conf. Great Lakes Res., 1970:15-46.

Howmiller, R.P. and M.S. Loden, 1976, Identification of Wisconsin Tubificidae and Naididae. Wis. Acad. Sci., Arts, Lett., 64:185-197.

Hrabě, S., 1967, Two new species of the family Tubificidae from the Black Sea, with remarks about various species of the subfamily Tubificinae. Publ. Fac. Sci. Univ. J.E. Purkyne, Brno., 485:331-356.

Hrabě, S., 1973, On a collection of Oligochaeta from various parts of Yugoslavia. Biolski Vestnik., 21:39-50.

Hunt, G.S., 1962, Water pollution and the ecology of some aquatic invertebrates in the lower Detroit River. Proc. 5th Conf. Great Lakes Res., Univ. Mich. Inst. Sci. Tech., Great Lakes Res. Div., Publ., 9:29-49.

Johnson, M.G. and R.O. Brinkhurst, 1971, Associations and species diversity in benthic macroinvertebrates of Bay of Quinte and Lake Ontario. J. Fish. Res. Bd. Can.,28:1683-1697.

Johnson, M.G. and D.H. Matheson, 1968, Macroinvertebrate communities of the sediments of Hamilton Bay and adjacent Lake Ontario. Limnol. Oceanogr., 13:99-111.

Judd, J.H. and J.G. Bocsor, 1975, Environmental changes in a portion of Lake Ontario following pollution abatement. Verh. Internat. Verein. Limnol., 19:1984-1989.

Kasprzak, K., 1972, Variability of Chaetogaster diaphanus (Gruithuisen). 1828 (Oligochaeta, Naididae) in different environments. Zoologica Poloniae., 22:43-51.

Kennedy, C.R., 1965, The distribution and habitat of Limnodrilus Claparede (Oligochaeta: Tubificidae). Oikos, 16:26-39.

Kenk, R., 1949, The animal life of temporary and permanent ponds in southern Michigan. Mis. Pub. Mus. Zool. Univ. Mich., 71:66 p.

Kinney, W.L., 1972, The macrobenthos of Lake Ontario. Proc. 15th Conf. Great Lakes Res., 1972:53-79.

Krecker, F.H. and L.Y. Lancaster, 1933, Bottom shore fauna of western Lake Erie: population study to a depth of six feet. Ecol., 14:79-93.

Langlois, T.H., 1954, The western end of Lake Erie and its ecology. J.W. Edwards, Inc., Ann Arbor., 479 pp.

Lauritsen, D., 1979, Distribution of oligochaetes in Lake Michigan and their use as indices of pollution. 22nd Conf. Grt. Lks. Res. (Abstract).

Loden, M.S., 1978a, A revision of the genus Psammoryctides (Oligochaeta: Tubificidae) in North America. Proc. Biol. Soc. Wash., 91:74-84.

Loden, M.S., 1978b, Life history and seasonal abundance patterns of aquatic Oligochaeta in four southeastern Louisiana habitats. Ph.D. dissertation, Louisiana State Univ., Baton Rouge, Louisiana. 132 p.

Loveridge, C.C. and D.G. Cook, 1976, A preliminary report on the benthic macroinvertebrates of Georgian Bay and North Channel. Fish. Mar. Serv. Tech. Rept., 610:46 p.

Maciorowski, A.F., E.F. Benfield, and A.C. Hendricks, 1977, Species composition, distribution, and abundance of oligochaetes in the Kanawha River, West Virginia. Hydrobiologia, 54:81-91.

Merna, J.W., 1960, A benthological investigation of Lake Michigan. M.S. thesis, Mich. State Univ., 74 p.

Michaelsen, W., 1900, Oligochaeta. Tierreich 10:1-575.

Michaelsen, W., 1926. Oligochaeten aus dem Ryck bei Greifswald und von benachbarten Meersgebieten. Mitt Zool. St. Inst. Hamb.,42:21-29.

Milbrink, G. and L. Nyman, 1973a, On the protein taxonomy of aquatic oligochaetes. Zoon 1:29-35.

Milbrink, G., 1973b, Protein taxonomy of aquatic oligochaetes and its ecological applications. Oikos 24:473-474.

Moore, G.M., 1939, A limnological investigation of the microscopic benthic fauna of Douglas Lake, Michigan. Ecol. Monogr., 9:537-582.

Moore, J.P., 1906, Hirudinea and Oligochaeta collected in the
    Great Lakes region. Bull. Bur. Fish., 25:155-171.
Mozley, S.C. and L.C. Garcia, 1972, Benthic macrofauna in the
    coastal zone of southeastern Lake Michigan. Proc. 15th
    Conf. Great Lakes Res., 1972:102-116.
Mozley, S.C. and R.P. Howmiller, 1977, Environmental status of
    the Lake Michigan region. vol. 6: zoobenthos of Lake Michigan.
    Argonne National Laboratory, Argonne, Ill., 148 p.
Mozley, S.C. and M.H. Winnell, 1975, Macrozoobenthic species
    assemblages of southeastern Lake Michigan, U.S.A. Verh.
    Internat. Verein. Limnol., 19:922-931.
Nalepa, T.F. and N.A. Thomas, 1976, Distribution of macrobenthic
    species in Lake Ontario in relation to sources of pollution
    and sediment parameters. J. Grt. Lks. Res., 2:150-163.
Nicholson, H.A., 1873, Contributions to a fauna canadensis;
    being an account of the animals dredged in Lake Ontario in
    1872. Can. J. Sci. Lit. Hist. H.S., 13:490-506.
Pliodzinskas, A.J., 1978, Aquatic oligochaetes in the open water
    sediments of Lake Erie's western and central basins. Ph.D.
    dissertation, The Ohio State University, Columbus, Ohio. 160 p.
Rains, J.H., 1971, Macrobenthos population dynamics in Indiana
    waters of Lake Michigan in 1970. M.S. thesis. Ball State
    Univ., Muncie, Ind., 78 p.
Reighard, J.E., 1894, A biological examination of Lake St. Clair.
    Bull. Mich. Fish Comm. No. 4, Append. 11th Bienn. Rep.,
    (1892-4):60 p.
Ricker, K.E., 1959, The origin of two glacial relict crustaceans
    in North America, as related to Pleistocene glaciation.
    Can. J. Zool., 37:871-893.
Shelford, V.E. and M.W. Boesel, 1942, Bottom animal communities
    of the island area of western Lake Erie in the summer of
    1937. Ohio J. Sci., 42:179-190.
Shrivastava, H.N., 1974, Macrobenthos of Lake Huron. J. Fish.
    Res. Bd. Can. Tech. Rep., 449:45 p.
Smith, F., 1919, Lake Superior lumbriculids, including Verrill's
    Lumbricus lacustris. Proc. Biol. Soc. Wash., 32:33-40.
Smith, F. and B.R. Green, 1916, The Porifera, Oligochaeta and
    certain other groups of invertebrates in the vicinity of
    Douglas Lake, Michigan. Rept. Mich. Acad. Sci., 17:81-84.
Smith, K.R., 1966, A comparison of benthic oligochaete fauna
    from different stations in the Put-in-Bay region of Lake
    Erie. M.S. thesis. Ohio State University. Columbus, Ohio.
Smith, S.I., 1874, Sketch of the invertebrate fauna of Lake
    Superior. Rep. U.S. Comm. Fish., 1872-73:690-707.
Smith, S.I. and A.E. Verrill, 1871, Notice of the invertebrate
    dredge in Lake Superior in 1871, by the U.S. Lake Survey,
    under the direction of Gen. C.B. Comstock, S.I. Smith,
    naturalist. Am.J. Sci., 2:448-454.
Spencer, D.R., 1977, A species of Pristina (Oligochaeta: Naidi-
    dae) new to Lake Erie. Ohio J. Sci., 77:24-25.

Spencer, D.R., 1978a, *Pristina acuminata* Liang, a naidid oligo-
    chaete new to North America. Trans. Am. Microsc. Soc.,
    97:236-239.
Spencer, D.R., 1978b, The Oligochaeta of Cayuga Lake, New York
    with a redescription of *Potamothrix bavaricus* and *P. bedoti*.
    Trans. Am. Microsc. Soc., 97:139-147.
Stimpson, K.S., J.R. Brice, M.T. Barbour and P. Howe, 1975,
    Distribution and abundance of inshore oligochaetes in Lake
    Michigan. Trans. Am. Microsc. Soc., 94:384-394.
Teter, H.E., 1960, The bottom fauna of Lake Huron. Trans. Am.
    Fish. Soc., 89:193-197.
Thomas, M.L.H., 1966, Benthos of four Lake Superior bays. Can.
    Field Nat., 80:200-212.
Timm, T., 1970, On the fauna of the Estonian Oligochaeta.
    Pedobiologia.,10:52-78.
Veal, D.M., 1968, Biological survey of the St. Mary's River.
    Ont. Wat. Resour. Comm., 53 p.
Veal, D.M. and D.S. Osmond, 1968, Bottom fauna of the western
    basin and near-shore Canadian waters of Lake Erie. Proc.
    11th Conf. Great Lakes Res., Univ. Mich. Great Lakes Res.
    Div. Publ., 15:151-160.
Walton, L.B., 1906, Naididae of Cedar Point, Ohio. Amer. Nat.,
    40:683-706.
Wiley, M.J. and S.C. Mozley, 1978, Pelagic occurrence of benthic
    animals near shore in Lake Michigan. J. Grt. Laks. Res.,
    4:201-205.
Wright, S., 1955, Limnological survey of western Lake Erie.
    U.S. Fish Wildl. Serv. Spec. Sci. Rep. Fish. No., 139:341 p.
Wright, S. and W.M. Tidd, 1933, Summary of limnological investi-
    gations in the western end of Lake Erie, 1929-30. Trans.
    Am. Fish. Soc., 63:271-285.

AQUATIC OLIGOCHAETA OF SOUTHERN ENGLAND

M. Ladle and
G.J. Bird

Freshwater Biological Association
River Laboratory
East Stoke
Wareham, Dorset, England

ABSTRACT

A brief account is given of the known distribution of aquatic oligochaetes in the waters of southern England. Four main water types are recognized on the basis of the geology, topography and usage of the areas concerned.

The small, acid streams of Devon, Cornwall and the New Forest are characteristically inhabited by *Rhyacodrilus coccineus* and *Stylodrilus heringianus*. In the highly calcareous streams of Wiltshire, Dorset and Hampshire a much more varied oligochaete fauna is present. Again *R. coccineus* and *S. heringianus* are important but *Limnodrilus hoffmeisteri*, *Tubifex ignotus*, *Aulodrilus pluriseta*, *Psammoryctides barbata* and the enchytraeid *Propappus volki* are all abundant and show characteristic longitudinal zonations like those of many other organisms. In the large river Thames, *Potamothrix moldaviensis* is predominant with *Psammoryctides barbata*, *Limnodrilus hoffmeisteri* and the naid *Stylaria lacustris*, also abundant. The slow flowing Norfolk broads have a community characterized by *Potamothrix hammoniensis*. A note is given of the occurrence of a form of *Limnodrilus hoffmeisteri*.

For the purposes of this paper the area described as Southern England is approximately south of a line drawn from the Wash in the East to the Bristol Channel in the West. The geology of the area is diverse, but several natural sub-divisions of the water-courses can be made.

In the South-West (Devon and Cornwall) Palaezoic granites and sandstones predominate and the streams consequently have stony substrata with nutrient-poor, acidic, soft water. In the New Forest area of Hampshire there are streams of a similar nature, but in contrast, they receive drainage from Tertiary gravels and sands.

Stretching north-eastwards from Dorset is a wide band of soft Mezozoic strata, mainly chalk and limestones. These formations often contain large natural reservoirs or aquifers, which are the source of many of the streams. These streams are fast-flowing, with stony or gravel substrata. The gravel in some stretches can be entirely composed of flints. The water is rich in dissolved inorganic salts, particularly those of calcium. Diversity of stream-bed, and hence micro-habitat for invertebrates, comes from the annual cycle of algal and higher plant growth (mainly *Ranunculus* and *Rorippa* species) and associated deposits of silt and sand. A notable feature of this chalk stream system is the temporary streams, or winterbournes, which dry up or flow only intermittently and weakly in summer. This is a rigorous environment for aquatic oligochaetes and lumbricid species are encountered in the upper, drier reaches.

The River Thames is a sole large river. Much of its initial catchment is in the limestone region but the presence of extensive industrial and domestic effluents, the large volume of water, and the extensive tidal region make it notably different in character from most other rivers in the South.

Finally, north of the Thames, in the low-lying district of East Anglia are the Norfolk Broads. These are areas of shallow, slow-flowing, silter watercourses, often spreading into wide, shallow eutrophic lakes, with dense summer plant growth.

DISTRIBUTION OF OLIGOCHAETA

Overall there is a general lack of information regarding distribution. Difficult taxonomy is not the sole reason. Records up until the present have been assessed primarily on a county basis, and some species, and whole groups, are poorly represented because of a lack of collections. The Tubificidae is the best recorded group and Table 1 summarizes their occurrence in the situations described.

The composition of the oligochaete community in each region is inadequately and unevenly known. The acid streams (Table 2) show a varying composition: the Oberwater Stream data is the most reliable as it is the result of a year's sampling of phreatic water (Ladle, 1971).

Table 1.   Distribution of tubificid species by stream type

	Ponds	Norfolk Broads	Thames	Chalk Streams	Winterbournes	Acid Streams
Limnodrilus hoffmeisteri	X	X	X	X	X	X
Psammoryctides barbata	X	X	X	X	X	X
Tubifex tubifex		X	X	X	X	X
Rhyacodrilus coccineus	X	X		X	X	X
Aulodrilus pluriseta	X		X	X	X	X
Limnodrilus udekemianus	X		X	X	X	X
L. claparedeianus				X	X	X
Tubifex ignotus			X	X	X	X
Potamothrix moldaviensis		X	X			
P. hammoniensis		X	X	X		
Peloscolex ferox		X		X	X	X
Limnodrilus cervix			X			
L. profundicola			X			
Tubifex newaensis			X			
Potamothrix bavaricus		X				
Branchiura sowerbyi			X	X		

Table 2. Species composition (%) in acid streams

	New Forest Stream	Oberwater	River Plym	River Dart
Rhyacodrilus coccineus	85	26	16	50
Psammoryctides barbata	8	41	--	--
Tubifex tubifex	3	1	--	--
Enchytraeidae	3	15	13	5
Nais elinguis	1	3	--	5
Nais alpina	--	2	--	11
Stylodrilus heringianus	--	12	71	29

The influence of substrate is important and also the particular reach of the river sampled. This complicates any attempt
at comparisons between different sites unless a standard technique
is applied. The most familiar stream-type, i.e. the Dorset chalk
streams, show this feature (Tables 3,4,5). It is notable that
Stylodrilus heringianus Clap. is the dominant species in nearly
every situation, especially in the gravel sediments.

Table 6a summarizes data from a series of airlift samples
from the middle reaches of the Thames (Furse 1978). In the
brackish tidal situation an entirely different community is
encountered, with *Tubifex costatus* Clap., *Peloscolex benedeni*
(Udekem), and *Clitellio arenarius* (Müller) common (Aston, pers.
comm.).

Table 3. River Frome - gravelbed species composition (%)

SOURCE	⟶			ESTUARY
Rhyacodrilus coccineus	77	8	6	9	9
Stylodrilus heringianus	13	83	52	19	9
Limnodrilus hoffmeisteri	3	5	6	16	17
Psammoryctides barbata	3	8	25	50	57
Tubifex tubifex	3	0	0	1	0
Tubifex ignotus	1	3	9	7	3

Table 4. Species composition of a chalk stream – Bere Stream, Dorset (all substrates)

Tubificidae	%	Naididae	%
Limnodrilus hoffmeisteri	19.7	Nais elinguis	0.3
Tubifex/Potamothrix	13.8	Nais communis	0.2
Tubifex ignotus	8.7	Stylaria lacustris	0.1
Aulodrilus pluriseta	7.5	Nais variabilis	0.08
Psammoryctides barbata	4.5	N. barbata	0.04
Tubifex tubifex	2.1	Pristina menoni	0.04
Limnodrilus udekemianus	0.4	Uncinais uncinata	0.02
L. claparedeianus	0.3	Lumbriculidae	
Rhyacodrilus coccineus	0.3	Stylodrilus heringianus	34.2
Peloscolex ferox	0.2	Lumbriculus variegatus	0.3
Peloscolex speciosus	0.08	Enchytraeidae	
Potamothrix hammoniensis	0.04	Propappus volki	5.6
		Others	1.6
		Haplotoxidae	
		Haplotaxis gordioides	0.02

Table 5. Species composition and order of dominance by substrate in Bere Stream.

GRAVEL BED	%		RANUNCULUS BED	%
Stylodrilus heringianus	56.8		*T. ignotus*	9.3
Propappus volki	16.0		*Psammoryctides barbata*	4.8
Limnodrilus hoffmeisteri	10.8		*Aulodrilus pluriseta*	4.6
Tubifex tubifex	7.5		Tubificidae	1.7
Psammoryctides barbata	7.1		Enchytraeidae	1.0
Tubifex ignotus	5.3		Naididae	0.3
Enchytraeidae	3.1		MARGINAL SAND/SILT	
Tubificidae (others)	1.6		*Stylodrilus heringianus*	23.0
Naididae (all)	1.3		*Limnodrilus hoffmeisteri*	21.3
RANUNCULUS BED			*Tubifex tubifex*	19.9
Stylodrilus heringianus	30.6		*Aulodrilus pluriseta*	19.5
Limnodrilus hoffmeisteri	27.0		*Tubifex ignotus*	12.0
Tubifex tubifex	19.6		*Psammoryctides barbata*	1.1
			Tubificidae	0.3
			Naididae	0.7

Table 6. Species composition

A. RIVER THAMES (MIDDLE REACH)	%	B. NORFOLK BROADS	%
Potamothrix moldaviensis	27	*Potamothrix hammoniensis*	51
Psammoryctides barbata	21	*Limnodrilus hoffmeisteri*	38
Limnodrilus hoffmeisteri	20	*Psammoryctides barbata*	6
Stylaria lacustris	12	*Potamothrix bavaricus*	2
Limnodrilus udekemianus	9	*Peloscolex ferox*	2
Potamothrix hammoniensis	4	*Rhacodrilus coccineus*	1
Stylodrilus heringianus	4		
Limnodrilus cervix	3		

NOTE ON INFREQUENT SPECIES

Rhyacodrilus falciformis Bretscher has been recorded from both acid and chalk regions, notably in a small temporary stream in south Dorset. It may be a characteristic species of such small headwater streams.

Peloscolex speciosus 'simsi' only recorded and described from a specimen collected in the Frome (Brinkhurst, 1966) has now been taken from the Bere Stream, a tributary of the River Piddle, flowing in a valley north of the Frome. It may have been over-looked elsewhere as in its immature stage is almost indistinguish-able from certain specimens of *Tubifex tubifex* and *T. ignotus*.

Limnodrilus cervix Brinkhurst is an introduced North American species found only in the south in the Thames, as is *Limnodrilus profundicola* (Verrill).

The brackish water forms *Tubifex costatus, Tubifex pseudo-gaster* (Dahl) *Monopylephorous rubroniveus* (Levinsen) and *Clitellio arenarius* are as yet recorded only from the Thames although, with-out doubt, they will be found elsewhere.

Unfortunately, the Naididae are even less well documented than the tubificids, for, in most instances, only one or two records have been obtained. *Nais elinguis* Müller, which is a pollution-tolerant form, has the widest distribution being reported from nine counties; *Nais simplex* Piguet and *Ophidonais serpentina* (Müller) have been found in four. Although not previously recorded in Britain (Brinkhurst, 1971) *Vejdovskyella intermedia* (Bretscher) seems to be quite common in chalk streams, whilst the related form *Vejdovskyella comata* (Vejdovsky) occurs in the acid streams of the New Forest and Devon. It may become apparent that differences in oligochaeṭa communities in these two stream types will be found in the ephemeral Naididae rather than the more catholic and robust Tubificidae.

A NOTE ON A VARIETY OF *LIMNODRILUS HOFFMEISTERI*

This species is widespread and common throughout the region. The general description is that of the form 'parvus' (Brinkhurst, 1971). However, several aberrant specimens have been found in the Frome and Bere Stream in Dorset. The penis tubes have a scalloped or lobate margin to the hood and they are somewhat stouter in appearance than the typical form (Figure 1 and Table 7). The bifid setae are also of a different form, having blunter teeth, with the upper tooth being as thick as the lower and slightly longer. The setae are also not as curved and sigmoid as the typical form.

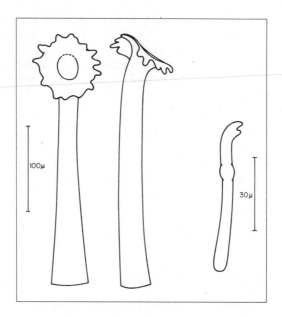

Figure 1. Penis tubes and anterior setae of *Limnodrilus hoffmeisteri* variant from River Frome and Bere Stream.

Table 7. Comparison of forms of *Limnodrilus hoffmeisteri*

Penis tube	Normal	Variant
length : max. width	8.6 – 12.5	6.6 – 8.4
(mean)	10.2	7.2
length	345 – 460μ	335 – 410μ
(mean)	370μ	370μ
base width	28 – 51	46 – 60
(mean)	38μ	53μ

Anterior setae		
relationship of distal to proximal teeth	shorter and thinner	longer and as thick
nature of teeth	"sharp"	"blunt"
shape of setae	distinctly sigmoid	weakly sigmoid

There are records of a similar type from the Great Lakes of North America (Hiltunen, 1967). In the present state of knowledge it is perhaps advisable to follow the view expressed by Hiltunen (1967) and Brinkhurst and Jamieson (1971) and regard it merely as a variety or ecomorph of *L. hoffmeisteri*. When more details of dates of maturity and breeding are collected, a different conclusion may be stated.

ACKNOWLEDGEMENTS

This work was supported in part by D.O.E. Contract number DGR - 480 - 33. The authors wish to thank Karen Astor, Ralph Brinkhurst, Mike Furse and Chris Mason for providing information and assistance.

REFERENCES

Brinkhurst, R.O., 1966, Taxonomical studies on the Tubificidae (Annelida, Oligochaeta). Supplement. Int. Revue ges Hydrobiol., 51, 727.
Brinkhurst, R.O., 1971, A Guide for the identification of British aquatic oligochaeta. Scient. Publs. Freshwat. Biol. Ass., 22 (2nd ed.).
Brinkhurst, R.O. and B.G.M. Jamieson, 1971, Aquatic Oligochaeta of the World. Oliver and Boyd, Edinburgh.
Furse, M.T., 1978, An ecological survey of the middle reaches of the River Thames, July to September 1977. A report to Thames Water Authority.
Hiltunen, J.K., 1967, Some oligochaetes of Lake Michigan. Trans. Amer. Microsc. Soc., 86:433-454.
Ladle, M., 1971, Studies on the biology of oligochaetes from the phreatic water of an exposed gravel bed. Int. J. Speleol., 3 : 311-316.

LIFE CYCLES OF MASS SPECIES OF TUBIFICIDAE

(OLIGOCHAETA)

T.L. Poddubnaya

Laboratory of Zoology
Institute of Biology of Inland Waters
Academy of Sciences of USSR
Borok, Nekovz
Jaroslavl, USSR

ABSTRACT

The life cycles of mass species of Tubificidae: *Isochaetides newaensis, Tubifex tubifex* and *Limnodrilus hoffmeisteri* have been studied in rivers, reservoirs, lakes and under experimental conditions. The length of the life cycles depend on abiotic and biotic factors of the environment. Thus, changes in the temperature regime, in the level of productivity of the mud, and in the population density result in changes in the time, duration, and intensity of reproduction of the worms causing transformations in the structure and productivity of the populations.

INTRODUCTION

Investigations of tubificid life cycles are necessary for solution of problems associated with the assessment of biological productivity and purification of freshwater ecosystems. Embryonic development of oligochaetes from fertilized eggs to completely formed worms proceeds in cocoons. The postembryonic period commences after emergence of an individual from the cocoon: worms grow to maturity, reproduce once or several times, and die. The whole life cycle in tubificids takes several years. Embryonic

development of oligochaetes was described in detail in the classic
works of the late 19th and early 20th centuries (Kowalevsky, 1871;
Penners, 1922; Svetlov, 1923, 1926; Meyer, 1931; and others).
Investigations of the postembryonic period started relatively late
(Poddubnaya, 1958, 1959, 1963; Timm, 1962, 1964; Brinkhurst, 1964a,
1964b; Brinkhurst and Kennedy, 1965; Kennedy, 1965, 1966a, 1966b).
As a result of these investigations a general impression of the
life cycles of Tubificidae was established, and directions in the
study of adaptive abilities of separate species determined. A
rather detailed knowledge of tubificid biology in various environments, and of the influence of environmental conditions on onto-
genesis has now been obtained (Poddubnaya, 1971, 1973, 1976; Archipova, 1976; Poddubnaya and Archipova, 1977; Timm, 1972, 1974;
Thorhauge, 1975, 1976; Johnson and Brinkhurst, 1971).

Three species of tubificids, *Isochaetides newaensis* Mich.
Tubifex tubifex (Mull), and *Limnodrilus hoffmeisteri* dominate
communities (up to 50-80% of biomass) in the majority of lakes,
rivers and reservoirs, take an active part in purifying sediments
of excessive buried organic matter, and serve as an important food
source for fish. Populations of these species living under various
ecological conditions were studied experimentally and in the field
for many years by the author and her assistants. This communication
summarizes their results.

Isochaetides newaensis

Isochaetides newaensis is a pelophil of rivers, backwaters, and
some lakes and reservoirs that inhabits various sediments: silt with
stones and pebbles; sandy and silted clays to most soft sapropel
muds; and grey sandy muds which it prefers. The greatest densities
of this species occur where at least feeble but steady currents are
observed. The restriction of *I. newaensis* to places with water
flow is explained by its greater oxygen requirements as compared
with other tubificids. Distinct morphological variability in the
characteristics of this species has not been revealed by morphological analysis of populations from different water-bodies; individuals from different populations may differ essentially in their
size-weight characteristics.

Embryonic development from egg to emergence of young, in the
temperature range 14-27°C, occupied 20-35 days (440-450 degree
days). Below 10°C the eggs only undergo fission and then die.
High mortality of early embryos (65%) was observed at temperatures
higher than 25-27°C. At optimum temperatures of 20-22°C, embryos
developed with the minimum mortality in 20-22 days (Poddubnaya,
1972).

Newly hatched worms weigh on the average 0.85 mg. and have a
mean length of 7 mm. The gonads (first testes and then ovaries)

are formed in the second month after emergence. In the following
period of low winter temperatures the genital system does not
develop further, but spermatogenesis takes place. Sexual organs
develop in 2 to 3 weeks when worms are 10-11 months old in the
spring of the next year; spermathecae develop last. Copulation of
the worms occurs a month before cocoons are layed (Fig. 1). Thus,
the young worms which attained sexual maturity at the age of 11-12
months start to reproduce first in the 13-14th month of their life.
at this time they weigh 80-120 mg (Poddubnaya, 1972).

In all the water-bodies investigated in Northwestern, central
and Southwestern regions of European USSR *I. newaensis* reproduced
once a year in May-July for 45-65 days. Laying of the coccons starts
at water temperatures above 12°C. The number of embryos in a cocoon
varies from 4 to 30 depending on water temperature; at $15-20^{\circ}$C the
mean number of eggs in a cocoon is 22+0.01, and at the temperatures
higher than 22° it is 12+0.05. In one reproduction season one
individual lays up to 60 eggs. After an individual has bred once
its sexual system begins to be resorbed; then it regenerates during
the second half of summer (July-August), and a second reproduction
starts after winter, in the third year of life. Worms reproducing
for the first and the second time can be distinguished by the condition of their gonads and sexual organs, and by the time of laying
of cocoons. Some individuals in the population, having repeated
resorbtion and regeneration of the sexual system, may reproduce for
a third time in the fourth year of life; such worms weigh 180-210
mg. Specific absolute fecundity, i.e., the number of eggs laid by
one individual during the whole life cycle, varies from 150 to
200.

Maximum growth rate occurs during the first 1.5-2 months of
life, when specific growth rate C_w=0.4 mg/day.

Tubifex tubifex

Tubifex tubifex is a euryoxybiotic pelophil occurring in
various habitats in water-bodies of any trophic condition. A comparative analysis of morphological and biological peculiarities of
populations of *T. tubifex* has shown high variability in the species
(Timm, 1974; Poddubnaya, 1976; Poddubnaya and Arkhipova, 1977) and
revealed the existence of two ecological forms in lakes, rivers and
reservoirs: a rheophilic one which is more eurybiotic; and a limnophilic one adapted to exist under relatively stable conditions.

The main habitats of rheophilic *T. tubifex* are places with a
poor water flow. These are: backwaters and pits in rivers;
estuaries and mouths of inflowing rivers in the head waters in
reservoirs; places adjacent to mouths of rivers in lakes. General
distinguishing features of these biotopes are: increased sedimentation compared to neighbouring parts; weak current acting mostly

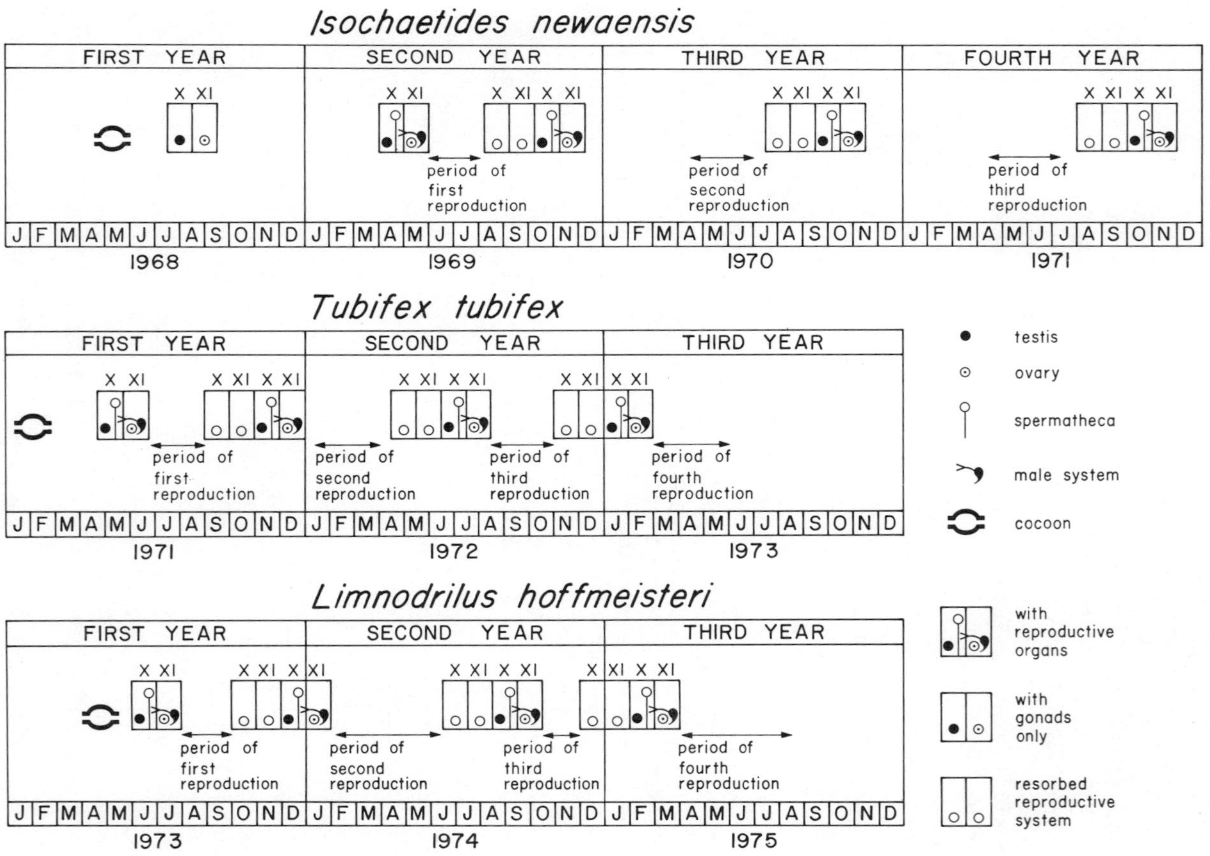

Figure 1. The scheme of life cycle of mass species of Tubificidae. A. *Isochaetides newaensis*, B. *Tubifex tubifex*, C. *Limnodrilus hoffmeisteri*. Roman numerals indicate segments. Symbols as on figure.

during spring and winter; greater depth than adjacent places; changeable oxygen regime but, as a rule, dissolved oxygen content not below 1 mg/L.

The limnophilic form of *T. tubifex* lives in flood-plain water-bodies that have lost their connection with the river, many lakes of glacial and karst origin, and shallow water-bodies fed by underground water. Its main habitats occur where currents are practically absent. As a rule, this is the profundal zone with little water exchange and soft sediments like grey muds and sapropel, and adjacents parts of the littoral. Organic matter in the sediments is mostly of autochthonous origin and the oxygen regime is always relatively favourable; a drop in the oxygen content below 4 mg/L is a rare phenomenon and may occur only at the end of winter.

The embryonic period in *T. tubifex* at various temperatures (2-30°) lasts from 12 to 60 days (240 degree-days). High mortality of embryos (60%) was observed at temperatures below 10 and above 20°C. Embryos at the earliest stages of development are most sensitive to changes in dissolved oxygen content within the temperature range of 2-7°. Normal development of embryos takes place within the temperature range 6-19°C at dissolved oxygen content 2.5-7.0 mg/L. At the optimum temperature (18-20°0 embryonic development lasts 12-15 days (Poddubnaya, 1976; Poddubnaya and Arkhipova, 1977; Aston, 1973).

Newly hatched worms weigh, on average, 0.08 mg and have a mean length 3 mm. The course of sexual maturation is influenced by water temperature and population density. At 20°C and a population density not greater than 20000/m^2, worms attain maturity within two months. At this temperature the gonads are formed at the end of the first month of life in the same way as in *I. newaensis*. During the next 2 weeks sexual organs are formed, the spermathecae being formed last. Copulation occurs about 20 days before cocoons are deposited. Low water temperature (2°) and high population density (above 70000/m^2) delay maturation by up to 10 months, but under any circumstances *T. tubifex* attains maturity in the first year of life.

Reproduction takes place at a wide temperature range of 0.5-27°C. The duration of the reproductive period in *T. tubifex* in different water-bodies varies from 4 to 12 months and, as is the case with development, it depends on water temperature, dissolved oxygen concentration, and population density. A simultaneous drop in water temperature to 0.3-1.0^{o}C and in dissolved oxygen content to 0.5-2.0 mg/L stops the process of reproduction (Poddubnaya and Arkhipova, 1977). At a very high population density (above 70000/m^2) reproduction slows down. The intensity of reproduction varies within a year. Mass laying of cocoons in spring and winter alternates with a sharp abatement to complete cessation of sexual

activity in summer and autumn. Each individual is capable of laying
eggs for 3-4 months without interruption, and resorbtion of the sexual system begins only after such a period of reproduction. Fig. 1b
shows the life cycle of *T. tubifex* in the river Latka, a tributary
of the Rybinsk reservoir.

From the cocoons laid in winter (at the end of January and
February) young emerge in early spring (April). They attain maturity
in 80 to 90 days and reproduce once during the first year. But
during the second year each individual reproduces twice. It is not
impossible that some worms of a given generation may reproduce for
a fourth time during the third year of life. The life cycle of
T. tubifex under the conditions of this river last 2-2.5 years
(Poddubnaya, 1976). The number of embryos in a cocoon varies from
2 to 20 and depends on temperature: with an increase in temperature
from 12 to 20° the number of eggs in the cocoon diminish. An individual lays from 35 to 120 eggs in one reproduction period, and
the absolute specific fecundity varies from 90 to 340.

In the rheophilic form of *T. tubifex* which inhabits more
variable habitats development, growth and maturation proceed more
intensively. The greatest rate of growth occurs during the first
month of life before maturation (specific growth rate - C_w= 0.04
mg/day); reproduction procees from January to August. The absolute
specific fecundity is higher (250 eggs on average) and the life
cycle is shorter than in the limnophilic form which lives at greater
depths and in relatively colder water where metabolism is retarded.
Limnophilic *T. tubifex* reaches sexual maturity by the age of six
months, and the specific growth rate is only 0.02 mg/day); the
reproductive period is extended for the whole year, and absolute
specific fecundity is almost half that of the rheophilic form.

Limnodrilus hoffmeisteri

Limnodrilus hoffmeisteri is a eurybiotic pelophil like
T. tubifex. It possesses a very wide ecological range, and inhabits
all types of waters with any sort of pollution (but it cannot withstand oxygen deficit). It competes with *T. tubifex* in heavily
polluted waters where it may or may not dominate the population
numerically. Its quantitative development is also associated with
the organic content of the sediments.

The embryonic period at various temperatures ($2-30^{\circ}$) lasts from
15 to 75 days (260 degree-days). Normal development of embryos
occurs within the temperature range of $10-25^{\circ}$ and at dissolved
oxygen content 2.5-9.6 mg/L. High mortality of embryos in cocoons
was observed at low ($2-5^{\circ}$) and high (above 30°) temperatures;
embryos, like those of *T. tubifex,* are most sensitive to variations
in dissolved oxygen content and to low temperature.

Maturation depends on water temperature and population density. At 18-20°C and a population density not more than 25000/m^2 the worms attain maturity by the end of the second month of life. Under these conditions the gonads and the whole sexual system are formed in the same time frame as was shown for *T. tubifex*. Production of cocoons starts by the end of the third month of life.

Low or high temperatures (1-4° or above 30°) and high population density (more than 35000/m^2) delay maturation by 4-6 months but do not stop reproduction (Timm, 1974; Poddubnaya, 1973). The reproduction period is extended and varies within a wide range but more often it lasts from May till October. *Limnodrilus hoffmeisteri*, like *T. tubifex*, attains maturity in the first year of its life.

Intensive laying of cocoons is observed twice a year, in May-June and in the second half of August-September. Each individual is capable of laying cocoons for two months without break, after which resorbtion of the sexual system and its subsequent regeneration begin. From the cocoons laid at the beginning of May the young emerge at the beginning of June (Fig. 1) in the Rybinsk reservoir. By the end of August the worms attain maturity and reproduce for the first time for 2 to 2.5 months (September-October). The second reproduction begins in spring, and the third one in the autumn of the second year of life. The whole life cycle of *L. hoffmeisteri* takes 2-3 years in the Rybinsk reservoir (Arkhipova, 1976). The number of embryos in a cocoon (from 2 to 10) increases with an increase in temperature (Poddubnaya, 1973; Timm, 1974; Aston, 1973); i.e. *L. hoffmeisteri* is a more thermophilic species. For one reproduction period one individual lays from 40 to 150 eggs, and absolute specific fecundity varies from 120 to 480 eggs.

Our observations on the biology of *L. hoffmeisteri* in the Upper Volga reservoirs and in experiments, Kennedy's (1966a,b) observations in English water bodies, and those of Timm in the Estonian waters indicates its great plasticity and dependence of the life cycle upon local conditions. Changes in the environment (temperature, productivity of the mud, and population density) result in changes in the time, duration, and intensity of reproduction of the worms, and cause transformations in the structure and productivity of the populations.

CONCLUSION

The tubificids considered here inhabit, in various combinations, all the main biotopes of inland waters. *Isochaetides newaensis* inhabits places with a relatively greater flow and has the greatest longevity among tubificids. It is the most stenobiotic, and does not suffer sharp fluctuations in numbers or shifts in phases of the life cycle. *Tubifex tubifex* is widespread in various biotopes, but it thrives on muds rich in organic matter.

Its numbers vary within a wide range and it is adapted to sharp changes in the environment. Life cycles differ in populations living under conditions of increased or decreased water flow and temperature. *Limnodrilus hoffmeisteri* also possesses a wide ecological range, and its quantitative development is closely associated with the amount of organic matter in the sediments. It competes with *T. tubifex* yielding to it only in heavily polluted waters, and like *T. tubifex* it lives on average 2 to 3 years.

Comparison of experimental data and field observations suggested the following common features in the life cycles of *T. tubifex* and *L. hoffmeisteri*: under close to optimal environmental conditions (considering temperature, population density, dissolved oxygen content) these species can reproduce for a long time without interruption; absolute specific fecundity averages 250 eggs; resorbtion and regeneration of the sexual system occurs once a year; minimum longevity is 12 to 13 months; a decrease in temperature and an increase in population density lead to an increase in the number of resorbtions and regenerations of the sexual system (up to three times per whole life cycle); absolute specific fecundity in the latter case diminishes to 150–180 eggs, while longevity increases to 2 to 3 years. Even under favourable conditions the population realises not more than 35% of the potential fecundity.

The greatest productivity is found in *T. tubifex*; some populations produce from 35 g/m^2 to 2 kg/m^2 of biomass annually, and production to maximum spring (May) biomass ratios (P/B) are from 1.5 to 5 (Poddubnaya, 1976; Poddubnaya et al, 1977). Populations of *L. hoffmeisteri* produce from 6 to 43 g/m^2, and P/B coefficients are 3.3–3.6 (Poddubnaya, 1973; Arkhipova, 1976). Production of *I. newaensis* is from 15 to 40 g/m^2 annually, and P/B varies from 2.6 to 5.7 (Poddubnaya, 1975). Production maxima are confined to periods of intensive reproduction, growth, and maturation.

REFERENCES

Arkhipova, N. R., 1976, Peculiarities of the biology and production in *Limnodrilus hoffmeisteri* Clap. (Oligochaeta, Tubificidae) on grey muds in the Rybinskoe reservoir. Tr. Inst. Biol. Inland waters Acad. Sci. USSR, 34:5-15 (In Russian).

Aston, R.J. 1973, Field and Experimental Studies on the Effects of a Power Station Effluent on Tubificidae (Oligochaeta, Annelida). Hydrobiologia,42:2-3, pp.225-242.

Brinkhurst, R. O., 1964a, Observations on the biology of the marine Oligochaete *Tubifex costatus*. J.Mar.Biol. Ass., 44:11-16., 1964b, Observations on the biology of lake-dwelling Tubificidae, Arch. Hydrobiol., 60:385-418.

Brinkhurst, R. O. and C. R. Kennedy, 1965, Studies on the biology of the Tubificidae (Annelida, Oligochaeta) in a polluted stream. J.Anim. Ecol., 34:429-443.

Johnson, M. G. and R. O. Brinkhurst, 1971, Production of bentic
     macroinvertebrates of Bay of Quinte and Lake Ontario. J. Fish.
     Res. Board Canada, 28:1699-1714.
Kennedy, C. R., 1965, The distribution and habital of *Limnodrilus*
     Claparede (Oligochaeta:Tubificidae). Oikos, 16:26-38.
Kennedy, C.R., 1966a, The life history of *Limnodrilus udekemianus*
     Clap. (Oligochaeta,Tubificidae). Oikos, 17:10-17.
Kennedy, C.R., 1966b, The life history of *Limnodrilus hoffmeisteri*
     Clap. (Oligochaeta:Tubificidae) and its adaptive significance.
     Oikos, 17:158-168.
Kowalevsky, A., 1871, Embriologische Studien an Würmern und Arthro-
     poden. Mem. Adac. Sci. St. Petersb., seria 7, 16:1-70.
Meyer, A., 1931, Cytologische Studien über die Gonoblasten in der
     Entwicklung von *Tubifex*. L. Morph.Okol. Tiere, 22:269-286.
Penners, A., 1922, Die Furschung von *Tubifex rivulorum* Lam.
     Zool. Jahrb., Anat., 43:323-368.
Poddubnaya, T.L., 1958, Some data on the reproduction of Tubificidae.
     Rep. Acad. Sci. USSR, 120:422-424. (in Russian).
Poddubnaya, T.L., 1959, On the population dynamics of Tubificidae
     (Oligochaeta, Tubificidae) in Rybinsk reservoir. Tr. Inst.
     Biol. Reserv. Acad. Sci. USSR,2:102-108. (in Russian).
Poddubnaya, T.L., 1963, Life cycle and growth rate of *Limnodrilus
     newaensis* Mich. (Oligochaeta, Tubificidae). Tr. Inst. Biol.
     Reserv. Acad. Sci. USSR,5:46-56. (in Russian).
Poddubnaya, T.L., 1971, Resorption and regeneration of the repro-
     ductive system in *Isochaetides newaensis* Mich, (Oligochaeta,
     Tubificidae). Tr. Inst. Biol. Reserv. Acad. Sci. USSR,
     22:81-90. (in Russian).
Poddubnaya, T.L., 1972, Quantitative changes and production of the
     population of *Isochaetides newaensis* Mich. (Oligochaeta).
     The Aquatic Oligochaeta. Materialy vtorogo vsesoyuzn. Simp.
     Borok 1972, 134-147. Jaroslavl Isdat. Acad. Sci. USSR,
     (in Russian).
Poddubnaya, T.L., 1973, Productivity of Tubificidae in some lowland
     dam reservoirs. In.: Sim.po ekologicheskoj fisiologii pres-
     novodnykh zhivotnykh. Leningrad, 20-22, Nov.,1973. (in Russian)
Poddubnaya, T.L., 1975, Determination of productivity of Tubificidae,
     p.192. In.: "Methods of investigating of biocenosis of inland
     waters". Edit. "Nauka". Moscow. (in Russian).
Poddubnaya, T.L., 1976, Peculiarities of biology and productivity
     of *Tubifex tubifex* (Mull.) in a polluted section of the tri-
     butary of the Rybinsk reservoir, p. 119. In.:"Biological and
     productive processes of the Volga basin". Edit. "Nauka"
     Leningrad. (In Russian).
Poddubnaya, T.L. and N.R. Arkhipova, 1977, The temperature-
     oxygen optimum for the development and surviving *Tubifex
     tubifex* in the embryonic period. p. 231 in Ecological and
     Physiological investigations in nature and experiments. Frunze.
Poddubnaya, T.L., T.V. Akinshyna and A.A. Tomilov, 1977, The role
     of oligochaeta *Tubifex tubifex* in the biomass and production

of the benthos of Bratskoye reservoir, p. 90 in: "Biological
    investigations of the East Siberian waters", Irkutsk.
Svetlov, P.G., 1923, The early stages of development of *Rynchelmis
    limosella*. Izv. Biol.nauchno-issled. In-ta Permsk.gos.univ-
    ta, 2, p.141-152. (in Russian).
    1926, Embryonic developement in the family Naididae. Izv.
    Biol. nauchno-issled.in-ta Permsk.gos.univ-ta, 4, 359-372.
    (in Russian).
Thorhauge, F., 1975, Reproduction of *Potamothrix hammoniensis*
    (Tubificidae, Oligochaeta) in Lake Esrom Denmark. A field
    and laboratory study. Arch. Hydrobiol., 76:449-474.
Thorhauge, F., 1976, Growth and life cycle of *Potamothrix hammo-
    niensis* (Tubificidae:Oligochaeta) in profundal of eutrophic
    Lake Esrom. Arch. Hydrobiol., 78:71-85.
Timm, T.E., 1962, Eesti NSV magevel-vähenarjasusside faunast,
    okologiast ja Avikust. - Riik. Toit. Univ. Tartu.,120:63-107.
    (in Russian).
Timm, T.E., 1964, The oligochaeta of Estonian waters (Faunistic
    and ecologycal revew). Cand.Thesis, Tartu, p.1-22 (in Russian).
Timm, T.E., 1972, Upon the methods of culture of water oligochaeta
    in p. 106-117. In.: "The water oligochaeta" Moskow. (In
    Russian).
Timm, T.E., 1974, On the life cycles of the aquatic Oligochaeta
    in aquaria. Acad. Sci. Estonian SSR Inst. Zool. Bot.
    Hydrobiol. Res.,6:97-118.

POPULATION DYNAMICS OF *TUBIFEX TUBIFEX*, STUDIED BY MEANS OF A NEW MODEL

G. Bonomi

C.N.R. Istituto Italiano di Idrobiologia
Pallanza, Italy

G. Di Cola

C.N.R. Istituto per l'Applicazione del Calcolo
Roma, Italy

INTRODUCTION

The difficulties connected with the study of tubificid population dynamics and production have long been recognized (Kennedy, 1966; Johnson and Brinkhurst, 1971; Jónasson and Thorhauge, 1976; Waters, 1977; Bonomi and Adreani, 1978). The problems encountered are mostly of a taxonomic nature, as it is often difficult, if not impossible, to describe the populations in terms of biologically meaningful compartments. Even in the unusual case of a single species population, the complicated biological cycle strongly limits the possibility of a suitable description; indeed, it is well known that after the cocoon laying period, the tubificids undergo a regression of the reproductive system (Černosvitov, 1930; Poddubnaya, 1971) that makes the regressed animals quite similar to the true young, that have not yet undergone any reproductive period. Only Poddubnaya and co-workers, as far as we know, have been able to account for the state of regression of the reproductive system in their work on the tubificids of Rybinsk Reservoir (Poddubnaya, 1972; Arkhipova,

185

1976). Moreover, the possibility of describing the cocoon (or egg)
population numbers has been either neglected or not taken into
consideration for the lack of description of the cocoons specific
characters.

INITIAL RESULTS FROM *TUBIFEX* AND *LIMNODRILUS* CULTURES

   Recently we had the opportunity of working on the profundal
benthos of a reservoir in South Italy (Pietra del Pertusillo
Reservoir), in which a two-species tubificid population - *Tubifex
tubifex* and *Limnodrilus hoffmeisteri* was present. This enabled us
to describe the populations in terms of egg numbers and to attempt
a first tentative approach to population dynamics of *T. tubifex*
by using development times of eggs and first maturation times,
that we obtained from some scattered information in the literature
(Bonomi and Adreani, 1978).

   Later, we started to culture *T. tubifex* and *L. hoffmeisteri*
from Pietra del Pertusillo Reservoir, in order to have experimental
measurements of the above mentioned development and maturation
times. The cultures were kept according to the method proposed
by Kosiorek for *Tubifex* cultures (Kosiorek, 1974), and the preliminary results from such cultures have already been published
(Adreani and Bonomi, 1979).

   As the results showed that the maturation time (time since
the hatching from the cocoon to the first cocoon laying) gave a
very high variance, we started to culture cohorts of both *Tubifex*
and *Limnodrilus* in order to investigate the source of this high
variability.

   An example of such cohort analysis is given in Figure 1,
describing the survival of 45 newborn *T. tubifex*, reared together
in the same jar. The sexual maturation process started at an age
of about 6.5 months and from that moment the immature, the mature
and the ovigerous (mature with evident oocytes in the oviducts)
individuals were counted separately, and the concomitant egg
production was measured. Since the moment indicated by an arrow
in Figure 1, we began to measure the individual wet weight of the
worms. The sequence of the weight frequency distributions is
shown in Figure 2. Note the ample weight range presented not
only by the immature, but also by the mature and ovigerous cohort-
ants. The same pattern is shown by the *Limnodrilus* cohorts
(Figure 3). It seems, therefore, that great caution should be used
when attributing polymodal size distributions in these animals to
co-existing generations (e.g., Arkhipova, 1976; Jónasson and
Thorhauge, 1976a, 1976b, 1976c). Furthermore, it appears that
the faster growing worms mature first (earlier than maturation
occurs in the smaller animals) and over a longer period of time.

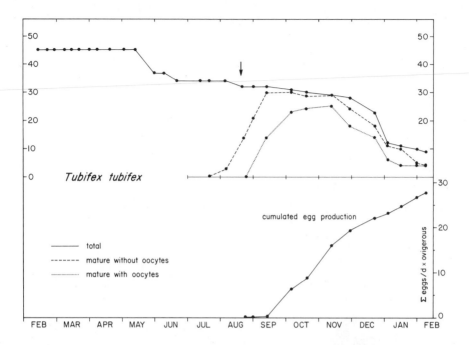

Figure 1. Survival and maturation of a cohort of 45 *T. tubifex*
           reared at 8°C. A cumulated eggs production curve is
           also given, the arrow indicates the date at which
           the individual weight measurements were started, that
           are shown in figure 2 (after Adreani e Bonomi, 1979).

The general tendency to mean weight decrease of the mature and
ovigerous individuals during the long cocoon deposition period
is also evident in the field *Tubifex* population (see below).
Ladle (1971) made similar observations on the *T. tubifex* and
L. hoffmeisteri populations from a Dorset chalk stream.

    As the time required for the first cocoon laying seems to
depend on the growth rate during the preceding period, it appears
that the application of maturation times in our previous population statistics computations (Bonomi and Adreani, 1978; Bonomi,
1979) requires revision. We have to use different "first maturation times" for different "first maturation weights". Our
results on cohorts clearly explain the high variance in the
"first maturation times" that we had measured.

    We also reared cohorts in order to describe cocoon and egg
production. Some results on clutch size variation during the
deposition period have already been published (Adreani and Bonomi,
1979). We now have new results on the effect of temperature and
of population density on the duration of the cocoon laying period

Tubifex tubifex

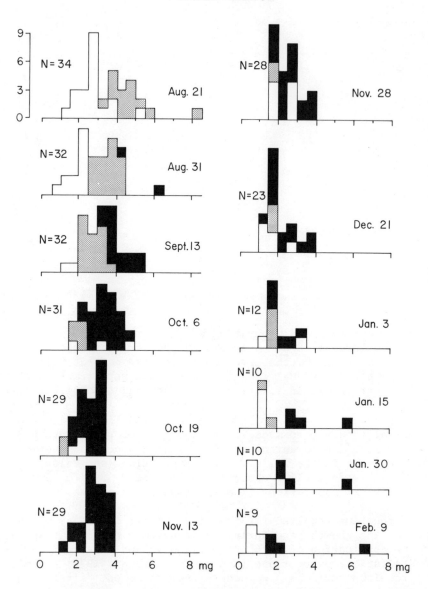

Figure 2.   Individual weight frequency distribution of a *Tubifex tubifex* cohort (see figure 1).   Mature with oocytes, mature without oocytes and immature worms are indicated by black, dashed and white areas respectively.   Total number of surviving individuals (N) and the dates of the observations are also given (after Adreani e Bonomi, 1979).

Limnodrilus hoffmeisteri

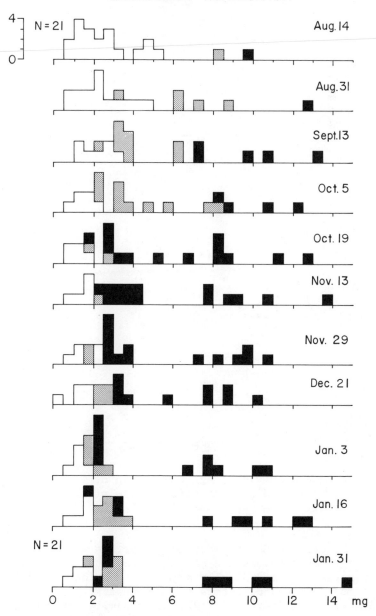

Figure 3. Individual weight frequency distribution of *Limnodrilus hoffmeisteri* cohort. For explanation, see figure 2 (after Adreani e Bonomi, 1979).

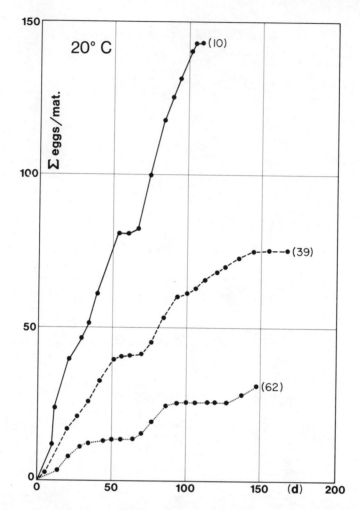

Figure 4.   Cumulated individual egg production of *Tubifex tubifex*
cohorts with different initial populations density (in
brackets).   Temperature: 20ºC.

and on egg production.   Cocoon and egg production appear positively correlated to temperature and clearly depend on the culture
crowding (Figures 4, 5 and 6). The deposition period is longer at
the lower temperatures and appears to be potentially without
interruption at 5ºC.   Thorhauge (1975) made similar observations
on *Potamothrix hammoniensis* from Lake Esrom (Denmark).

THE MODEL

     The set of results obtained from tubificid cultures suggested

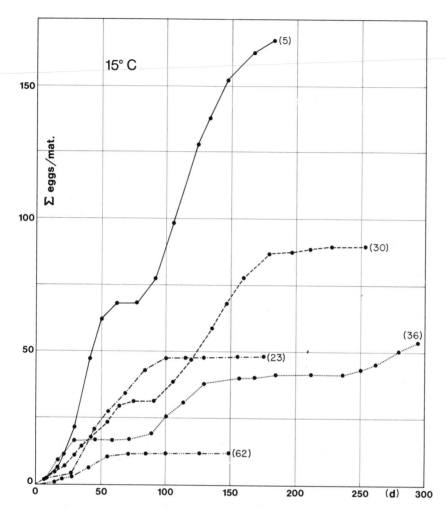

Figure 5.   Cumulated individual egg production in *Tubifex tubifex*
            cohorts with different initial population density (in
            brackets).   Temperature: 15°C.

a new model for *T. tubifex* that could incorporate some of our
observations and allow at least a preliminary application to field
material.   With this aim we subdivided the population into six
age classes: E_1, E_2, y, Y, M, O; and into six weight classes (mg):
(0-0), (0-1), (1-3), (3-6), (6-10), (10-15). E_1 and E_2 correspond to successive developmental stages inside the cocoon, E_2
lasting from a stage of an evidently elongated embryo to a
completely developed worm.   The y compartment includes the free
living immature up to 1 mg; the class is to be considered an age

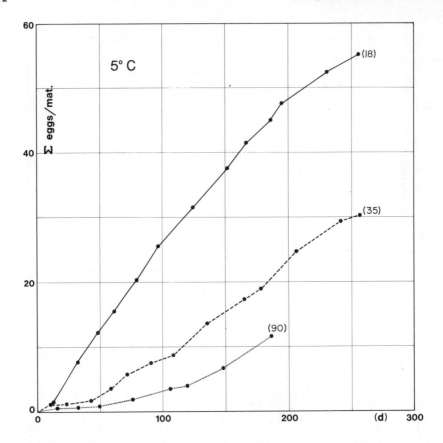

Figure 6. Cumulated individual egg production of *Tubifex tubifex*
           cohorts with different initial population density (in
           brackets). Temperature: 5°C.

class on the assumption that it is highly improbable that an
individual resulting from the sexual regression process be
included in it. The Y, M and O classes represent the subsequent
young, mature, and oocyte bearing stages, interconnected as
illustrated in the scheme of Figure 7. The (0-0) weight class
includes E_1 and E_2; the (0-1), the already defined y young.
The (1-3), (3-6) and (6-10) classes cover 1-10 mg weight range
that approximately correspond, according to our observation on
T. tubifex cultures, to the one covered by the first $Y \to M \to O \to M \to Y$
cycle (Figure 7). The (10-15 mg) class should result by a second
(third?) cycle, but up to now we have been unable to describe the
processes that generate it.

    The birth, growth, maturation and regression processes are
represented in the scheme of Figure 7. The dynamics of these

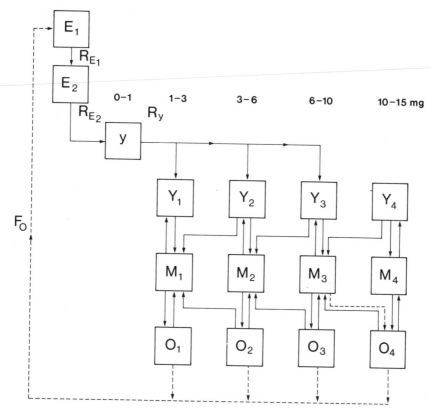

Figure 7. Provisional bi-dimensional model describing egg development, growth, maturation, egg deposition and regression. For explanation see text.

processes is described by the system of ordinary differential equations given in Table 1.

This age-weight population model seems to be a good descriptor of the real processes involved, and satisfies the realism and generality requisites. However, the model represents a complex structure and requires a set of experimental data (maturation times, regression times, a.s.o.) not yet completely available to us. Nevertheless it remains a useful conceptual tool and a guide for future investigation and for the design of next experiments.

As an intermediate approach to the Pertusillo Reservoir data analysis and production estimation, we have formulated a more simple model, considering only a unidimensional age structure. The main processes are represented in Figure 8, and the population dynamics described in Table 2.

Table 1. Mathematical description of birth-growth-death process
          in age-weight structured population.

$$\frac{dE_1}{dt} = F_0 - R_{E_1}$$

$$\frac{dE_2}{dt} = R_{E_1} - R_{E_2} - D_{E_2}$$

$$\frac{dy}{dt} = R_{E_2} - R_Y - D_Y$$

$$\frac{dy_j}{dt} = a_j R_y - R_{Y_j} + B_{M_j} - D_{Y_j}$$

$$\frac{dM_1}{dt} = C_j R_{Y_j} + b_{m+1} R_{Y_{j+1}} - R_{M_j} - B_{M_j} + q_j B_{0_j} + P_{j+1} B_{0_{j+1}} - D_{M_j}$$

$$\frac{dO_j}{dt} = R_{M_j} - B_{0_j} - D_{0_j}$$

where $F_0 = \Sigma\, f_m\, O_j$ = fertility rate

$R_{X_j} = \dfrac{X_j}{T_{X_j}}$ = developmental rate of compartment X_j

X_j = population density of compartment X_j

B_{X_j} = regression rate

T_{X_j} = developmental time

D_{X_j} = mortality rate

with the following constraints

$a_j = 1$

$b_j + C_j = 1$

$P_j + q_j = 1$

Table 2. Mathematical description of birth-growth-death process in age structured population.

$$\frac{dE_1}{dt} = F_0 - R_{E_1}$$

$$\frac{dE_2}{dt} = R_{E_1} - R_{E_2} - D_{E_2}$$

$$\frac{dy}{dt} = R_{E_2} - R_y - D_y$$

$$\frac{dY}{dt} = R_y - R_Y + B_M - D_Y$$

$$\frac{dM}{dt} = R_Y - R_M + B_0 - B_M - D_M$$

$$\frac{d0}{dt} = R_M - B_0 - D_0$$

where

$$Y = \Sigma \, Y_j$$

$$M = \Sigma \, M_j$$

$$0 = \Sigma \, 0_j$$

The rates R_X, B_X, D_X are given as weighted means.

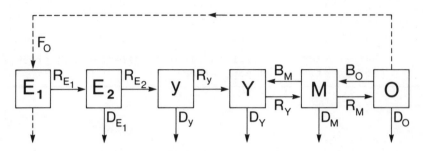

Figure 8.   Simplified linear model.   For explanation see text.

DATA ANALYSIS

A total of 2711 *T. tubifex* from Pertusillo Reservoir, covering all the sampling dates from March 1, 1976 to March 14, 1977,
and corresponding to a set of 30 Ekman grab samples, have been
individually weighed (Figure 9) and positioned according to the
y, Y, M and O classes.   The corresponding population densities
(N^o/m^2) of Table 3 were obtained by proportioning the previously
estimated "Young" (y+Y) and "adult" (M+O) numbers (Bonomi and
Adreani, 1978).   The data have been submitted to a smoothing
procedure and interpolated at five days time intervals.

The fertility and mortality rates (Table 4) have been estimated by means of the differential equations of Table 2, which
fundamentally represent a density balance on the assumption that
any change in the population density is the resultant of the
opposed processes due to recruitment, development, regression and
elimination.

The production estimation was obtained after the method
proposed by Petrovich et al.   (1964).

DISCUSSION

The analysis of *T. tubifex* data from Pietra del Pertusillo
Reservoir seems to have supplied fairly acceptable results.   The
temporal sequence of fecundity and death rates shown in Table 4
is indeed reasonable and the scattered negative values in the
computer output may be neglected.   The D_{E_2}, that may be considered
an intracocoon mortality rate, attains a first peak value at 195
days, in coincidence with the sharp decrease of the E_2 population
density (Table 3) and a second one at 250–260 days, in a moment
when the egg number is at the annual minimum. D_y shows an ana-

Table 3.   Pietra del Pertusillo reservoir.   Experimental data used
           with the linear model.   p.t. - progressive time, in days.

p.t.	E_1	E_2	y	Y	M	O
Compartment densities ($N^o.m^{-2}$)						
0	6545	4035	3076	1903	405	1162
35	10576	5736	2405	1598	235	842
62	19533	6398	964	1519	119	728
94	11870	6921	1146	1486	276	844
126	4206	7443	2027	1453	381	958
153	600	6065	3131	1267	480	917
195	130	45	4000	1185	168	1028
217	67	20	1709	1760	308	379
252	599	61	1136	776	299	459
322	6745	2349	4125	1562	386	669
349	6086	3759	5635	586	106	848
Residence times (d)						
0	20	30	129	105	51	317
35	19	29	119	104	48	285
62	18	27	108	91	43	252
94	17	25	106	82	46	251
126	15	23	103	85	48	247
153	15	26	103	87	51	251
195	18	27	103	88	59	252
217	16	24	76	68	50	185
252	12	19	84	73	42	203
322	17	26	116	90	50	287
349	17	26	108	76	49	256
Individual biomasses (mg)						
0	0.00	0.00	0.38	3.05	6.27	4.60
35	0.00	0.00	0.33	2.36	5.02	5.07
62	0.00	0.00	0.43	2.91	4.28	6.90
94	0.00	0.00	0.38	3.04	3.57	5.87
126	0.00	0.00	0.32	3.24	3.58	5.21
153	0.00	0.00	0.31	2.76	3.67	4.53
195	0.00	0.00	0.32	2.57	3.35	4.06
217	0.00	0.00	0.49	2.06	2.37	4.04
252	0.00	0.00	0.55	2.51	3.90	5.01
322	0.00	0.00	0.33	3.72	5.00	4.84
349	0.00	0.00	0.28	4.74	4.69	6.32

Table 4. Fecundity rate (F_0) and mortality rate (D_{E_2}, D_y, D_Y, D_M, D_0) computed at five day intervals.

(p.t.)	F_0	D_{E_2}	D_y	D_Y	D_M	D_0
20	0.4909	0.0426	0.0585	0.0080	0.0483	0.0168
25	0.5829	0.0430	0.0654	0.0080	0.0569	0.0160
30	0.7370	0.0468	0.0757	0.0076	0.0670	0.0151
35	0.9702	0.0551	0.0913	0.0064	0.0790	0.0141
40	1.2444	0.0686	0.1131	0.0046	0.0930	0.0128
45	1.4866	0.0850	0.1412	0.0025	0.1079	0.0113
50	1.6667	0.1019	0.1780	0.0001	0.1208	0.0095
55	1.7508	0.1168	0.2257	-0.0021	0.1262	0.0075
60	1.7045	0.1272	0.2812	-0.0041	0.1177	0.0053
65	1.5207	0.1308	0.3268	-0.0054	0.0948	0.0031
70	1.2961	0.1274	0.3460	-0.0062	0.0702	0.0015
75	1.0643	0.1183	0.3326	-0.0065	0.0508	0.0005
80	0.8458	0.1052	0.2984	-0.0063	0.0379	0.0000
85	0.6578	0.0895	0.2603	-0.0059	0.0305	-0.0001
90	0.5136	0.0728	0.2288	-0.0053	0.0272	0.0000
95	0.4225	0.0565	0.2069	-0.0046	0.0273	0.0003
100	0.3586	0.0422	0.1917	-0.0038	0.0277	0.0009
105	0.3034	0.0298	0.1796	-0.0027	0.0272	0.0018
110	0.2544	0.0193	0.1687	-0.0015	0.0255	0.0030
115	0.2093	0.0105	0.1580	-0.0001	0.0225	0.0045
120	0.1655	0.0030	0.1466	0.0015	0.0182	0.0066
125	0.1207	-0.0033	0.1342	0.0034	0.0125	0.0092
130	0.0751	-0.0085	0.1205	0.0056	0.0070	0.0119
135	0.0344	-0.0125	0.1057	0.0081	0.0043	0.0134
140	0.0000	-0.0150	0.0901	0.0111	0.0042	0.0135
145	-0.0255	-0.0160	0.0742	0.0145	0.0065	0.0119
150	-0.0392	-0.0149	0.0583	0.0185	0.0115	0.0083
155	-0.0394	-0.0112	0.0429	0.0233	0.0193	0.0030
160	-0.0324	-0.0058	0.0296	0.0278	0.0278	-0.0011
165	-0.0224	0.0015	0.0191	0.0313	0.0365	-0.0032
170	-0.0117	0.0115	0.0112	0.0331	0.0458	-0.0035
175	-0.0019	0.0258	0.0059	0.0326	0.0558	-0.0021
180	0.0062	0.0480	0.0027	0.0290	0.0655	0.0010
185	0.0119	0.0880	0.0019	0.0221	0.0724	0.0058
190	0.0149	0.1854	0.0035	0.0123	0.0712	0.0128
195	0.0141	1.0918	0.0086	0.0006	0.0567	0.0234
200	0.0107	-0.2459	0.0162	-0.0085	0.0397	0.0366
205	0.0066	-0.0585	0.0243	-0.0120	0.0335	0.0501
210	0.0015	0.0525	0.0325	-0.0114	0.0354	0.0635
215	-0.0050	0.6017	0.0383	-0.0071	0.0418	0.0722
220	-0.0106	-0.1938	0.0367	0.0004	0.0491	0.0656
225	-0.0069	-0.0758	0.0294	0.0076	0.0517	0.0471
230	0.0137	-0.0300	0.0165	0.0147	0.0502	0.0231
235	0.0500	0.0217	-0.0001	0.0219	0.0455	0.0032
240	0.0954	0.1201	-0.0158	0.0287	0.0381	-0.0078
245	0.1463	0.3215	-0.0266	0.0338	0.0287	-0.0108
250	0.2036	0.5988	-0.0312	0.0339	0.0180	-0.0087
255	0.2692	0.6957	-0.0309	0.0255	0.0075	-0.0037
260	0.3363	0.6219	-0.0290	0.0138	0.0001	0.0005
265	0.4032	0.5264	-0.0262	0.0020	-0.0042	0.0038
270	0.4682	0.4478	-0.0231	-0.0071	-0.0058	0.0066
275	0.5299	0.3857	-0.0198	-0.0127	-0.0053	0.0089
280	0.5865	0.3356	-0.0167	-0.0151	-0.0035	0.0107
285	0.6363	0.2940	-0.0138	-0.0151	-0.0006	0.0121
290	0.6776	0.2586	-0.0111	-0.0135	0.0031	0.0130
295	0.7087	0.2280	-0.0087	-0.0108	0.0072	0.0135
300	0.7279	0.2013	-0.0066	-0.0073	0.0118	0.0133
305	0.7335	0.1776	-0.0047	-0.0029	0.0168	0.0126
310	0.7241	0.1563	-0.0031	0.0023	0.0225	0.0113
315	0.6987	0.1372	-0.0016	0.0086	0.0291	0.0093
320	0.6572	0.1196	-0.0002	0.0167	0.0369	0.0066
325	0.6019	0.1033	0.0011	0.0271	0.0464	0.0036
330	0.5443	0.0881	0.0023	0.0394	0.0574	0.0007
335	0.4889	0.0742	0.0034	0.0548	0.0710	-0.0018

Tubifex tubifex

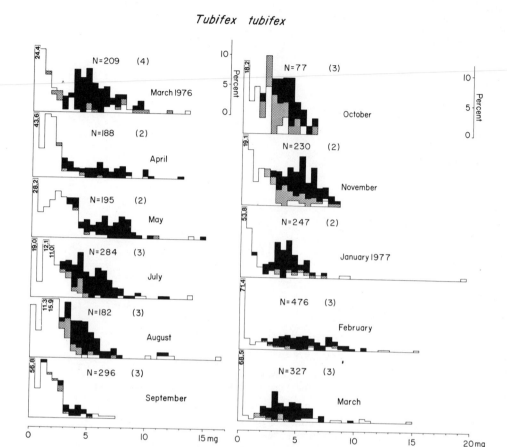

Figure 9.   Pietra del Pertusillo Reservoir. Weight frequency
distribution of *T. tubifex* in some (number in brackets)
Ekman samples. N = total number of individually weighed
worms. Black, shaded and white areas as in figures
2 and 3.

logues trend. It is important to point out that the high young
mortality reflects also the inevitable loss of the smallest young
worms that pass through the sieve during the field sieving procedure (Jónasson, 1955). Nethertheless, it is clear that it is
actually the resultant of comparatively very high fecundity, that
seems to be a real demographic strategy in this species. In the
cultures, not only *Tubifex* exhibits an higher fecundity rate than
Limnodrilus does (Adreani and Bonomi, 1979), but the same holds
for mortality, as shown in Figures 2 and 3.

Our production estimates are shown in Figure 10 and given
in Table 5 as annual values, together with the average biomass

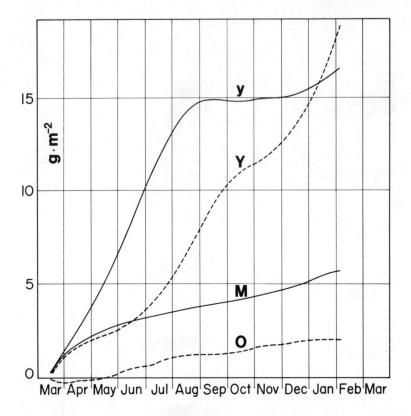

Figure 10. Pietra del Pertusillo Reservoir. Cumulated production
          of youngs (y and Y), mature (M) and ovigerous (O) of
          T. tubifex.

values and annual P/B ratios. A total production of 48.9 $g \cdot m^{-2} \cdot a^{-1}$
(formalin wet weight) has been estimated, largely performed by the
y and Y compartments. A previous computation of P in this population (Bonomi, 1979) was admittedly overestimated, due to the
inadequacy of the population dynamics model available at that
time.

    We obtain a value of 5.25 for the overall annual P/B ratio;
the P/B for the y compartment is much higher, but probably a
little overestimated. Indeed the production of the y compartment
includes the egg production, as we attributed a zero weight to
the eggs (Table 3), while the contrary holds for its B value due
to the sieving losses already mentioned. A comparison with the
few tubificid production estimates that may be found in the

Table 5 -   Pietra del Pertusillo Reservoir. Average biomass,
            annual production and annual P/B ratio for the
            y, Y, M, and O compartments of the *T. tubifex*
            population.

	(unit)	y	Y	M	O	Σ
$\overline{B}$	$(g\ m^{-2})$	0.63	3.59	1.13	3.91	9.32
P	$(g\ m^{-2}a^{-1})$	18.89	21.31	6.46	2.26	48.92
$\dfrac{P}{\overline{B}}$	(a^{-1})	27.4	5.9	5.7	0.6	5.25

literature is difficult because of the different approaches and
assumptions adopted by the authors. A P/B ratio perfectly comparable with ours, was calculated by Arkhipova (1976) for *L. hoffmeisteri* of the Rybinsk Reservoir, but Potter and Learner (1974)
in the "multivoltine" *Limnodrilus* spp. of the Eglwys Nunydd
Reservoir, in S. Wales, U.K., obtained a value of 12.5. The
calculations of Jönasson and Thorhauge (1976) supply very low P
and P/B values for *Potamothrix hammoniensis* in Lake Esrom. No
such estimates for *T. tubifex* seem to be available in the
published literature.

In conclusion, we believe that the possibility of keeping
cultures and cohorts of these oligochaeta for long term observations proves to be very profitable, as it allows a good understanding of their population biology, that stimulates the development of suitable population dynamics models. We also wish to
stress the importance of accounting for the cocoon and egg numbers
in the field samples. A good perspective before us seems to be
offered by the use of culture observations for the quantitative
estimate of the biotic factors that may be responsible of distribution and succession in the tubificids. We are now beginning a
set of laboratory experiments with the aim of testing some pre
and post-eutrophication tubificid species for their specific
fecundity. Our hypothesis is that eutrophication may favour the
shift to species with shorter life cycle and higher egg production
in these bottom animals.

REFERENCES

Adreani, L. and G. Bonomi, 1979, Elementi per una dinamica di
     popolazione dei tubificidi (Anellida, Oligochaeta). 3° Con-
     gresso dell'Associazione Italiana di Oceanologia e Limnologia:
     Sorrento, (in press).
Arkhipova, N.R., 1976, Peculiarities of the biology and production
     in *Limnodrilus hoffmeisteri* Clap. (Oligochaeta, Tubificidae)
     on grey muds in the Rybinskoe Reservoir Proc. Reservoir Biol.
     Inst., 34 (37):5-15 (In Russian).
Bonomi, G., 1979, Production of *Tubifex tubifex* Müller and *Lim-
     nodrilus hoffmeisteri* Claparede (Oligochaeta, Tubificidae),
     co-existing in the profundal zone a reservoir. Boll. Zool.,
     (in press).
Bonomi, G. and L. Adreani, 1978, Significato adattative della
     struttura comunitaria e della dinamica di popolazioni nel
     macrobenton profondo di un lago artificiale. In: "Il Lago
     di Pietra del Pertusillo: definizione delle sue caratteris-
     tiche limno-ecologiche". Ed. Ist.Ital.idrobiol., 133-201.
Černosvitov, L., 1930, La regression physiologique des organes
     genitaux du *Tubifex tubifex*. Bull.Biol. France et Belgique,
     64:211-248.
Johnson, M.G. and R.O. Brinkhurst, 1971, Production of benthic
     macroinvertebrates of Bay of Quinte and Lake Ontario.
     J. Fish Res. Bd. Can., 28:1699-1714.
Jónasson, P.M., 1955, The efficiency of sieving techniques for
     sampling bottom fresh water fauna. Oikos, 6:183-207.
Jónasson, P.M., and F. Thorhauge, 1976a, Population dynamics of
     Potamothrix hammoniensis in the profundal of Lake Esrom with
     special reference to environmental and competitive factors.
     Oikos, 27:193-203.
Jónasson, P.M., and F. Thorhauge, 1976b, Production of *Potamothrix
     hammoniensis* in the profundal of eutrophic Lake Esrom.
     Oikos, 27:204-209.
Jónasson, P.M., and F. Thorhauge, 1976c, Growth and life cycle
     of *Potamothrix hammoniensis* (Tubificidae, Oligochaeta) in
     the profundal of eutrophic Lake Esrom. A field and labora-
     tory study. Arch.Hydrobiol., 78:71-85.
Kennedy, C.R., 1966, The life history of *Limnodrilus hoffmeis-
     teri* Claparède (Oligochaeta, Tubificidae) and its adaptive
     significance. Oikos,17:158-168.
Kosiorek, D., 1974, Development cycle of *Tubifex tubifex* Müller
     in experimental culture. Pol.Arch.Hydrobiol., 21:411-422.
Ladle, M., 1971, The biology of oligochaeta from Dorset chalk
     streams. Freshwater Biol., 1:83-97.
Petrovich, P.G., E.A. Shushkina and G.A. Pechen, 1964, Computation
     of zooplankton production. Dokl.Akad.Nauk SSSR, 139,
     5:1235-1238 (In Russian).
Poddubnaya, T.L., 1971, Rezorbtsiya i regeneratsiya polovoi
     sistemy u tubifitsid naprimere *Isochaetides newaensis* Mich.

(Oligochaeta, Tubificidae). Tr.Inst.Biol.Vnutr.Vod.Akad. Nauk SSSR, 22 (25):81-90.

Poddubnaya, T.L., 1972, Quantitative dynamics and production of the population of *Isochaetides newaensis* Mich., (Oligochaeta) in the Rybinsk Reservoir. Proc. 2nd. SSR Congress on the Oligochaete Annelids, Borok (In Russian).

Potter, D.W.B. and M.A. Learner, 1974, A study of the benthic macroinvertebrates of a shallow eutrophic reservoir in South Wales with emphasis on the Chironomidae (Diptera); their life histories and production. Arch.Hydrobiol., 74:186-226.

Thorhauge, F., 1975, A field and laboratory study on the reproduction of *Potamothrix hammoniensis* (Tubificidae, Oligochaeta). Arch.Hydrobiol., 76:449-474.

Waters, T.F., 1977, Secondary production in inland waters. Adv. Ecol.Res., 10:91-164.

THE PRODUCTION BIOLOGY OF THE TUBIFICIDAE

(OLIGOCHAETA)

R.O. Brinkhurst

Ocean Ecology Laboratory
Institute of Ocean Sciences
P.O. Box 1700, Sidney,
British Columbia, Canada

ABSTRACT

Few estimates of production of aquatic oligochaetes have
been made. Those that have vary considerably in methodology,
the units employed, the values obtained where these can be
compared, and sometimes in the concept of production used. Laboratory studies on pure populations may be unreliable.

INTRODUCTION

There have been few studies of production in aquatic oligochaetes, but most of those that have been carried out involve
the Tubificidae. This is probably due to their abundance in
silty sediments and hence their potential importance in bioturbation, as fish food and as pollution indicators. The methods
used vary, but generally exclude those based on the recognition
of age classes or cohorts, as these cannot often be distinguished
in oligochaetes with any confidence, though Thorhauge (1976) was
able to do this in Lake Esrom populations of *Potamothrix hammoniensis* that seem unusually long-lived.

REVIEW

Erman and Erman (1975) note the modifications of the field
methods employed, from the summation of the products of instantaneous growth rates (determined in laboratory culture) and mean

205

biomass for a time interval (Johnson and Brinkhurst 1971) to mean
annual biomass multiplied by turnover ratio,(estimated at 5 for
oligochaetes in the Thames by Berrie 1972) and the methods of
Standen (reported elsewhere in this volume). Erman and Erman
employed the Hynes/Coleman-Hamilton method in their studies of
fens in which oligochaetes were important components of the fauna,
especially the lumbriculid *Kincaidiana freidris*. Johnson and
Brinkhurst (1971) showed that instantaneous growth rates at two
of four stations in the Bay of Quinte, Lake Ontario, were related
to temperature, and those of the other two stations were calculated. Production and respiration at each station shared the
same pattern throughout the years, with pronounced summer maxima.
These maxima differed in the two succeeding years in which this
small section of the St. Lawrence Great Lakes was studied. Production of the various oligochaete species involved (which
differed from site to site) varied from site to site and year to
year, ranging from 94.0 to 1.8 Kcal m^{-2}day^{-1}. The commonest
species were *Limnodrilus hoffmeisteri*, *Tubifex tubifex*, and the
lumbriculid *Stylodrilus heringianus* where production was lowest
(in the open Lake Ontario). Turnover rates were estimated at
12.3 to 1.8.

Erman and Erman (1975) studied seven minerotrophic peatlands
or fens in California at nearly 2000 m elevation. Production was
estimated in g wet m^{-2} rather than Kcals (though total production
was converted to Kcals using the data in Brinkhurst (1970) and
Johnson and Brinkhurst (1971). The values obtained for oligochates ranged from 7.4 to 58 g wet wt., within the range of 2.3
to 118.7 g obtained by Johnson and Brinkhurst (1971) and 43 g
for *T. tubifex* populations studied by Tilly (1968). The early
estimates by Teal (1957) for *L. hoffmeisteri* were much higher
(229g) but few of the details of that study are available for
comparison. The turnover ratio formula derived by Johnson and
Brinkhurst (1971) did not apply in the California study where
the average rate was 6.7 as opposed to 5.6, a similar 6.7 by
Teal (1957) but 28.7 by Tilly (1968).

In a detailed study of *Limnodrilus newaensis* (now variously
known as *Isochaetides* or *Tubifex*) by Poddubnaya (1963) "Actual
production" (observed plus losses due to all mortalities) is cited
as 9.5 g·m^{-2}, the "coefficient of actual production" as 3.1. This
author only has access to an incomplete translation of this paper,
and is aware that many more recent studies of this important
species have been made by Poddubnaya, which are referred to elsewhere in this volume. The work was done in the Rybinsk reservoir
of the Volga river.

While Thorhauge (1976) found growth of *P. hammonieinsis* to
be slow at low temperatures and negative at low oxygen concentra-

tions in late summer, Mason (1977), studying the same species, found the larger specimens were lost in late summer but that recruitment continued throughout the year. In this study length/ weight and length/respiration regressions were obtained, and production was estimated by the formula of McNeil and Lawton (1970) and was cited as 3.90 and 9.79 KJm^{-2} in Upton Broad in two years, but 73.35 and 216.50 in Alderfen Broad in Britain.

Aston (1968) studied growth in relation to temperature in *Branchiura sowerbyi* and respiration in the same species (1966), but did not couple these studies with measures of ingestion and egestion to obtain production estimates. This method was pursued by this author with a series of collaborators. Instead of simply achieving production estimates for all possible combinations of the three species that dominate Toronto Harbour *(T. tubifex, L. hoffmeisteri* and *Peloscolex multisetosus)* and four temperatures, these studies revealed that the various parameters used for pro- duction estimates were affected not only by temperature but by interactions between the species such that respiration was reduced, growth, feeding rate, and growth efficiency increased in mixed populations as compared with the same species, under identical conditions, in pure culture. Hence production estimates obtained by this method would produce values as variable as those observed in the field studies already discussed. Assimilation estimates based on estimated respiration and observed growth for cultures of *T. tubifex* and *L. hoffmeisteri* maintained in separate vials but in common aquaria (which should result in them acting the same way as mixed populations according to the experiments of Chua and Brinkhurst,(1973) ranged from 9.28 to 10.75 calories/g dry wt./day, estimates based on ingestion and egestion of the same worms gave values of 7.38 to 15-48 calories (Brinkhurst and Austin, 1979). Fluid losses and secretions may amount for a significant amount of assimilated calories, but these were not estimated here. Earlier estimates of assimilation based on long- term growth data and short-term respiration data yielded assimila- tion values of 232 calories/g dry wt./day for mixed species, 9.8 for the sum of values for pure populations of the two species. The comparable values here varied from 77 to 181 for separate populations in single aquaria, but the signal/noise ratio inherent in the method for obtaining ingestion values makes a statistical comparison of the data invalid. Attempts to convert data published by Standen (1973) for an enchytraeid gave results greatly at variance with those obtained for the tubificids.

SUMMARY

The variations in methods used, aims of the study, and even the units in which production is expressed vary so widely that no clear pattern can be determined so far. There are several detailed

studies of populations of tubificids in progress, at least two
of which will be discussed elsewhere in this volume, and we can
only hope that the factors that determine production rates will
be further explored. Ideally, the measurement of worm production
should be assessed in relation to other elements of the biota, so
that the relative importance of oligochaetes in material and
energy flows in ecosystems can be determined. Most of the studies
referred to here were set in such a context, although the oligochaetes concerned were often the dominant, if not sole, group of
organisms present. Laboratory studies of single species are
suspect now that the apparent cooperation between tubificid species
in exploiting their food resources has been recognised.

REFERENCES

Aston, R.J., 1966, Temperature relations, respiration and
    burrowing in *Branchiura sowerbyi* Beddard (Tubificidae,
    Oligochaeta). PhD. thesis, Univ. Reading, U.K.
Aston, R.J., 1968, The effect of temperature on life cycle,
    growth and fecundity of *Branchiura sowerbyi* (Oligochaeta,
    Tubificidae) J. Zool. Lond., 154:29-40.
Berrie, A.D., 1972, Productivity of the River Thames at Reading.
    Symp. Zool. Soc. Lond., 29:69-86.
Brinkhurst, R.O., 1970, Distribution and abundance of tubificid
    (Oligochaeta) species in Toronto Harbour, Lake Ontario.
    J. Fish. Res. Bd. Canada,27: 1961-1969.
Brinkhurst, R.O., and M.J. Austin, 1979, Assimilation by aquatic
    Oligochaeta. Int. Revue ges. Hydrobiol.,63:863-868.
Chua, K.E. and R.O. Brinkhurst, 1973, Evidence of interspecific
    interactions in the respiration of tubificid oligchaetes.
    J. Fis. Res. Bd. Canada, 30:617-622.
Erman, D.C. and N.A. Erman, 1975, Microinvertebrate composition
    and production in some Sierra Nevada minerotrophic peatlands.
    Ecology, 56:591-603.
Johnson, M.G. and R.O. Brinkhurst, 1971, Production of benthic
    macroinvertebrates of Bay of Quinte and Lake Ontario.
    J. Fish. Res. Bd. Canada, 28:1699-1714.
Mason, C.F., 1977, Populations and production of benthic animals
    in two contrasting shallow lakes in Norfolk. J. Anim. Ecol.,
    46:147-172.
McNeil, S. and J.H. Lawton, 1970, Production and respiration in
    animal populations. Nature, Lond., 225:472-474.
Poddubnaya, T.L., 1963, Life history and growth rate of the Newa
    Limnodrilus (*Limnodrilus newaensis* Mich. Oligochaeta,
    Tubificidae). Trudy. Inst. Reservoir Biol. Acad. Sci.
    USSR,5:46-56 (In Russian)
Standen, V. 1973, The production and respiration of an enchytraeid
    population in a blanket bog. J. Anim. Ecol., 42:219-244.
Teal, J.M., 1957, Community metabolism in a temperate cold spring.
    Ecol. Monogr., 27:283-302.

Tilly, L. J., 1968,   The structure and dynamics of Cove Spring.
    Ecol. Monogr., 38:169–197.
Thorhauge, F., 1976,   Growth and life cycle of *Potamothrix
    hammoniensis* (Tubificidae, Oligochaeta) in the profundal
    of eutrophic Lake Esrom.   A field and laboratory study.
    Arch. Hydrobiol., 78:71–85.

THE PRODUCTION BIOLOGY OF TERRESTRIAL ENCHYTRAEIDAE

(OLIGOCHAETA)

V. Standen

Department of Zoology
University of Durham
South Road, Durham DHI 3LE
England

ABSTRACT

The Enchytraeidae are an abundant and widespread group of
Oligochaetes; most species are terrestrial but a few are commonly
found in freshwater, some are confined to or are most abundant
in wet peatlands and bogs and a number are marine and littoral.
Some species commonly occur in large numbers and populations up
to 300,000 m^{-2} have a significant effect on processes such as
decomposition, especially in soils where earthworms are absent.

The most appropriate method for measuring production is
considered to be the summation of the production of successive
growth stages. To estimate production of *C. sphagnetorum* which
is extremely abundant and widespread in temperate peatlands and
which reproduces by fragmentation, the weights of individuals
dying over a given time period were summed. The ratio of annual
production to average biomass is also often used but different
meanings are attached to it by different authors and several
obstacles attached to its use are considered further. It may be
possible, however, to establish patterns of life history type in
relation to production. Data on life histories, longevity,
growth and reproductive rate are given for several species and
this, together with field estimates of density, indicates the
scale of production by enchytraeids in certain temperate grassland
and peatland habitats.

INTRODUCTION

The Enchytraeidae are a widespread group of oligochaetes which reach their maximum abundance in soils of the temperate zone. Probably the greatest species diversity is found in grasslands. Most species are terrestrial but a few are found commonly in freshwater; some are confined to wet peatlands and bogs and a large number are littoral or marine (Nielsen and Christensen, 1959).

The mean number of enchytraeids in a variety of habitats is shown in Table 1. A maximum density of 300,000 m^{-2} was found by Springett (1967) in blanket bog dominated by *Juncus squarrosus*, and over 100,000 m^{-2} was recorded from *Calluna* (Nielsen, 1955), Douglas Fir plantation (O'Connor, 1957) and wet meadow tundra (MacLean et al. 1977). Lower numbers of enchytraeids were found in grasslands although these populations often were composed of larger species (Springett, 1967).

Recent studies on the distribution of enchytraeids at moorland sites in Britain confirm that some species, such as those of the genus *Fridericia* are confined to mineral soils, while certain species of *Cernosvitoviella* and *Mesenchytraeus* are confined to peat, and others, *Cognettia sphagnetorum* (Vejdovsky) for example, occur in both peat and mineral soils. But whereas in peat the latter species is always dominant and abundant, it may be scarce where it occurs with *Fridericia* spp. in mineral soil.

The distribution of enchytraeid species between the moorland sites is correlated with soil pH, vegetation, drainage and rainfall. *Fridericia* species are correlated with the same factors as many earthworm species (Standen, in press). *Cognettia sphagnetorum* was dominant at all the peat sites; it is one of few enchytraeids that has evolved the ability to live under acid oligotrophic conditions, and in the absence of severe competition it exists in large numbers.

THE USE OF THE P/B RATIO IN PRODUCTION STUDIES

Production studies are of great interest when a population is an important source of food to a higher trophic level, or is numerically abundant and therefore a significant participant in energy flow through the community; however, estimates of production are difficult to obtain.

In the absence of direct data on the productivity of animal species and groups, the prospect of a short-cut method of calculating production from some more easily measured parameter of the population is attractive. The ratio of annual production to average

Table 1. The Abundance and Biomass of Enchytraeidae in Terrestrial Habitats

Habitat	Number 1000m^{-2}	$\bar{x}$ Biomass g·m^{-2}	Country	Authority
Meadow				
dry	8–14		USSR	Moszynski, 1930
	5–37		Norway	Solhoy, 1975
wet	20–35		USSR	Moszynski, 1930
	12–69		Norway	Solhoy, 1975
tundra	44–50		Alaska	MacLean et al., 1977
polygon	11–95		Alaska	MacLean et al., 1977
limestone	59	26.0	England	Springett, 1967
tropical	0.5–7	0.14	India	Thambia & Dash, 1973
Bog				
Calluna	41–202	7.0	Denmark	Nielsen, 1955
Calluna	20–90	13.0	England	Standen, 1973
Juncus	170	24.0	England	Springett, 1967
Woodland				
Spruce	2–54		Finland	Nurminen, 1967
Spruce	8–85		Norway	Abrahamsen, 1969
Pine	1–37		Finland	Nurminen, 1967
Pine	3–55		Norway	Abrahamsen, 1969
Fir	42–250	10.8	Wales	O'Connor, 1962
Aspen	2–22	3.0	Canada	Dash & Cragg, 1972
Tropical	2–23		Malaya	Chiba et al., 1976
Sewage	2,390	1150.0		Williams et al., 1969

biomass - P/$\overline{\text{B}}$ seems potentially useful, but Winberg (1968) stated
that it should be used only where a population has a stable age
structure and a constant level of biomass. Where these conditions
are not met, the ratio depends on several conflicting factors such
as the type of developmental life history of the organism and the
survivorship of the population. Therefore the ratio is examined
further.

    The first question to be decided is which measurement of
production and biomass should be used. Waters (1977) considers
that P/B derives its main significance from its expression as a
cohort P/$\overline{\text{B}}$, which is the total production of a cohort divided by
the mean standing crop measured over its whole life span. However
P/$\overline{\text{B}}$ is most often expressed as annual P/$\overline{\text{B}}$ and while this is its
most useful form in comparative studies there are certain restric-
tions to its use.

    There is less difficulty in using annual P/$\overline{\text{B}}$ predictively for
longlived species or species with co-existing populations where
average biomass is likely to be more constant and therefore more
accurately measure. Many enchytraeid species are of this type
but others are synchronised and annual, and in these cases it is
necessary to know what type of development the species undergoes
because where development is synchronised and discontinuous, this
may greatly affect the calculation and use of $\overline{\text{B}}$.

    For example a species like *Cognettia cognettii* (Issel), which
hatches and grows to full size within a relatively short period
but then matures much later in the year, has a high average
biomass compared with *Achaeta eiseni* (Vejdovsky) which spends a
long period in the cocoon but matures as soon as it is full grown
(Figure 1). In these cases biomass must be measured over the
whole growth period or else average B is misleading. In the two
examples cited, average biomass and therefore P/$\overline{\text{B}}$ are different
although total production is the same.

    In addition to this effect in synchronized species, annual
P/$\overline{\text{B}}$ depends on two main factors. 1) The type of mortality
suffered by the population; if heaviest mortality occurs early in
the life cycle there will be a large proportion of small fast
growing individuals in the population which will thus have a low
average biomass and high annual P/$\overline{\text{B}}$. This is in contrast to
species with low juvenile mortality and therefore a large proportion
of large slow growing individuals in the population which is thus
unproductive in relation to its average biomass (low annual P/$\overline{\text{B}}$).
2) Longevity; where a species takes more than one year to complete
its life cycle the same production is spread out over two or three
years but the biomass of co-existing generations is summed; therefore, the average biomass in each year is high and the annual P/$\overline{\text{B}}$

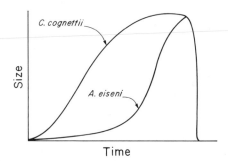

Figure 1.   The effect of different types of developmental life
             history on average biomass (size) of annual,
             synchronised species.

is low. Conversely in species which complete more than one generation in a year (e.g. microorganisms; a few enchytraeids) the
production of several generations is summed and compared with the
average of successive biomasses; annual $P/\overline{B}$ for such species is
therefore high (Mann, 1967).

In nature there is a spectrum of species with different
combinations of characteristic developmental type and population
survivorship. The extremes of growth in biomass of populations of
synchronised annual species showing discontinuous growth are
represented in Figure 2. They are:-
  i)   Heavy juvenile mortality; prolonged egg stage; therefore,
       high annual $P/\overline{B}$.
  ii) Heavy juvenile mortality; prolonged adult stage; therefore,
       intermediate annual $P/\overline{B}$.
  iii) Heavy mortality late in the life cycle; prolonged egg
       stage; therefore, intermediate annual $P/\overline{B}$.
  iv) Heavy late mortality; prolonged adult stage; therefore,
       low annual $P/\overline{B}$.

A second use of the ratio suggested by Heal and MacLean (1975)
is that, as $P/\overline{B}$ reflects the efficiency of production of a population it could be used to compare the rate of functioning of one
population with another. But such comparisons are valid only if
the effects of discontinuous growth on the calculation of average
biomass of synchronized species are taken into account.

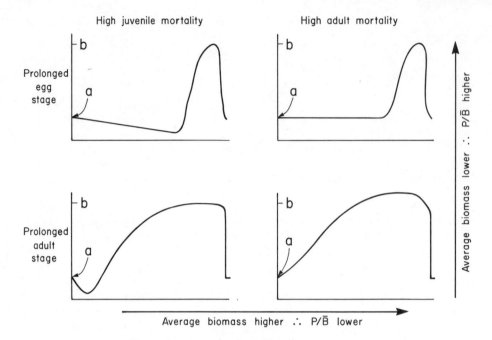

Figure 2. The effect of developmental type and population
          mortality on $P/\bar{B}$ where growth is discontinuous.

          In summary, a high annual $P/\bar{B}$ results when the mean biomass
measured over a year is low due to: a prolonged egg stage in
synchronised species; or to a short life cycle and therefore
several generations in one year; or because most mortality occurs
early in the life cycle and there is a high proportion of fast
growing juveniles in the population.

METHODS OF CALCULATING PRODUCTION

          One frequently used estimate of population production is
given by the product of the biomass of each age group and the
relative growth rate of each (Petrusewicz and Macfadyen, 1970).

$$P_g = \sum_{s=1}^{s=m} v'_s \cdot \bar{B}_s \cdot T$$

Where v'_s is the relative weight gain of each stage $g \cdot g^{-1} \cdot d^{-1}$, B_s
is the average weight of each stage, m = number of stages and T =
duration of each stage.

A second method is to sum the weight of individuals dying over a given time period.

$$P_{gT_1} = \Delta N_T \cdot \Delta \frac{W_T}{2} + N_{T_2} \cdot \Delta W$$

The second method was used to estimate production of *Cognettia sphagnetorum*, the dominant enchytraeid in many northern temperate organic soils. This species reproduces by fragmentation and the decline in numbers of the larger worms is due to fragmentation as well as death. Similarly the number of fragments decreases due to mortality and also to regeneration into whole worms. Therefore, a simple mathematical model was used which predicted population structure and mortality from field data used in conjunction with rates of growth and fragmentation established in laboratory cultures (Standen, 1973).

The estimates of production for two years were 12.44g m^{-2} and 10.04g m^{-2} (53754 and 43430 J m^{-2}y^{-1}). The ratio of annual P/B was:-

1. 12.44/10.78 = <u>1.15</u> - 1968

2. 10.04/12.89 = <u>0.78</u> - 1969

It is suggested that by reproducing by fragmentation *C. sphagnetorum* conserves minerals and energy and that this species is adapted to live under cold, mineral-poor conditions.

LIFE HISTORIES

Most of the enchytraeid species whose life history has been studied so far appear to produce cocoons in winter and early spring which then hatch and grow to full size during summer, and over-winter as adults; e.g. *Enchytraeus coronatus* Nielsen and Christensen and *Cernosvitoviella briganta* Springett. Juvenile mortality is often low and most mortality seems to occur in late autumn and early winter; e.g. *Cognettia sphagnetorum* and *C. glandulosa* (Michaelsen). This type of development leads to a high average biomass and annual low P/B ratio. In most species the cocoons hatch within a month e.g. *Enchytraeus albidus* Henle, *C. briganta*, *Enchytraeus buchholzii* Vejdovsky, and a few larger species of *Fridericia* and *Mesenchytraeus* may take two years to develop; both these life history characteristics also lead to a high average biomass and low annual P/B. *Cognettia glandulosa* seems to be typical of many species showing this type of development and its annual P/B ratio was calculated to be 2.6 (Standen, 1973).

Conversely *Achaeta eiseni* shows discontinuous growth with a prolonged egg stage, and the growth rate of some other species such

as *E. albidus* and *Cernosvitoviella atrata* (Bretscher) indicates
that they could complete their development within a year; in both
cases average biomass will be low and thus annual $P/\bar{B}$ will be
higher at approximately 3 or 4. Maclean (pers. commun.) estimated
$P/\bar{B} = 2.89$ for *C. atrata*.

   Cognettia sphagnetorum has an exceptionally low annual $P/\bar{B}$
ratio (approximately = 1) because it reproduces by fragmentation,
but only a few other species reproduce in this way e.g. *Buchholzia
appendiculata* (Buchholz).

BREEDING BIOLOGY

   Enchytraeids are hermaphrodite and therefore all mature animals
produce eggs. The number of eggs per cocoon and the number of
cocoons produced by each worm varies between species. There is a
range of reproductive strategy from those which have large numbers
of eggs per cocoon but produce few of them (i.e. semelparous), to
others which have a small number of eggs per cocoon and a larger
number of cocoons produced over a long period of time (i.e. intero-
parous).

   Observations on a limited number of species indicate considerable variation between individuals of the same species. *Lumbricillus
rivalis* has the largest number of eggs per cocoon (up to 18 according to Williams et al. 1969) and this species also produces the
largest number of cocoons per worm (up to 10 were estimated from
field observations). Some other species produce only 3 to 6 cocoons
per worm each containing only 1 or 2 eggs e.g. *Achaeta eiseni,
Enchytraeus buchholzii*, and *Marionina clavata* N. and C. (Springett,
1970). *Lumbricillus rivalis* and two other species abundant in
sewage bacterial beds all have a large number of eggs per cocoon
(Reynoldson, 1947). *Enchytraeus coronatus* has 1 - 7 (Williams
et al. 1969) and *Enchytraeus albidus* has 1 - 35 with an average of
10 (Ivleva, 1953). It seems possible that only those species which
produce large numbers of eggs in one batch are able to survive in
a habitat such as sewage beds where worms may be washed out of the
system at any time. The same characteristics of breeding biology
are apparent in species inhabiting decaying wrack beds which are
also of an unstable nature and subject to disturbance (Giere, pers.
commun.). In contrast to these species most other enchytraeids
appear to have a low reproductive rate and to produce a few eggs
over a long period of time. This may reflect the fact that they
live in stable environments with a low rate of predation.

THE LEVEL OF PRODUCTION IN DIFFERENT HABITATS

   When detailed information on population production of each
species in a community is not available an estimate of total
production can be made by using the average biomass and an annual

$P/\overline{B}$ ratio. The average biomass can be calculated from population
density using the relationship given by Abrahamsen (1973), between
length and volume, and volume and weight for six categories of
enchytraeids. The $P/\overline{B}$ ratio has been calculated for only a few
species but the existing information on life history types, growth
rates, and longevity suggest that an annual $P/\overline{B}$ of 2.5 is suitable
for the majority of species in stable habitats. Phillipson (1973)
used an annual $P/\overline{B}$ of 3 for invertebrates generally.

Production has been calculated for enchytraeids in four
habitats (Table 2) using published information on biomass and a
$P/\overline{B}$ of 2.5. The production of enchytraeids in blanket bog is also
shown, but in this habitat the enchytraeid population was dominated
by *Cognettia sphagnetorum* which has a low $P/\overline{B}$ ratio of 1.0. The
type of life history shown by the enchytraeids found in sewage
bacteria beds suggests that they have a higher than average $P/\overline{B}$
ratio and that the very high level of production calculated in
Table 2 may be higher still.

THE INFLUENCE OF ENCHYTRAEIDS ON ENERGY FLOW

Several studies of enchytraeid communities give the standing
crop of the populations and their energy costs due to respiration.
These figures are shown in Table 2 together with estimates of
production. It can be seen that the population respiration is
high. The proportion of assimilation which is used for production
is given by the ratio $P/A \times 100$ and, apart from that calculated
from Springett's data (1967), all are low (9.4 - 19.5) by comparison with McNeil and Lawton's (1970) predictions for invertebrates
which vary from 25 - 62. Their generalisations were based on data
for relatively large invertebrates, whereas most species show a
weight specific respiration rate such that small animals like
Enchytraeidae will have high respiratory costs and consequently
low population production efficiencies. The bulk of the populations
studied by Springett (1967) in limestone grassland were the larger
Fridericia species and this may explain the high $P/A \times 100$ ratio
recorded by her.

The overall contribution by Enchytraeidae to energy flow
through an ecosystem is difficult to assess. The significance of
this group depends on its 'activity' with respect to other saprovores and soil microorganisms and also to the quality and quantity
of dead organic matter (D.O.M.) on which they feed (Coulson and
Whittaker, 1978). In habitatswhere herbivory is high, such as
grassland grazed by sheep, a large proportion of D.O.M. is faeces
and a smaller proportion is unchanged dead leaf litter. Earthworms
are often an important element of the decomposer fauna in these
situations and enchytraeids probably complement their activity.
Where herbivory is low, as in tundra, moorlands and woodlands, the

Table 2.   Annual Production and Respiration of Enchytraeid Population

in Different Habitats

	Average Biomass $g \cdot m^{-2}$	Estimated Production $g \cdot m^{-2}$	Estimated Production $Kcals \cdot m^{-2} \cdot y^{-1}$	Respiration $Kcals \cdot m^{-2} \cdot y^{-1}$	$\dfrac{P}{A \times 100}$	
Meadow	26.0	65	67	76	46.8	Springett, 1967
Heath	10.5	26	27	154	14.9	Nielsen, 1961
Conifer forest } P/B =2.5	10.8	27	28	151	15.6	O'Connor, 1962
Sewage	1150	2875	2973	15600	16.0	Williams et al.,1969
Blanket bog } P/B =1.0	11.3	11.3	13	62	17.2	Standen, 1973

input of unchanged dead organic matter is relatively high. Deciduous woodlands support large populations of earthworms and other large saprovores, but the soils of tundra, moorland and coniferous woodland tend to be acid and earthworms are few, and in these situations Enchytraeidae are often the most important group of decomposer organisms. Large standing crops of Enchytraeidae are found in these soils and it has been shown that they accelerate the rate of decomposition of dead plant material by feeding on it directly and also by their influence on microorganisms (Standen, 1978). This was demonstrated by enclosing various types of leaf litter in nylon mesh bags containing a known number of enchytraeids and measuring the weight loss of litter and its respiration rate compared with control litter without animals.

In conclusion, where herbivory is high, enchytraeids and other groups of decomposer animals usually process most of the annual input of dead organic matter so that little remains. Where herbivory is low, the nature of the dead organic matter is such that even very large populations of enchytraeids plus their stimulatory effect on microorganisms cannot assimilate the annual input of dead plant material and so organic debris accumulates as peat on moorlands or as humus in coniferous woodland and tundra. The life histories of most enchytraeids indicates that the ratio of annual $P/\overline{B}$ is relatively low and 2.5:1 is suggested for use when an estimate of production has to be made from average biomass only. Production by enchytraeids is high in some habitats. Sewage bacteria beds supported 200 times greater production than blanket bog peat.

REFERENCES

Abrahamsen, G., 1969, Sampling design in studies of population densities of Enchytraeidae. Oikos,20:54-66.
Abrahamsen, G., 1973, Studies on body volume, body surface-area, density and live weight of Enchytraeidae. Pedobiologia,13: 6-15.
Chiba, S., T. Abe, M. Kondoh, M. Shiba and H. Watanake, 1976, Studies on the productivity of soil animals in the Pason Forest Reserve, W. Malaysia. Science Rep. Hirosaki Univ.,23: 74-78.
Coulson, J. C. and J.B. Whittaker, 1978, Ecology of moorland animals. In Ecological Studies 27, (Ed. by O.W. Heal and D.F. Perkins) Springer Verlag Berlin.
Dash, M. C. and J.B. Cragg, 1972, Selection of microfungi by Enchytraeidae and other members of the soil fauna. Pedobiologia,12:282-286.
Heal, O. W. and S. F. MacLean, 1975, Comparative productivity in ecosystems - secondary production. In: Unifying concepts in ecology. Eds. W. H. van Dobben and R. H. Lowe-Connell, Junk, Hague.

Ivleva, I. V., 1953, The influence of temperature and humidity on
    the distribution of Enchytraeidae. Trudy Latvijskogo
    Otdelenija VNIRO v,1. (Russian).
MacLean, S. F., G. K. Douce, E. A. Morgan and M. A. Skeel, 1977,
    Community organisation of the soil invertebrates of Alaskan
    arctic tundra. In: Soil organisms as components of ecosystems.
    Ecological Bulletin 25. Eds: Lohm, U. and T. Persson,
    Stockholm.
McNeil, S. and J. H. Lawton, 1970, Annual production and respiration
    in animal populations. Nature,225:472-474.
Mann, K. H., 1967, The cropping of the food supply. In: The biolo-
    gical basis of freshwater fish production 243-257. E.S.P.
    Gerking I.B.P. Symposium Reading. Wiley N.Y.
Moszynski, A., 1930, Repartition quantitative des Enchytraeidae
    dans different milieux. Ann. Mus. Zool. Polon,9:65-127.
Nielsen, C. O., 1955, Studies on Enchytraeidae 2. Field studies.
    Natura Jutlandica,4-5:1-58.
Nielsen, C. O., 1961, Respiratory metabolism of some populations
    of enchytraeid worms and free living nematodes. Oikos,12:
    17-37.
Nielsen, C. O. and B. Christensen, 1959, The Enchytraeidae, a
    critical revision and taxonomy of the European species.
    Natura Jutlandica,8-9:1-160.
Nurminen, M., 1967, Ecology of enchytraeids in Finnish coniferous
    forest soils. Ann. Zool. Fenn.,4:147-157.
O'Connor, F. B., 1957, An ecological study of the enchytraeid worm
    population of a coniferous forest soil. Oikos,8:162-198.
O'Connor, F. B., 1962, Oxygen consumption and population metabolism
    of some populations of enchytraeidae in North Wales. In:
    Soil organisms, Ed: Doeksen, E.J. and J. van der Drift, North
    Holland Publishing Co.
Petrusewicz, K. and A. Macfadyen, 1970, Productivity of terrestrial
    animals. I.B.P. Handbook 13. Blackwells, Oxford.
Phillipson, J., 1973, Biological efficiency of protein production
    by grazing and other land-based systems. In: Biological
    efficiency of protein production. Ed. J.G.W. Jones C.U.P.
Reynoldson, T.B., 1947, An ecological study of the enchytraeid
    worm population of sewage bacterial beds. Field investigaion.
    J. Anim. Ecol.,16:26-37.
Solhoy, T., 1975, Dynamics of Enchytraeidae populations on Hardan-
    gervidda. Ecological studies: analysis and synthesis 17,
    55-58, Fennoscandia Tundra Ecosystems, Ed: Wielgolaski, F. E.
Springett, J.A., 1967, An ecological study of moorland Enchytraei-
    dae. Ph.D. Thesis, Durham University.
Springett, J.A., 1970, Distribution and life histories of some
    moorland Enchytraeidae. J.Anim. Ecol., 39:725-737.
Standen, V., 1973, The production and respiration of an enchytraeid
    population in blanket bog. J. Anim. Ecol., 42:219-245.
Standen, V., 1978, The influence of soil fauna on decomposition by
    microorganisms in blanket bog litter. J. Anim. Ecol., 47:25-38.

Standen, V., in press. Factors affecting the distribution of
    lumbricids in associations in peat and mineral soils in
    northern England. Oecologia:
Thambi, A.V. and M.C. Dash, 1973, Seasonal variation in numbers
    and biomass of Enchytraeid populations in tropical grassland
    soil for India. Tropical Ecology,14:228-237.
Waters, T.F., 1977, Secondary production in inland waters. Rec. Ad.
    Ecol., Vol. 10.
Williams, N.V., J.F. de L.G. Solbe and R.W. Edwards, 1969, Aspects
    of the distribution, life histories and metabolism of the
    enchytraeid worms *Lumbricillus rivalis* and *Enchytraeus
    coronatus* in a percolating filter. J. Appl. Ecol., 6:171-183.
Winberg, G.G., 1968, The dependance of rate of development on
    temperature. In: Metody opredelenija produkcii vodnyk
    zivotnyh izd 'Vysejsaja' Minsk (Russian).

ECOLOGICAL STUDIES OF AQUATIC OLIGOCHAETES IN THE USSR

A. Grigelis

Department of Ichthyology and Hydrobiology
Institute of Zoology and Parasitology
Academy of Sciences, Vilnius K Pozelos, 54
Lithuanian SSR

ABSTRACT

Research on the ecology of aquatic oligochaetes in fourteen different ecological regions of the USSR, such as West Siberia, the Volga reservoirs, the Black Sea estuaries, the Caspian and Aral seas, Central Asia, Lake Baikal and the Far East is reviewed.

INTRODUCTION

Literature on the aquatic oligochaetes of the USSR is extensive and varied, having both a systematic – faunistic and biological – ecological as well as applied character. The European region is the best known, early publications being those of Grimm (1877), Braun (1884), Kojevnikov (1890), Michaelsen (1903), Andrusov (1914), Lastochkin (1918), Svetlov (1924, 1925, 1926), and Malevich (1925, 1927, 1929). Extensive work was begun after the second World War, and we are now able to divide the USSR into fourteen ecological regions.

WEST SIBERIA

Michaelsen (1903) recorded *Limnodrilus udekemianus* Clap., *Peloscolex ferox* (Eisen), *Haplotaxis gordioides* (Hartman) and *Pelodrilus* Mich. Other contributions by Malevich (1926), Pirozhnikov (1929), Michaelsen (1935) and Svetlov (1946) added further information on the lower Ob and Lake Sartlan among others.

More recent studies by Semernoi (1970) and Zaloznyi (1973, 1974, 1976) identify sixty-four species in the region, which can now be subdivided into five districts.

Upper Ob

The headwaters flow across mountainous and woody steppe and taiga zones. Low-water flows are 0.3 to 0.5 m s^{-1}, and up to 2 m s^{-1} at high flows. In soft sediments *T. tubifex* and *L. hoffmeisteri* in particular, with *L. udekemianus*, contribute to a benthic biomass of 0.8 to 2.85 gm^{-2}. In the Novosibirsk reservoir, built in 1957, the two former species predominate.

Middle Ob

This large swampy basin contains taiga rivers bearing much colloidal and dissolved organic matter which causes oxygen depletion. The water is slightly acid (pH 6.0 to 6.9) and dissolved inorganic salts do not exceed 200 mg L^{-1}, and there is little vegetation. *Potamothrix hammoniensis* is added to the three species of the previous district, with populations of 80-420 m^{-2} at a biomass of 0.06 to 0.3 gm^{-2}.

Irtish basin

The tributaries of the Tobol, Ishim and other rivers have 500-1000 mg L^{-1} dissolved inorganics, and while sediments of the Irtish are fine sandy clays, sand occupies a large proportion of the river bed. In 1965-1970 samples of the species list was extended by the addition of *L. claparedeianus*.

Lower Ob

A large number of lakes and sloughs are found in this part of North West Siberia. The river water is a yellow colour, and is rich in oxygen (40-50 mg L^{-1}, Alekin, 1949). Most worms are found in the silts, where they make up 31% of the benthos by numbers, 1.3% by weight. The lakes of the bottomland are very abundant and support a varied zoobenthos in the littoral. Biomass averages 10.2 gm^{-2}, of which 0.47 gm^{-2} is oligochaetes. Common species are *Stylaria lacustris, Chaetogaster limnaei, Ch. langi, Nais variabilis, T. tubifex, L. hoffmeisteri, P. ferox,* and *Lumbriculus variegatus* (Zaloznyi, 1974).

Extreme North

On the Gydansk and Jamalsk peninsula in the permafront zone, there is almost no vegetation in the large lakes. Up to 1200 m^{-2} specimens of *T. tubifex, L. hoffmeisteri* and *Lumbriculus variegatus* make up the bulk of the oligochaetes, with *Ch. diastrophus, Ch. langi, N. communis, N. variabilis* in addition. Zaloznyi (1974) found 20 to 110 *Alexandroria oregensis* and 10 to 200 *Peloscolex oregonensis* per square meter. Twenty-eight species are found in the rivers, thirty-one in the tributaries and thirty-eight in the lowland lakes. The mean oligochaete biomass is 4.1 gm^{-2} (1485 m^{-2}), in the oligochaete – poor upland lakes as little as 0.7 gm $^{-2}$ (180 m^{-2}). The species were divided into the following categories by Zaloznyi (1972):

Region	Species	Percent
cosmopolitan	7	11.1
2-3 zoogeographical regions	14	28.8
holarctic	13	20.3
palearctic	23	35.9
european-siberian	4	6.2
siberian	3	4.6

NORTHERN EUROPEAN WATERS

This area includes the Murmansk, Karelia, Komi ASSR, and the Leningrad regions. Popchenko (1972) summarized work in Karelia, listing 106 species belonging to 7 families, among them five Lumbriculids (*Lumbriculus alexandrovi, L. Karelicus, L. tetraporophorus, Trichodrilus aporophorus, Tatriella longiatriata*) and two forms (*Nais bretscheri* f. *variedentatus, Lamprodrilus acsaetus* f. *palearcticus*) being new to science. Of the whole list of species, 98 are found in the lakes, 53 in reservoirs, 64 in rivers and 22 on the shore of the White Sea. The dominant species are, as usual, *T. tubifex, L. hoffmeisteri, L. udekemianus, L. claparedeianus* and *P. hammoniensis*.

In the oligotropic and mesotrophic lakes of Karelia the worms make up from 5 to 20 percent of the total number, with up to 198 m^{-2}, and 8 to 20 percent of the biomass (up to 0.69 gm^{-2}). In most Karelian lakes oligochaetes make up from 55 to 92% of the number, 65 to 70 per cent of the biomass, but in dystrophic laes the figures are 20-110 m^{-2} and 0.12 gm^{-2}. In Lake Vozhe worms are 14% of the biomass, whereas in Lake Ladoga they are 20 to 40 per cent, by number and 50 percent of the biomass, which increases to 60-70% and 70-80% respectively when the chironomids emerge. In sandy parts of the lake 200-500 m^{-2} are found. *Alexandroria oregensis* was thought to be endemic (but see the section on the lower Ob, above) and thirty-three species are listed by Slepukhina (1972). Detailed studies of the River Neva (Tsvetkova, 1970) show that the large number of worms (350,000 m^{-2}) is caused by rapid decomposition of organic matter together with good aeration.

THE UPPER VOLGA

The early work in this area was by Michaelsen (1923), Malevich
(1926, 1940, 1947, 1949, 1956), Lastochkin (1936), Malevich and
Zevina (1958) and Poddubnaya (1958 to 1976). The latter made intensive studies of the Rybinsk reservoir, the nutrition of tubificids (1961) and naidids (1965) and reproduction in tubificids,
expecially *I. newaensis* (1958, 1959, 1963) and the production of
T. tubifex in polluted areas (1976). The abundance of *T. tubifex*
varied from 3,120 to 260,000 m^{-2} (21.2 - 540.2 gm^{-2}, $\bar{x}$ 146 gm^{-2})
with an annual production of 2,377 gm^{-2}.

BALTIC REPUBLICS

The fauna of Estonia, Latvia, Lithuania and Byelorussia has
been investigated since the early publications by Braun (1884),
Kojevnikov (1890) and Schneider (1908). The species found in
Estonian lakes were listed by Riikoja (1955), who found sixteen
species, whereas twenty-six were listed for Lithuania by Sivickis
(1933, 1934, 1940). The more recent works by Timm (1962 et seq.)
accounted for 52 lacustrine species, 61 in rivers, including the
new species *Peipsidrilus pusillus* described from Lake Peipsi-
Pihkva in 1977. A total of 76 species from 7 families has been
found in Estonia.

Latvian lakes and rivers have been studied by Kachelova and
Parele (1972) and Parele (1975) as part of a water pollution investigation. In the Lielupe river there are 44 species, in the Dau-
guva they make up 65.2 to 94.2% of the entire benthos. The same
species are found throughout Lithuania and Byelorussia according
to Grigelis (1961 to 1975) and Sokolskaya (1953, 1956), the common
ones being *Stylaria lacustris, N. elegans, T. tubifex, L. hoffmeisteri, L. claparedeianus, P. hammoniensis* and *Lumbriculus variegatus*.

UKRAINE AND MOLDAVIA

This area has not been so well investigated as those already
considered. Twenty-eight species are known from Moldavian ponds
according to Chokyrlan (1960), where numbers reach 480-5640 m^{-2},
biomass 0.68 to 18.84 gm^{-2}. *Dero obtusa, Ch. limnaei, L. hoffmeisteri* (73%, 1920 m^{-2}),*P. hammoniensis* (40%, 1040 m^{-2}), *L. udekemianus* (39%, 840 m^{-2}), and *T. tubifex* (14%, 420 m^{-2}) predominate. In
the psammophilic biocenoses of the Dubosarski reservoir *Aulodrilus
limnobius, I. newaensis,* and the enchytraeid *Propappus volki* are
common, and in the pelophylis biocenoses the usual *Limnodrilus -
P. hammoniensis* assemblage is abundant.

In the River Prut (0.4-1.45 ms^{-1}, 1.5 to 3.3 in deep, dissolved
inorganics 103-770 mg L^{-1} pH 7.3 to 8.6, 1.5° C winter, 25.6° C
summer) and the Lapushiskoje reservoir studied by Mushchinskii (1970)

the *I. newaensis* - *P. ferox* - *P. volki* fauna of the oligosaprobic river sections was replaced by *T. tubifex* - *L. hoffmeisteri* - *P. hammoniensis* in the reservoir.

In the steppe ponds (including Dnepropetrovsk and Kakhovsk reservoirs) Lubyanov (1953, 1955) and Gaidash (1970) found *T. tubifex*, *L. hoffmeisteri*, *L. udekemianus*, *L. caparedeianus* and *P. hammoniensis* at 9 to 45 m in the profundal zone making up 97 to 98% of all zoobenthos (1320 to 6400 m^{-2}, 7.8 to 24.6% of biomass). Where sewage from industrial sources adversely affects the environment *T. tubifex* and *L. hoffmeisteri* (10,100 m^{-2}) dominate, the biomass being 848.0 gm^{-2}, but these cannot be utilised by fish because of the environmental conditions.

The fauna of the Dnieper includes 33 worm species of which *Aeolosoma travancorense* (14,392 m^{-2}) and *P. moldaviensis* (847 m^{-2}) were common in silty-sand and six species were added to the species list (*A. travancorense, Haemonais waldvogeli, Dero dorsalis, Aulophorus piguete, Haplotaxis gordioides, Potamothrix vejdovskyi*) by Fomenko (1963). The rheophil fauna was divided into polyrheophilic, α mesorheophylic, β-mesoreophilic and limnophilic, based on various running-water conditions (Fomenko, 1972).

PONTO-CASPIAN ESTUARIES

The species of this unusual brackish-water region were divided by Finogenova (1972) as follows:

holoeurihalinic
  Enchytraeus albidus
  Lumbricillus lineatus

stenohaline-salt
  Peloscolex svirenkoi
  P. euxinicus

euryhaline salt
  Paranis frici
  P. simplex
  Amphicaeta sannio
  Potamothrix bavaricus
  P. mrazeki
  P. danubialis
  P. deserticola
  T. tubifex blanchardi

marine - weak euryhaline
  Tubifex costatus

Marine - strong euryhaline
  Paranais litoralis

Freshwater - euryhaline
  Psammoryctides barbatus
  Nais elinguis
  N. communis
  Ch. langi
  Uncinais uncinata
  Aeolosoma hemprichi
  Potamothrix hammoniensis
  P. heuscheri
  P. moldaviensis

Freshwater - stenohaline
  Vejdovskyella comata
  Nais barbata
  Limnodrilus hoffmeisteri
  L. udekemianus
  L. claparedianus
  Isochaeta newaensis

Moroz (1977) recorded 105 species and forms in the estuaries of the north-west part of the Black Sea, where there are 13 Ponto-Caspian endemics. Seven species predominate (*Psammoryctides deserticola* 51%, *P. moldaviensis* 49%, *Isochaeta michaelseni* 40%, *Potamothrix caspicus* 40%, in the Dniestr estuary, *P. hammoniensis* 43%, *I. michaelseni* 42.5%, *P. deserticola lastockini* 27%, *P. barbatus* 27% in the Dnieper - Bug). *Isochaeta michaelseni* (48-64%) and *P. danubialis* (24-36%) inhabit the main channel and tributaries of the Danube, where densities fluctuate from 20-180,000 m^{-2}, 0.02 to 135 gm^{-2}.

THE VOLGA

The main study in the sector has been made in the Kuibyshev and Tsimlyanskoye reservoirs. In the Kamsk spur of the former thirty-two species are named by Ekaterininskaya (1964), the common ones being *T. tubifex, L. hoffmeisteri, P. hammoniensis* and *Uncinais uncinata* with the rare species including *Rhynchelmis limosella, Limnodrilus helveticus* and *L. aurostriata* (sic.). Fourty-four species were listed by Lyubin (1972, 1974) from 1958-1959 collections, with *Propappus volki* (152,000 m^{-2}), 8.4 gm^{-2}) *Isochaeta newaensis* (up to 5,760 m^{-2}) and *I. michaelseni* (480 m^{-2}).

In the first year of existence of the Tsimlyanskaya reservoir Ioffe (1956) found twenty-four species including *Stylaria lacustris, I. michaelseni* and *I. newaensis*. The species list then increased, until Miroshnichenko (1971) identified fifty-eight species; with up to 19,000 m^{-2} (19.2 gm^{-2}) in silty-sand, 5,841 m^{-2} (7.36 gm^{-2}) in non-silt, and 1,790 to 3,629 m^{-2} (4.67 to 5.41 gm^{-2}) in sand. The usual *Tubifex - Limnodrilus - Potamothrix* assemblage dominated.

THE CAUCASUS

Many studies of this zone have been published from Malevich (1930, 1947) and Chekanovskaya (1959) to Pataridze (1957 to 1970). The north Caucasus was investigated by Slepukhina (1970, 1975), where worms are not abundant because the rivers are very turbid. In the ponds (transparency 5-25 cm) predominant species are *T. tubifex* (up to 5,000 m^{-2}) and *L. udekemianus* (up to 6,000 m^{-2}). Forty-four species have been found in Georgia, and at 570-630 m^{-2} they are the most abundant element in the zoobenthos of Chromski reservoir. In lake Paravani there are 15,348 m^{-2} (10.1 gm^{-2}) and in Lake Rica the minimum biomass is found in summer (26 gm^{-2}) the maximum of 67 gm^{-2} in winter.

CASPIAN SEA

The average salinity in this unusual water body is 13°/oo. The first oligochaete survey (Grimm, 1877) revealed five species. More recent studies by Hrabě (1950) and Finogenova (1972, 1976a, b) added other species, of which *Aktedrilus svetlovi, Potamothrix manus,* and *Marionina micula* are new to science. At depths of greater than 200 m, worms may be as abundant as 30–190 m^{-2}, the common species being *P. caspicus* (<1,145 m^{-2}) and *P. deserticola, P. grimmi, P. moldaviensis mitropolskiji* are endemic (from 50–570 m^{-2} at 103 m).

ARAL SEA

The origin and biology of the water body are of interest, with its very turbid rivers with low suspended organic load and and average salinity in the lake of 10.4°/oo. The oligochaete fauna is poor, only four species (*Paranais simplex, Lumbricillus lineatus, Potamothrix bavaricus* and *Psammoryctides albicola*) being named, with the last predominating (Gavrilov, 1972).

CENTRAL ASIA

The oligochaetes of Lake Chalkar and Lake Isyk-kull were listed by Hrabě (1935), Hrabě and Černosvitov (1929), and Chekanovskaya (1952) described those of Fergana and the seventeen species from West Kazakhstan. The steppe and desert zones of Uzbekistan and Kazakhstan, which include rivers, freshwater, brackish and saline lakes have been studied by Sbarach (1968), Mamilova (1970), Akhmetova (1975), Semernoi and Belozub (1979). Akhmatova found nineteen species in south and central Kazakhstan of which eight (including *A. hemprichi, S. lacustris, N. variabilis* and *L. hoffmeisteri*) are predominant in rivers, and three are rare (*A. tenebrarum, S. fossularis, L. variegatus*). These are found in the freshwater lakes but *L. variegatus* is missing from the brackish lakes. The tubificids are present in ponds and silting waters, but oligochaetes are absent from the saline lakes.

The zoobenthos of the shallow freshwater lakes of Biilikul and Akkul were investigated by Sbarach (1968) who found five species with *L. hoffmeisteri* and *P. bavaricus* being widely distributed. Mamilova (1970) studied the shallow lake Kara-Kul in which worms were second in abundance (0.215 gm^{-2}) with an average production of 0.731 gm^{-2}. Semernoi and Belozub (1979) found 24 species and 2 forms in rice fields.

BAIKAL AND ZABAIKAL

This is a unique zone, with Lake Baikal being a huge natural laboratory for the formation of new species. Verescshagin (1940) identified 57 forms, 53 being endemic. Izosimov (1962) described 7 new species of the family Lumbriculidae, while Sokolskaya (1962) described 11 naidid species, 5 of which were new endemics. Noskova (1967) investigated the shallow Selenga district, finding 13 widespread species and 35 endemics among 50 species. The greatest

population densities recorded were for *Peloscolex inflatus* and
L. baicalensis (84 and 328 m^{-2} respectively). The smallest number
of worms was found in clean sand (1,060 m^{-2}). The current list
of species in Lake Baikal has now reached 70, with both baicalian
and siberian elements. The water-bodies beyond Baikal were divided
into three provinces by Semernoi (1972), finding a total of 59
species including 9 newly-described species.

In these cold regions, survival mechanisms have been observed.
The Aeolosomatidae and Naididae lay wintering coccons and then die,
but specimens of *Rhyacodrilus coccineus, R. sibiricus,* and *T.*
tubifex and the lumbriculid *L. variegatus* overwinter in frozen
sediments at temperatures down to -11° C, forming slime capsules.
The enchytraeids survive in frozen sediment. Ice cover persists
for seven months in the Zabaikal, the ice reaching 160–180 cm. The
central parts of the lakes of the region contain *T. tubifex* and
L. hoffmeisteri (Shapovalova, 1974).

KAMCHATKA, SAKHALIN, LOWER AMUR

Eleven species are known from the freshwaters of this area
(Sokolskaya, 1967, 1972). *Ilyodrilus templetoni* is common in the
Amur River, and the endemics *Peloscolex nikolskyi* Last, and *P.*
apapillatus Last. are abundant. Eighteen species were found in
the Sakhalin and twenty-eight in Kamchatka, including the endemic
species *Lumbricillus kamtschatikanus* (Mich.), *Alexandrovia ringulata*
(Sok.), *Peloscolex kamtschaticus* Sok., *P. kurenkovi* Sok., *Stylos-*
colex opisthothecus Sok., *Kurenkovia magna* Sok., and *Propappus*
arhynchotus Sok.; *Limnodrilus hoffmeisteri* and *L. claparedeianus*
are absent, which makes this area unique, though with some genetic
contacts with north Europe, the Amur and east Asia.

Shurova (1977, 1978) investigated the littoral enchytraeids
of the far-eastern seas. Twenty-six species were described, 19
of them being newly discovered. Fourteen percent of the species
were found in the Kuril Islands and Kamchatka, with twelve species
in the upper and mid littoral, six in the lower littoral and only
Grania pacifica in the sublittoral. In the north-western part of
the Sea of Japan *Lumbricillus corallinae, L. nipponicus, L. ignotus,*
Enchytraeus cryptosetosus, Marionina limpida and *M. spicula* are
abundant.

CHUKOTKA

Only eight species are known of which five were described as
new by Sokolskaya (1976, 1977).

REFERENCES

Akhmetova, B. A., 1975, On the aquatic oligochaete fauna of central
    and southern Kazakhstan. Helminths of Birds and Fishes of Ka-
    zakhstan and Their Intermediary Hosts [K faune vodnykh oligo-
    khet (Oligochaeta) Tsentral'nogo i Yuzhnogo Kazakhstana. Gel'-
    minty ptits i ryb Kazakhstana i ikh promezhutochnye khozyaeva].
    Alma-Ata, 29-35.
Alekin, O. A., 1949, General Hydrochemistry [Osnovy gidrokhimii].
    Gidrometeoizdat, Leningrad.
Andrusov, L., 1914, Data on the oligochaete fauna in the vicinity
    of Kiev [Materialy dlya fauny Oligochaeta okrestnostei g.
    Kieva]. Zap. Kievsk. o-va estestvoispytat., 23: 91-95.
Braun, M., 1884, Physikalische und biologische Untersuchungen im
    westlichen Theile des Finnischen Meerbusens. Arch. Naturk.
    Liv., Est- und Kurlands, 2. Ser., 10(1).
Chekanovskaya, O. V., 1952, On the oligochaete fauna of the West
    Kazakhstan oblast [K faune maloshchetinkovykh chervei (Oligo-
    chaeta) zapadno-Kazakhstanskoi oblasti]. Trudy zoologicheskogo
    instituta Akademii nauk SSSR, 11: 293-299.
Chekanovskaya, O. V., 1959, On the oligochaete fauna of the northern
    Caucasus [K faune maloshchetinkovykh chervei Severnogo Kavkaza].
    Trudy zoologicheskogo instituta Akademii nauk SSSR, 26: 347-354.
Chokyrlan, V. K., 1960, On the oligochaete fauna in Moldavian ponds
    [Materialy po faune oligokhet prudov Moldavii]. Sbornik rabot
    molodykh uchenyikh. Kishinev, 181-182.
Ekaterinskaya, N. G., 1964, The oligochaete fauna of Kamsk bay in the
    Kuibyshev reservoir in 1960-1962 [Fauna maloshchetnikovykh cher-
    vei Kamskogo otroga Kuibyshevskogo vodokhranilishcha po materialam
    1960-1962 g]. Trudy Tatarskogo otdeleniya gosudarstvennogo
    nauchno-issled. Instituta ozernogo i rechnogo rybnogo khozyai-
    stva, 60: 133-141.
Finogenova, N. P., 1972, The oligochaete fauna of brackish waters in
    the Ponto-Caspian basin. The aquatic Oligochaeta [Oligokhety
    v solonovatykh vodakh basseinov Ponto-Kaspiya. Vodnye malo-
    shchetnikovye chervi]. Trudy vsesoyuznogo gidrobiologicheskogo
    obshchestva, 27: 65-74.
Finogenova, N. P., 1976a, Oligochaetes of the central and southern
    Caspian [Maloshchetnikovye chervi Srednego i Yuzhnogo Kaspiya].
    Gidrobiol. issled. samoochishcheniya vodoemov, 141-151.
Finogenova, N. P., 1976b, New species of Tubificidae (Oligochaeta)
    from the Caspian Sea [Novye vidy maloshchetinkovykh chervei
    semeistva Tubificidae iz Kaspiiskogo morya]. Zoologicheskii
    zhurnal 55: No. 10, 1563-1566.
Fomenko, N. V., 1963, Quantitative data on the oligochaetes in the
    waters of the lower Dnieper. The hydrochemical and hydrobiolog-
    ical regimen of the lower Dnieper after the construction of the
    Kakhovsk reservoir [Materialy do kil'kiskoi kharakteristiki
    oligokhet u vodoemakh ponizzya Dnepra. Gidrokhimicheskii to

biologicheskii rezhim ponizzya Dnepra posle sporuzhennya
    Kakhovs'kogo vodoemishcha]. No. 29, Kiev, 81–98.
Fomenko, N. V., 1972, On ecological groups of oligochaetes of the
    Dnieper. The aquatic Oligochaeta [Ob ekologicheskikh gruppakh
    oligokhet (Oligochaeta) reki Dnepra. Vodnye maloshchetinkovye
    chervi]. Trudy gidrobiologicheskogo obshchestva SSSR, 27: 94–
    10ố.
Gaidash, Y. K., 1970, The effects of flood-control runoff and indus-
    trial pollution on the feeding base of fish-benthophages of the
    Dnieper reservoir [Vliyanie zaregulirovannogo stoka i promysh-
    lennogo zagryazneniya na kormovuyu bazu ryb-bentofagov Dneprov-
    skogo vodokhranilishcha]. Voprosy ikhtiologii, 10: 530–536.
Gavrilov, G. B., 1972, Collections of oligochaetes made in the Aral
    Sea during the cruises of 1964 and 1965. The aquatic Oligochaeta
    [Sbory oligokhet v Aral'skom more za navigatsii 1964 i 1965 gg.
    Vodnye maloshchetinkovye chervi]. Trudy vsesoyuznogo gidrobiolo-
    gicheskogo obshchestva, 17: 82–86.
Grigelis, A. I., 1961, Oligochaete fauna and dynamics of number and
    biomass of *Ilyodrilus hammoniensis* Mich. and *Psammoryctes barb-
    batus* (Grube) in Lake Disnai [Fauna oligokhet i dinamika *Ilyo-
    drilus hammoniensis* Mich. i *Psammoryctes barbatus* (Grube) v
    ozere Disnai]. Trudy Akademii nauk Litovskoi SSR, 3: 145–152.
Grigelis, A. I., 1966, Invertebrates in the soils of the Spit of
    Kurshyu-Nyariya. 4. Enchytraeids in the soil and oligochaetes
    in the surrounding waters [Pochvennye bespozvonochnye kosy
    Kurshyu-Nyariya. 4. Pochvennye enkhitreidy i vodnye oligokhety
    kosy i okaimlyayushchikh vod]. Trudy Akademii nauk Litovskoi
    SSR. Ser. V, 2: 169–175.
Grigelis, A. I., 1974, Population structure of the dominant benthic
    organisms of Lake Dusya (I. *Potamothrix hammoniensis* Mich., 1969–
    1971) [Struktura populyatsii dominiruyushchikh bentosnykh
    organizmov oz. Dusya (I. *Potamothrix hammoniensis* Mich. 1969–
    1971 gg.)]. Trudy Akademii nauk Litovskoi SSR. Ser. V,
    4: 77–87.
Grigelis, A. I., 1975, Number and biomass of oligochaetes of the
    species *Potamothrix hammoniensis* Mich. in Lakes Dusya, Galstas,
    Obyaliya, and Shlavantas in 1968–1971 [Chislennost' i biomassa
    oligokhet vida *Potamothrix hammoniensis* Mich. v ozerakh Dusya,
    Galstas, Obyaliya i Shlavantas v 1968–1971 gg]. Trudy Akademii
    nauk Litovskoi SSR, Ser. V. 2: 53–59.
Grimm, O. A., 1877, The Caspian Sea and its fauna [Kaspiiskoe more
    i ego fauna]. Trudy Aralo-Kasp. ekspeditsii 1876–1877, 2: 108–
    112.
Hrabě, S., 1935, Oligochaetes of Lake Issyk-Kul. Proc. Kirghiz
    Compl. Exped. 3.
Hrabě, S., 1950, Oligochaeta Kaspiskoho jezera. Prace Moravsko-
    slezske akademic ved Prirodnich 22: 251–290.
Hrabě, S., and Černosvitov, L. V., 1929, Oligochaetes of Lake
    Chalkar. Russk. Gidrobiol. Zh. 8–9: 211–218.

Ioffe, T. I., 1956, The bottom feeding resources in Tsimlyanskaya
    reservoir during its first year [Dannye komovye resursy Tsim-
    lyanskogo vodokhranilishcha v pervyi god ego sushchestvovaniya].
    Izvestiya VNIORK 34: 1-2.
Izosimov, V. V., 1962, Oligochaetes of the family Lumbriculidae.
    Oligochaetes and Turbelaria of Lake Baikal [Maloshchetnikovye
    chervi semeistva Lumbriculidae. Maloshchetinkovye chervi i
    planarii ozer Baikal], 3-126.
Kachalova, O. L., and E. A. Parele, 1972, New materials on the
    freshwater oligochaete fauna of Latvia [Novye materialy po faune
    presnovodnykh oligokhet Latvii]. Izvestiya Akademii nauk
    Latviiskoi SSR. No. I(294): 27-32.
Kojevnikov, G., 1890, La fauna de la mer Baltique orientale et les
    problèmes des explorations prochaines de cette faune. 1-26.
Kuzin, A. M., 1953, Isotopes in Biochemistry [Izotopy v biokhemii].
Lastochkin, D. A., 1918, Data on the aquatic oligochaete fauna of
    Russia. I. Register of species found in Petrograd and its
    environs [Materialy po faune vodnikh Oligochaeta Rossi. I.
    Spisok vidov, naidenhikh v Petrograde i ego okrestn]. Trudy
    Petrogr. o-va estestv. 49: 57-63.
Lastochkin, D. A., 1936, Hydrobiological research of the Rivers
    Volga and Mologa [Gidrobiologicheskie issledovaniya rr. Volgi
    i Mologi]. Trudy Ivanovskogo sel'sk. in-ta, 2.
Lepneva, S. G., 1930, Study of the bottom fauna of the upper Ob
    [K izucheniyu donnoi fauny verkhnei Obi]. Zapiski GGI, III
    Leningrad.
Lepneva, S., 1931, Ergebnisse der Erforschung des Teleckoje-Sees.
    Sonder-Abdruck aus dem Archiv für Hydrobiologie, 33.
Lubyanov, I. P., 1955, Seasonal changes of the bottom fauna of
    Dnepropetrovsk reservoir after its restoration [Sezonnye
    izmeneniya donnoi fauny Dnepropetrovskogo vodokhranilishcha
    posle ego vosstannovleniya]. Vestnik Dnepropetrovskogo
    nauchno-issled. in-ta gidrobiologii, 2: 83-101.
Lubyanov, I. P., and I. A. Fed'ko, 1953, The bottom fauna of ponds
    of the Ukrainian steppe in relation to environmental conditions
    [Donnaya fauna prudov stepnoy Ukrainy v svyazi s usloviyami ee
    sushchestvovaniya]. Vestnik Dnepropetrovskogo nauchno-issled.
    in-ta gidrobiologii, 123-152.
Lyubin, V. A., 1972, Changes in the oligochaete fauna of the
    Kuibyshev reservoir during 1958-1967. The aquatic Oligochaeta
    [Izmeniya v faune oligokhet Kuibyshevskogo vodokhranilishcha
    vo vremya 1958-1967 gg. Vodnye maloshchetinkovye chervi].
    Materialy vtorogo Vsesoyuznogo simposiuma v Barke, 91-93.
Lyubin, V. A. 1974, Changes in the composition of the oligochaete
    fauna in the Kuibyshev water storage basin [Izmeneniya v sos-
    tave fauny maloshchetinkovykh chervei Kuibyshevskogo vodo-
    khranilishcha]. Gidrobiologicheskii zhurnal, 10: 47-52.
Malevich, I. I., 1925, Notes on the Oligochaeta of the Shatursk
    lakes [Zametki ob Oligochaeta Shaturskikh ozer]. Trudy
    Kosinskoi Biolog. stantsii, 3.

Malevich, I. I., 1926, Notes on the oligochaete fauna of the USSR [Zametki po faune Oligochaeta SSSR]. <u>Russk. gidrobiol. zhurnal</u>, 5: (10-12) 223-225.

Malevich, I. I., 1927, Oligochaeta of Kosinsk waters [Oligochaeta Kosinskikh vodoemov]. <u>Trudy Kosinskoi Biol. stantsii</u>, 5.

Malevich, I. I., 1929, Oligochaetes of Meshchersk lowland waters [Oligochaeta vodoemov Meshcherskoi nizmennosti]. <u>Trudy Kosinskoi gidrobiol. stantsii</u>, 9: 41-63.

Malevich, I. I. 1930, On the freshwater oligochaete fauna of the Caucasus [K faune presnovodnikh Oligochaeta Kavkaza]. <u>Raboty Severo-Kavk. gidrob. stantsii</u>, 3.

Malevich, I. I., 1940, Data on the oligochaete fauna of rivers of the USSR [Materialy po faune Oligochaeta rek SSSR]. <u>Byull. Mosk. o-va ispyt. prirody, otd. biol.</u> 49: 181-185.

Malevich, I. I., 1947, Oligochaetes of the caves of the Caucasus [Oligokhety pescher Kavkaza]. <u>Byull. Mosk. o-va ispyt. prirody, otd. biol.</u> 52, 4.

Malevich, I. I., 1949, On the oligochaete fauna of Teletskoye Lake [K faune oligokhet Teletskogo ozera]. <u>Trudy zoologicheskogo instituta AN SSSR</u>, 4.

Malevich, I. I., 1956, Oligochaetes of the Moscow district [Maloshchetinkovye chervi (Oligochaeta) Moskovskoi oblasti]. <u>Uchenye zapiski Mosk. gor. ped. inst. im. Potemkina</u>, 61: 403-439.

Malevich, I. I., and G. B. Zevina, 1958, Data on the oligochaete fauna of Rybinsk reservoir [Materialy po faune maloshchetinkovykh chervei Oligochaeta Rybnogo vodokhranilishcha]. <u>Trudy biol. stan. "Borok" AN SSSR</u>. 3: 399-496.

Mamilova, R. Kh., 1970, Biomass and production of macrozoobenthos of Lake Karakul and Lake Sorkul (Ili River basin) [Biomassa i produktsiya makrozoobentosa ozer Karakul' i Sorkul' (basseina r. Ili)]. 1-26.

Michaelsen, W., 1903, Neue Oligochaeten und neue Fundorte altbekannter. <u>Mitt. Zool. Inst. Hamb.</u> XIX.

Michaelsen, W., 1923, Die Oligochaeten der Wolga. <u>Proc. Wolga biol. station</u>, 1-2.

Michaelsen, W. 1935, Oligochaetes of lakes of the Central Altai [Maloshchetinkovye chervi ozer tsentral'nogo Altaya]. <u>Issledovaniya ozer SSSR</u>, 8.

Miroshnichenko, M. P., 1971, Intensity of the development, distribution, and dynamics of chironomid larvae and oligochaetes in the Tsimlyanskoye reservoir [Intensivnost' razvitiya, respredeleniya, dinamika lichinok khironomid i oligokhet v Tsimlyanskom vodokhranilishche]. <u>Trudy Volgogradskogo otdeleniya GosNIORK</u>, 5: 51-73.

Moroz, T. G., 1977, Oligochaetes from river mouths of the northwestern Black Sea area [Oligokhety ust'evykh oblastei rek severo-zapadnogo Prichernomor'ya]. <u>Gidrobiologicheskii zhurnal</u>, 13: 20-25.

Mushchinskii, V. G., 1970, The oligochaete fauna of the River Prut
    [O faune oligokhet r. Pruta]. Gidrobiologicheskie i rybokho-
    zyaistvennye issledovaniya vodoemov Moldavii, I: 114-115.
Noskova, A. A., 1967, Oligochaetes of the Selenga District of Lake
    Baikal [Oligokhety Selenginskogo raiona ozera Baikal].
    Kazaks., 1-24.
Parele, E. A., 1975, Oligochaetes of the Daugava and Lielupe estu-
    aries and their importance in sanitation-biological assays
    [Maloshchetinkovye chervi ust'ebykh raiona rek Daugava i Lielupe,
    ikh znacheniye v sanitarnobiologicheskoi otsenke], 1-24.
Pataridze, A. I., 1957, The oligochaete fauna of the Tbilisi reservoir
    during the first three years of its existence [Fauna oligokhet
    Tblisskogo vodokhranilishcha v pervye tri gody ego sushchest-
    vovaniya]. Soobshcheniya AN GSSR, 19: 217-224.
Pataridze, A. I., 1959, The oligochaete fauna of the Paldoisk reser-
    voir and River Iori [K izucheniyu oligokhet fauny Paldoiskogo
    vodokhranilishcha i reki Iori]. Soobshcheniya akademii nauk
    Gruzinskoi SSR, 22: 201-207.
Pataridze, A. I., 1962, The oligochaetes of Lake Paravini [Materialy
    k poznaniyu oligokheto fauny ozera Paravini]. Soobshcheniya
    Akademii nauk Gruzinskoi SSR, 29: 203-207.
Pataridze, A. I., 1963, Oligochaetes of the Tbilisi reservoir
    [Maloshchetinkovye chervi (Oligochaeta) Tbilisskogo vodokhran-
    ilishcha]. Trudy instituta zoologii, 19: 163-204.
Pataridze, A. I., 1969, Vertical distribution of oligochaetes in the
    sediments of the Tbilisi reservoir [Vertikal'noe raspredelenie
    oligokhet v ilovykh gruntakh Tblisskogo vodokhranilishcha].
    Voprosy biologicheskoi produktivnosti vnutrennikh vodoemov
    Gruzii, 122-132.
Pataridze, A. I., 1970, A freshwater oligochaete of Georgia
    [Presnovodnye maloshchetinkovye chervi Gruzii]. Biologicheskie
    protsessy v morskikh i kontinental'nykh vodoemakh, 290-291.
Petkevich, A. N., and B. G. Ioganzen, 1958, Perspectives of the
    fishery industry of the upper Ob in relation to hydrotechnical
    construction [Perspektivy rybnogo khozyaistva verkhei Obi v
    svyazi s gidrostroitel'stvom]. Izvestiya VNIORK, 44, Moskva.
Pirozhnikov, P. L., 1929, Lake Sartlan [K poznaniyu ozera Sartlan].
    Trudy Sibirsk. nauchn. rybokhoz. stantsii, 4: 1-117.
Poddubnaya, T. L., 1958, Some data on the reproduction of Tubifi-
    cidae [Nekotorye dannye po razmnozheniyu tubifitsid]. Doklady
    Akademii nauk SSSR, 120: 422-424.
Poddubnaya, T. L., 1959, Autotomy and regeneration of Tubificidae
    [Ob autotomii i regeneratsii u tubifitsid]. Byulleten'
    Instituta biologii vodokhranilishch. No. 5, 15-16.
Poddubnaya, T. L., 1961, Data on the nutrient mass species of
    Tubificidae in the Rybinsk reservoir [Materialy po pitaniyu
    massovykh vidov tubifitsid Rybinskogo vodokhranilishcha].
    Trudy Instituta biologii vodokhranilishch, 4: 219-231.

Poddubnaya, T. L., 1963, Life cycle and growth of the Neva Limno-
    drilus (*Limnodrilus newaensis* Mich., Oligochaeta, Tubificidae)
    [Zhiznenyi tsikl i temp rosta nevskogo Limnodrila (*Limnodrilus
    newaensis* Mich., Oligochaeta, Tubificidae)]. Trudy Instituta
    biologii vodokhranilishch, 5: 46-56.
Poddubnaya, T. L., 1965, Nutrition of *Chaetogaster diaphanus* Gruit
    (Naididae, Oligochaeta) in the Rybinsk reservoir [Pitaniye
    Chaetogaster diaphanus Gruit. (Naididae Oligochaeta) v Rybinskie
    vodokhranilishchie]. Biologicheskie protsessy vo vnutrennikh
    vodoemov, 178-190.
Poddubnaya, T. L., 1976, Peculiarities of biology and production of
    Tubifex tubifex (Müll.) in the polluted region of the Rybinsk
    reservoir [Osobennosti biologii i produktsiya *Tubifex tubifex*
    (Müll.) v zagryaznenom uchastke pritoka Rybinskogo vodokhran-
    ilishcha]. Biologicheskie produktsionnye protsessy y basseine
    Volgi, 119-127.
Popchenko, A. I., 1972, Aquatic oligochaetes of Karelia [Vodnye
    oligokhety Karelii (ekologofaunicheskii obraz)], 1-26.
Riikoja, H., 1955, Eesti NSV salegrootute fauna uurimisc kusimusi.
    LUS Aastaraamat 48.
Sbarakh, T. I., 1968, Macrozoobenthos of the Assinsk lakes (Dzambulsk
    oblast, Kaz. SSR) and its importance to feeding fish [Makro-
    zoobentos Assinskikh ozer (Dzhambul'skaya obl., Kaz. SSR) i
    ego znachenie v pitanii osnovnykh promyslovykh ryb], 1-35,
    Alma-Ata.
Schneider, G., 1908, Der Obersee bei Reval. Berlin.
Semernoi, V. I., 1970, Oligochaetes of some huminic waters of Tyumen
    oblast [Oligokhety nekotorykh gumifitsirovannykh vodoemov
    Tyumenskoi oblasti]. Izvestiya Sibirskogo otdel. Akademii
    nauk SSSR. No. 5, 1, 80-85.
Semernoi, V. I. 1972, Oligochaetes of the Gusino-Ubukunsk lake group
    [Maloshchetinkovye chervi (Oligochaeta) Gusino-Ubukunskoi
    ozernoi gruppy]. Biologia vnutrennikh vod., 13: 17-23.
Semernoi, V. I., and L. G. Belozub, 1979, Oligochaetes of the rice
    fields of Uzbekistan [Oligochaeta risovykh polei Uzbekistana].
    Gidrobiologicheskii zhurnal, 23: 31-35.
Shapovalova, I. M., 1974, The zoobenthos of the Ivano-Arachleisk
    lakes [Zoobentos Ivano-Arkheiskikh ozer (Zabaikal'e)], 1-25.
Shurova, N. M., 1977, New species of the littoral oligochaete
    Lumbricillus [Novye vidy litoral'nykh oligokhet roda *Lumbri-
    cillus*]. Biologiya morya 1: 57-62.
Shurova, N. M., 1978, Oligochaetes of littoral Eastern Kamchatka
    [Maloshchetinkovye chervi (Oligochaeta) litorali vostochnoi
    Kamchatki]. Litoral' Beringova morya i yugo-vost. Kamchatki,
    98-106.
Sivickis, P., 1933, The freshwater summer fauna of northwestern
    Lithuania [Siaures rytu Lietuvos geluju vandenu fauna vasaros
    metu]. V.D.U. Matematikos-gamtos fakulteto darbai, 9.

Sivickis, P., 1934, Studies of the fauna of Sventosios harbor in the
    summer of 1934 [Sventosios uosto faunos tyrinejimai 1934 m.
    vasare]. Kosmos, 15: 380-398.
Sivickis, P., 1940, The fauna of Lake Luodis at the end of summer
    [Luodzio ezero fauna vasaros gale]. Kosmos, 21: 106-114.
Slepukhina, T. D., 1970, The paucity of the invertebrate fauna in the
    reservoirs of the Northern Caucasus and its causes [Bednost'
    fauny bespozvonochnikh v vodokhranilishchakh Predkavkaz'ya
    i ee pricheny]. Biologicheskie protsessy v morskikh i
    kontinental'nykh vodoemakh, 337-338.
Slepukhina, T. D., 1972, On the oligochaete fauna of the littoral
    of Lake Onega [K faune oligokhet litorali Onezhskogo ozera].
    Materialy vtorogo Vsesoyuznogo simposiuma v Barke, 53-59.
Slepukhina, T. D., 1975, The ecological importance of the turbid
    water in the ponds of the Northern Caucasus [Ekologicheskoe
    znachenie mutkosti vody na primere prudov Severnogo Kavkaza].
    Ekologiya, 5: 94-96.
Sokolskaya, N. L., 1953, Oligochaetes of the lakes of Byelorussia
    [Materialy po faune Oligochaeta ozer Belorussii]. Uchenye
    zapisi Belorusskovo universiteta, 17: 88-95.
Sokolskaya, N. L., 1956, Data on the aquatic oligochaete fauna
    of Pripet-Polesye [Materialy po faune vodnikh maloshchetin-
    kovykh chervei Pripyatskogo poles'ya], 189-196.
Sokolskaya, N. L., 1962, New species of Naididae (Oligochaeta)
    from Lake Baikal [Novye vidy Naididae (Oligochaeta) iz ozera
    Baikal]. Zool. zhurnal, 41: 660-665.
Sokolskaya, N. L., 1967, Data on the aquatic oligochaete fauna of
    the Fast East of the USSR [Materialy po faune vodnikh Oligo-
    chaeta Dal'nego Vostoka]. Simpozium po vodnym maloshchetinkovym
    chervyam, 31-33.
Sokolskaya, N. L., 1972, On the aquatic oligochaete fauna of the
    Far East of the USSR. The aquatic Oligochaeta [K faune
    vodnykh Oligochaeta Dal'nego Vostoka SSSR. Vodnye maloshchet-
    inkovykh chervei]. Trudy vsesoyuznogo gidrobiologicheskogo
    obshchestva, 27: 59-65.
Sokolskaya, N. L., 1976, Data on the oligochaete fauna of the waters
    of Chukotka peninsula [Materialy po faune maloshchetinkovykh
    chervei (Oligochaeta) vodoemov Chukotskogo poluostrova].
    Trudy Biologopochvennogo Instituta Dal'nevostochnogo nauchn.
    tsentra AN SSSR, 36: 89-101.
Sokolskaya, N. L., 1977, A new species of Eclipidrilus (Oligochae-
    ta, Lumbriculidae) from Chukotka peninsula [Novyi vid Eclipi-
    drilus (Oligochaeta, Lumbricidae) s Chukotskogo poluostrovo].
    Zoologicheskii zhurnal, 56: 296-3001
Svetlov, P. G., 1924, Observations on oligochaetes of the Perm
    district [Nablyudeniya nad Oligochaeta Permskoi gub]. Izv.
    biol. nauchno-issled. instituta pri Permsk gos. univers. 3.

Svetlov, P. G., 1925, Some data on the oligochaete fauna near Cherdyn
    [Nekotorye dannye o faune Oligochaeta Cherdynskogo kray].
    Izv. biol. nauchno-issled. instituta pri Permsk. gos. univers.
    III, 10: 471-475.
Svetlov, P. G., 1926, On the oligochaete fauna of the Samarsk
    district [K faune Oligochaeta Samarskoi gub.]. Izv. biol.
    nauchno-issled. instituta pri Permsk. gos. univers., 249-256.
Svetlov, P. G., 1946, The oligochaete fauna of Tomsk oblast [K
    faune oligokhet Tomskoi oblast]. Trudy Tomskovo gos. univer-
    siteta, 97.
Timm, T. E., 1962, On the fauna, ecology, and distribution of
    freshwater oligochaetes in the Estonian SSR [Eesti NSV magevee-
    vaheharjassusside faunast, ökoloogiast ja levikust]. Zoologija-
    alaseid toid, 2: 63-106.
Timm, T. E., 1964, Oligochaetes of the waters of the Estonian SSR
    [Maloshchetinkovye chervi vodoemov Estonii]. Faunistiko-
    ekologicheskii obzor. 1-22, Tartu.
Timm, T., 1977, Peipsidrilus pusillus gen. n.sp.n. (Oligochaeta,
    Tubificidae). Eesti NSV teaduste Akademia toimestised, Koide
    Biologia 26: 279-283.
Tsvetkova, L. I., 1970, The distribution of oligochaetes in the
    River Neva in relation to its pollution [Rasprostranenie
    oligokhet v r. Neve v svyazi s ee zagryazneniem]. Biologiche-
    skie protsessy v morskikh i kontinental'nykh vodoemakh, 390-391.
Vereshchagin, G. Yu., 1940, Genesis and history of Lake Baikal, its
    fauna and flora [Proiskhozhdenie i istoriya Baikala, ego fauny
    i flory]. Trudy Baikal'skoi limnologicheskoi st. 10.
Zaloznyi, N. A., 1972, The aquatic oligochaete fauna of the basin
    of the Middle Ob. [K faune vodnykh maloshchetinkovykh chervei
    basseina Srednei Obi]. Vodnye Maloshchetinkovye chervi (material
    II simpoz) 33-42.
Zaloznyi, N. A., 1973, Studies of oligochaetes of the lower Tom
    River [K izucheniyu fauny presnovodnykh maloshchetinkovykh
    chervei basseina nizhnego techeniya reki Tomi]. Problemy
    ekologii, 3: 135-138.
Zaloznyi, N. A., 1974, Oligochaeta and Hirudinea of the waters of
    Western Siberia [Maloshchetinkovye chervi i piyavki vodoemov
    Zapadnoi Sibiri]. Tomsk, 1-20.
Zaloznyi, N. A., 1976, The aquatic fauna of Oligochaeta and Hirudi-
    nea of Western Siberia [Fauna vodnikh oligokhet i piyavok Zapad-
    noi Sibiri]. Problemy ekologii, 4: 97-112.

AQUATIC OLIGOCHAETA OF THE RHONE-ALPES AREA:

CURRENT RESEARCH PRIORITIES

J. Juget

Département de Biologie Animale et Zoologie
Université Claude-Bernard (LYON I)
43, Bd du 11 Novembre 1918
69621 - VILLEURBANNE (FRANCE)

ABSTRACT

This review describes long-term changes in the fauna of lake Leman, the systematics and ecology of the fauna of the Rhône as well as studies on *B. sowerbyi* and *N. variabilis* in a pond.

OLIGOCHAETA AND LAKE POLLUTION

During the last twenty years lake Leman has been the main focus of research works on lacustrine Oligochaeta in the region of Rhône-Alpes. Since 1957, these works have been connected to those of the International Control Commission for the protection of the water of lake Leman against pollution.

Impact of the Eutrophication of lake Leman on the Oligochaeta

The impact of the eutrophication of lake Leman on the structure and the evolution of the annelid fauna has been thoroughly studied by Lang (Fauna Conservation, Lausanne), (1974, 1978a, b). The use of multivariate analysis has enabled this worker to classify the lacustrine species of Tubificidae and Lumbriculidae along a pollution gradient as indicators of a given level of pollution and to carry out a critical comparative analysis of chemical and biological criteria of pollution.

The reduction, in large proportions, of the specific diversity of benthic communities in the central plain of Leman (-309 m)
during the last ten years reflects the growing disquilibrium
between the epilimnion and the hypolimnion in this big subalpine
lake and the deoxygenation of the deepest water layers. Since
1967, the Lumbriculidae *Stylodrilus heringianus* and *Stylodrilus
lemani* have disappeared from this sector threatened by asphyxia.
The Tubificidae *Peloscolex velutinus*, *Limnodrilus profundicola*
and *Potamothrix heuscheri* have also disappeared from this zone
where only *Tubifex tubifex* and *Potamothrix hammoniensis* survive.

Techniques of Restoration of Lakes : Impact of Hydrogen Peroxide
on Oligochaeta

        In the context of perfecting the techniques of restoration of
lakes, an analysis of the impact of hydrogen peroxide (H_2O_2) on
sediment and benthos (Tubificidae) has been undertaken in the
laboratory (Soares de Assis, 1977). The works undertaken, which
concern samples of mud from Leman, point out the deferred toxicity
of this oxidizing agent responsible for releasing toxic substances
such as ammoniacal nitrogen and copper sulphate in water in contact
with the treated sediment, in quantities which can be lethal to
the benthic fauna. Present experiments undertaken in the laboratory
aim at determining the stages of the evolution of the water-sediment
complex after treatment with hydrogen peroxide from the synecological point of view of the interrelations among the physicochemical
components of the aquatic environment, the microflora (sulphato-
reducing bacteria) and the burrowing macrofauna (Tubificidae).

SYSTEMATICS AND ECOLOGY OF OLIGOCHAETA IN THE RHONE AND ITS
APPENDICES

        A general survey of Oligochaeta in the Rhône has been recently
established by Lafont and Juget (1976). It reveals a high degree
of similitude between the fauna of lake Leman and the fauna of the
Rhône and between the fauna of the upstream sector and the downstream
sector of the river: this tends to confirm the alpine character of
the Rhône fauna which is related to that of the Rhine and the Danube.
The most diversified populations belong to the Upper Rhône (before
Lyons) and to the lower parts of the river (after the junction of
the Isère). *Limnodrilus hoffmeisteri* is at the same time, the
most constant and the most abundant species (25% of the total number,
excluding the underground interstitial fauna).

        Since 1975, the structure and dynamics of the ecosystems in
the French Upper Rhône (between Seyssel and Lyons) have been submitted to concerted studies carried out by members of the Research
Unit Associated with the C.N.R.S. (E.R.A. no. 849) at the Department of Animal Biology and Zoology of Claude Bernard University
(Lyon I). Those works form part of a large program which aims at

defining the interrelations between the main river on the one hand, and the arms corresponding to ancient meanders cut off from the river and the phreatic waters surrounding the Rhône, on the other hand. They must also provide a reference inventory of the river and of its appendices in anticipation of future arrangements on this part of the river (building of hydroelectric dams, establishment of new nuclear power stations, establishment of the industrial harbour zone of the Ain plain, evacuation of waste waters from towns bordering the big subalpine lakes, etc.).

The Ancient Meanders of the Rhône

One of the specificities of the upper course of the Rhône, between lake Leman and the lyonese urban centre, consist in the existence of ancient arms or "lônes" which result from the evolution of ancient meanders (Ain et al, 1973; Dorgella, 1973; Juget et al, 1975). Their origin is approximately situated between the XVIII century and now. Some of them have lost all direct link with the Rhône and have only kept an indirect contact with the river through the under-flow; others, with a more recent age, are still in direct communication with the main course. The conjunction of different factors, responsible for the modelling and the more or less long term evolution of those ancient arms (historic factors in relation to the relative age of the "lônes", topographic factors connected to the configuration of the ancient meanders with regard to the present course of the Rhône, pedological factors related to the degree of filling up of the "lônes" by the riparian vegetation and to silting, hydrological factors linked to the way in which the arms are fed by the river itself and (or) the under-flow, human factors corresponding to the arrangement of the main course (dams) and to its repercussions on "lônes") create a complex border situation ("effet de lisiere") within such ecosystems at the frontier of the phreatic and fluvial sphere (Amoros et al. 1978; Juget et al., 1979). Those peripheral effects influence the structure and the dynamics of the populations, the inventory of which has shown their richness and diversity. Numerous zoological groups, including Oligochaeta, are characterised by the coexistence and the imbrication, in those ancient meanders, of ecologic types having different affinities, including species of phreatic and fluvial origin, species from running and stagnant waters, species liking vegetation related to the faune of peat-bogs and limivorous species related to those settled in river silts or lacustrian fine sediments. Their study is now under way.

The Underground Fauna of the Alluvial Plain of the Rhône

Another characteristic of the alluvial plain of the Rhône, east of Lyons, between the junction Ain-Rhône and Saône-Rhône, lies in the existence of considerable phreatic waters whose supply is independent of the under-flow of the Rhône. Of an average thickness of 30 m, this layer of underground waters has a flow rate evaluated at 470 m^3/hr for a one kilometre-front and an approximate speed of 0.50 m/hr and follows a general east-west direction, convergent towards the river (David, 1967, Pinchard; 1975; Gibert et al., 1977). Samples obtained by using manual strainer pump (Bou-Rouch method) (Bou, 1974; Bou and Rouch, 1967) from underground interstitial water are collected from between -50 to -90 cm below the surface of the ground. In some stations, the water samples collected come from the infiltration of the main Rhône (under-flow); other pumping stations are chosen because of their close relation to the "lônes"; in a third category of stations, the underground interstitial water does not show any visible relation with surface water.

It is, right now, possible to draw some conclusions about Oligochaeta:

- Among the species of underground interstitial water recorded, some are of a rare or new type; among those, we can mention notably *Potamodrilus fluviatilis, Dorydrilus michaelseni, Trichodrilus leruthi* and two new species of the genus *Rhyacodrilus* which cohabit with the species *Rhyacodrilus balmensis* first described from the Karst (French southern Jura, Juget, 1959).

- The diversity and the richness of the interstitial fauna collected near the piezometrical level are due to the coexistence, at this level, of epigean species and phreatobiont and troglobiont species (Gilbert et al., 1977) Oligochaeta represent with Crustacea (Niphargids, Copepoda, ...) one of the numerically and gravimetrically predominant zoological groups. Densities of more than 300 individuals per 100 L of interstitial water obtained from pumps have been recorded, though the average population density is around 50 individuals, with the predominance, varying from one station to the other, of Lumbriculidae, Naididae, Tubificidae or Enchytraeidae.

- The discovery of the tubificid *Rhyacodrilus balmensis* in the group of dominant species living in interstitial water coming from the infiltration of the Rhône (under-flow) which is also found in the caves of the southern part of the Jura (grotte de la Balme, puits de Rappe) (Juget, 1959; Juget and Lavont, 1979) shades off the frontiers conventionally drawn to distinguish the phreatobiont species from the troglobiont species. Under this circumstance, a factor of a trophic nature can be put forward to explain this distribution. Like many underground Oligochaeta, *Rhyacodrilus balmensis* exploits the fine part of the sediments. An adjustment in the absorption capacity of the digestive tract allows, even in species of small size as is the case here, to increase the variety of their food up to the whole group of fine elements (including coarse silts). I have shown, about the underground limivorous species, the existence of a negative correlation between the "specific absorbance index" (defined as the average value of the ratio between the maximum calibre of grains ingested and the maximum diameter of the alimentary canal) and the size of the species considered (Juget, 1979). As a matter of fact, clays and silts which constitute the essential source of food for the under-ground limivorous species are never absent from the sediments of the "underground Rhône".

- The good transmissivity of aquiferous alluvions which are intersected by the Rhône and its ancient meanders hinders the detection of species related to fluvial under-flow (or para-flow) and of species related to the phreatic waters independent of the river. The differences observed in different stations are more quantitative than qualitative. The similarity matrices, established from the calculation of inter-species association indices, however reveal, in the present state of research, affinity nuclei connected to the characteristics of the underground run-off (speed and flow rate). For example, *Potamodrilus fluviatilis* has been found only in the under-flow and *Propappus volki* in the under-flow and para-flow of the Rhône (advance of Rhône ground waters flowing parallel to the river). Those two species, considered as representative of the underground rheopsammon also show a great affinity with *Marionina argentea*, *Pristina foreli* and *Rhyacodrilus balmensis*. On the contrary, *Pristina idrensis* and *Enchytraeus buchholzi* preferably colonize alluvial waters without any close connection with the river. The two species *Pristina foreli* and *Pristina idrensis* have been observed together only in 2% of the samples analysed. *Pristina idrensis* itself never cohabits with either *Potamodrilus fluviatilis* or *Propappus volki*. In respect to the first results obtained, we shall have to check whether such affinities (or incompatibilities) are confirmed or invalidated when we widen the horizon explored by pump sampling, at each station, by including deeper aquiferous alluvions.

RESEARCH ON BENTHOS AND PERIPHYTON IN AN AGROECOSYSTEM, THE PISCICULTURAL POND OF THE DOMBES (AIN).

The Dombes is a slightly undulating plateau of about 100 000 ha, at about thirty km north of Lyons, of an average altitude of 290 m, dominating in the south, the east and the west, the valleys of the Rhône and the Saône with hill ranges, 50 to 150 m high. Washing by rain water of contemporary morainic mounds from the shrinkage of quaternary glaciers has brought about the accumulation (in the hollows) of impermeable silts and clays favorable to the establishment of ponds. Arranged from the beginning of the XIII th century, the ponds of the Dombes, of an average surface of 30 ha, are characterised by a traditional triennial mode of exploitation: such a system undertakes alternately two years of water filling for fish breeding (chiefly carp) and one year of drainage, most often together with cereal crops. This traditional shift system is now

disturbed in its duration (extension of the drainage period), but
not in its principle, because of the requirements of maize cultivation which has displaced that of oats. The influence of the
triennial rotation on the fertility of the soil's (which are related
to pseudogley and on productivity of the waters in those agroecosystems and the dynamics of physico-chemical parameters of the
aquatic environment in relation to the soils), flora and fauna
have been dealt with by numerous works and publications during
the last ten years. The works concerning Oligochaeta are mainly
connected with synecological studies on benthos and periphyton,
in relation to pisciculture and the exploitation mode of ponds
(Juget and Amoros, 1970, 1973; Juget et al., 1972; Fanget, 1972;
Uribe-Zamora, 1975; Yi, 1977).

Only certain aspects of the ecophysiology of two common
species, one belonging to the pond benthos (*Branchiura sowerbyi*),
the other to the periphyton (*Nais variabilis*) will be briefly
examined here.

Growth, Reproduction, Resistance to Dessication of *Branchiura sowerbyi*

The average density of this species amounts to about 3000
individuals/m^2 in the ponds of the Dombes. Samples collected from
several ponds reveal the coexistence of two distinct cohorts. The
life cycle of this species seems to be of two years; the egg laying
period is itself limited to May and June, thus coinciding with the
plankton-eating phase of the alimentary system of the carp in the
ponds of the Dombes.

There is, for the specimens found in the Dombes, a relation
between the weight of the worms and the number of caudal gills
which is the following (Giauffret, 1976).

$$Y = 0.0079 \ X^{2,21} \qquad (r = 0.91 \ ; \ n = 148)$$

with Y = wet weight (in mg)
     X = number of branchial segments
     95 % confidence limits of r : 0.93 ... 0.88.

The relative growth of the number of branchial segments
with respect to time and temperature (from +5 to +20° C)
has also been measured in the laboratory. If Bi represents the
average number of pairs of gills, at the beginning of the experiment, for a group of 10 individuals and Bf the average number
of pairs of gills for those 10 worms at later observation, the
relative growth Bf/Bi, with respect to temperature, can be
expressed in the form of logY = a + bX. The equation of the line
of regression, calculated for the worms found in the Dombes, is
the following (T in °C):

$$\log \frac{Bf}{Bi} \text{ (I month)} = -0.042 + 0.013 \text{ T} \qquad (r = 0.82)$$

(95 % confidence limits of r : 0.93 ... 0.54).

For the sexually mature worms, growth is slowed down proportionally to the increase in the egg laying rate : the metabolism of the worm is redirected from growth (gill formation) to reproduction (production of ovigerous cocoons).

The regeneration of the branchial region has also been studied in the laboratory at different temperatures. For the individuals having undergone an amputation of the caudal region situated before the first pair of gills, the complete regeneration of the initial number of gills requires 10 to 14 days at 20° C, about 40 days at 15° C, at least 6 months at 10° C. At 5° C, as a rule, no blastema will appear within 6 months; only branchial anlages will appear in the segments anterior to the section ("spare" gills) which will normally regress with the development of blastema. The regeneration rate evolves in the same way in respect to time, for the different age groups. However, the regenerating power of gills decreases from the tail to the front extremity : at 20° C, the total regeneration of the worms having undergone an amputation at the thirty-fifth segment requires a month and a half. The regeneration power also diminished progressively with successive amputations. Regeneration is, nevertheless, not incompatible with egg laying.

The fragmentation, followed by regeneration, can be an active or passive process which, according to situations, can compensate predation effects or unfavorable environment conditions. In the case of drying or freezing of the sediments, the worm reduces its water content (which increases from the front to the back of the body) by fragmenting itself and by penetrating into the soil where it becomes inactive. The populations from the Dombes resist the dessication of the sediment (undertaken experimentally in laboratory) better than those of the same species originated from the Rhône: specimens from the Dombes split up earlier and die later. Actually, soil drainage is periodic in the Dombes.

Those elements confirm those put forward by Aston (1967) concerning populations of *Branchiura sowerbyi* collected from the Thames and the Avon.

Ecology and Production of Zooperiphyton. Ecophysiological aspects of the Reproduction of *Nais variabilis*

Naididae form an important part of fauna living on macrophytes growing in the ponds of the Dombes (Juget, 1976; Juget and Rostan, 1973). During summer, some ponds are overgrown with water caltrop

Trapa natans whose floating leaf rosettes constitute a true
floating plant cover on the water surface. This plant shelters
numerous species of Naididae, chiefly *Nais variabilis, Dero digitata,
Dero obtusa, Pristina longiseta, Ophidonais serpentina, Ripistes
parasitica,* and *Chaetogaster langi.* During the summer vegetation
development, the centre of gravity of the zooperiphyton moves from
the lower zone of immersed plants to the upper stratum formed by
the inferior face of the floating leaf rosettes. At the end of
summer, before the autumn sedimentation phase of plants at the
bottom of the pond, 99% of the *Nais variabilis* populations are
concentrated at this level. This migration is influenced at the
same time by the chemical stratification and depletion of dissolved
oxygen during summer under the floating vegetation and the increase
in the biovolume of the phytoperiphyton and of the detritus stuck
at the lower face of the leaf rosettes of the plant.

The formula of Galkovskaya (in Winberg, 1971):

$$\frac{I}{D} \, \Delta t \, \frac{N_1 + N_2}{2}$$

has been used in the estimation of the production of *Nais variabilis* during the asexual reproduction phase (the sexual reproduction being exceptionally observable in less than 1% of the
specimens examined). In this formula, D represents the generation
time, i.e. the time necessary, in given experimental conditions,
for a group of n zooids, to give through scissiparity without
mortality 2n zooids characterised, like the previous ones, by
complete regeneration of eyes. N_1 and N_2 represent the number of
zooids (isolated or in series) recorded on the plants per surface
unit at the beginning and at the end of a given time t (in days).
The measure of D is undertaken for group of 10 or 20 zooids picked
up from pond and cultivated in glass enclosures which are immersed
in situ at the same location and which are held near the surface
of the pond by means of floats. The nutritive medium used is
composed of the water of the pond added to periphyton obtained
by squeezing out the floating leaves of *Trapa natans* in pond
water, filtrating the suspension thus obtained through filter
with meshes 0.07 mm, and adjusting the concentration of the
filtrate to 500 mg/1. The maintenance of an air reserve trapped
between the growing medium and the neck of the container, whose
lid is formed by a thin waterproof but gas-porous nylon gauze,
favours exchanges through gas diffusion between the pond water
and the medium as well as the oxygenation of the latter. The
picking out of zooids on fresh medium and the control of the absence of eventual predators was undertaken every 8 to 10 days.

For example, after the experiments carried out during the summer from the 22nd of August to the 12th of September 1973 in the pond "Sainte Anne" (Biological Reserve of Villars-les-Dombes), the average value of I/D was estimated for *Nais variabilis* at 0.21 ± 0.04, at an average incubation temperature of 20.2° C (average of daily maxima : 23.7° C; average of daily minima : 16.8° C). The average daily production, expressed in mg dry weight for the same period, was evaluated at 3.12 mg per surface unit (or per floating leaf rosette of *Trapa natans* of an average diameter of 26 cm). The least generation time (D) observed during the summer exponential phase of growth of *Nais variabilis* is 3.33 days. The growth rate is affected by the temperature and for a given temperature can vary in important proportions in relation to the quantity as well as the quality of food available.

REFERENCES

Ain, G., B. Gillot, N.C. Neuburger, G. Pautou and J. Tetard, 1973, Etude écologique des anciens lits du Rhône entre le confluent du Guiers et le confluent de l'Ain. Univ. Sc. Med., Grenoble, C.N.R., E.I.D., Ain, Isère, Rhône, Savoie. 75 p.

Amoros, C., J. Juget, D. Levet, J.L. Reygrobellet, M. Richardot, Ph. Richoux, C. Roux and B.O.J. Yi, 1978, Une lône du Rhône, zone humide en position de lisière dans l'espace et dans le temps. Rapport final du contrat n° 76-55, avec le comité scientifique "Faune et Flore" du Ministere de l'Environnement et du Cadre de vie. 165 p.

Aston, R.J., 1967, The effect of temperature on the life cycle, growth and fecundity of *Branchiura sowerbyi* (Oligochaeta, Tubificidae). Journ. Zool. London, 154:29 - 40.

Bou, C., 1974, Recherches sur les eaux souterraines. 25 : les méthodes de récolte dans les eaux souterraines intersti-tielles. Ann. Spéléol., 29:611 - 619.

Bou, C. and R. Rouch, 1967, Un nouveau champ de recherches sur la faune aquatique souterraine. C.R. Acad. Sci. Paris, 265 (D):369 - 370.

David, L., 1967, Formations glaciaires et fluvio-glaciaires de la région lyonnaise. Doc. Labo. Géol. Fac. Sci., Lyon, 22.159p.

Dorgelo, J., 1973, Etude de la végétation dans les anciens lits du Rhône et des moustiques qui lui sont liés, de Lyon au confluent de l'Ain. D.E.S. Université Claude-Bernard, LYON I.

Fanget, R., 1972, Contribution à l'écologie des étangs pisci-coles de la Dombes: sur le régime alimentaire de la carpe à miroirs (*Cyprinus carpio* L.). Thèse Dr. 3° cycle, Lyon. 70 p.

Giauffret, C., 1976, Ecophysiologie de *Branchiura sowerbyi*: croissance, régénération, résistance à la dessication. D.E.A. (Ecologie fondamentale et appliquée des eaux continentales), Université Claude-Bernard, LYON I. 30 p.

Gibert, J., R. Ginet, J. Mathieu, J.L. Reygrobellet and A. Seyed-
    Reihani, 1977, Structure et fonctionnement des écosystèmas
    du Haut-Rhône francais. IV. Le peuplement des eaux phréa-
    tiques. Premiers résultats. Annls. Limnol., 13:83 - 97.
Juget, J., 1959, Recherches sur la faune aquatique de deaux
    grottes du Jura méridional francais: la grotte de la Balme
    (Isère) et la grotte de Corverssiat (Ain). Ann. Spéléol.,
    14:391 - 401.
Juget, J., 1976, Les Oligochètes du zoopériphyton des herbiers
    à Trapa natans. Ecologie et production. C.R. Ann. Hydrobiol.,
    7:43 - 45.
Juget, J., 1979, La texture granulométrique des sédiments et le
    régime alimentaire des Oligochètes limnicoles. Hydrobiologia,
    in press.
Juget, J. and C. Amoros, 1970, Données préliminaires sur la faune
    planctonique et benthique de deux étangs piscicoles de la
    Dombes (Ain). Annls. Limnol., 6:215
Juget, J. and C. Amoros, 1973, Les communautés d'Invertébrés des
    étangs piscicoles de la Dombes: quelques exemples de régula-
    tion biologique. Ann. Hydrobiol., 4:125.
Juget, J., C. Amoros, R. Fanget and J.C. Rostan, 1972, Les
    communautés d'Invertébrés des étangs piscicoles de la Réserve
    Biologique de Dombes: données préliminaires sur leur répar-
    tition quantitative, leur évolution, leur importance dans
    l'alimentation de la carpe. Bull. Soc. Nat. Archeol. Ain,
    86:11 - 41.
Juget, J., C. Amoros, D. Gamulin, J.L. Reygrobellet, M. Richardot,
    R. Richoux and C. Roux, 1975, Structure et fonctionnement
    des écosystemes du Haut-Rhône francais. II. Etude hydrolo-
    gique et écologique de quelques bras morts; premiers résultats.
    Bull. Ecol., 7:479 - 492.
Juget, J., C. Amoros, J.L. Reygrobellet, M. Richardot, Ph. Richoux,
    C. Roux and B.O.J. Yi, 1979, Structure et fonctionnement des
    écosystemes du Haut-Rhône francais. VII. Le complexe hydro-
    graphique de la Lône des Pecheurs, un ancien méandre du Rhône.
    Schweiz. Z. Hydrol., (in press).
Juget, J. and M. Lafont, 1979, Description de Peloscolex turquini
    n. sp. et redescription de Peloscolex moszynskii Kasprzak,
    1971 (Tubificidae, Oligochaeta) avec quelques remarques sur
    la répartition du genre Peloscolex dans les eaux douces
    francaises. Bull. Soc. Linn., Lyon, 48:75-118.
Juget, J. and J.C. Rostan, 1973, Influence des herbiers à Trapa
    natans sur la dynamique d'un étang en période estivale.
    Annls. Limnol., 9:11 - 23.
Lafont, M. and J. Juget, 1976, Les Oligochètes du Rhône. I.
    Relevés faunistiques généraux. Annls. Limnol., 12:253-268.
Lang, C., 1974, Macrofaune des fonds de cailloux du Léman.
    Schweiz. Z. Hydrol., 36:301 - 350.

Lang, C., 1978a,   Factorial correspondence analysis of Oligochaeta communities according to entrophication level. Hydrobiologia, 57:241-247.

Lang, C., 1978b, Approche multivariable de la détection biologique et chimique des pollutions dans le lac Léman (Suisse).   Arch. Hydrobiol., 83:158-178.

Pinchant, R.,1975, Protection des nappes alluviales: orientations et conclusions des études du site alluvial de l'est lyonnais. Bull. Inform. Comité et Agence de Bassin Rhône-Mediterranée-Corse, 9:34-50.

Soares de Assis, L.F., 1977, Etude préliminaire de la toxicité du peroxyde d'hydrogène (H_2O_2) sur *Peloscolex ferox* et *Potamothrix* sp. (Oligochaeta, Tubificidae). Rapport D.E.A. (Ecologie fondamentale et appliquée des eaux continentales), Université Cl. -Bernard, LYON I. 36p.

Uribe-Zamora, M., 1975, Sélection des proies par le filtre branchial de la carpe-miroir (*Cyprinus carpio L.*). Thèse Dr. 3° cycle, Lyon. 127 p.

Winberg, G.G., 1971, Methods for the estimation of production of aquatic animals. Academic Press, London. 175 p.

Yi, B.J., 1977, Contribution à l'écologie des étangs piscicoles de la Dombes (Ain). Recherches sur le périphyton et son environnement par la technique du substrat artificiel. Thèse Dr. 3° cycle, Lyon. 114 p.

EFFECTS OF TUBIFICID OLIGOCHAETES ON PHYSICAL AND

CHEMICAL PROPERTIES OF LAKE ERIE SEDIMENTS

P. L. McCall
J.B. Fisher

Department of Earth Sciences
Case Western Reserve University
Cleveland, Ohio
U.S.A. 44106

ABSTRACT

Laboratory experiments and field observations show that
tubificid oligochaetes have important effects on sediment properties. Tubificids pelletize the surface sediment of the western
and central basins of Lake Erie and increase the median sediment
grain size and settling velocity by two orders of magnitude. The
thickness of the pelletized layer in the lake is about 1 - 2 cm.
The critical entrainment stress of pelletized sediment in box
cores from Lake Erie is about twice that of unpelletized sediment.
The difference in entrainment rate of sediment may be even greater.
In sandy sediments from the Vermilion River, tubificids have the
opposite effect: ultimate grain size in the upper 1 cm of sediment
is decreased, water content is increased, and surface sediments
are more easily eroded than non-pelletized sediments because the
tubificids feed selectively on clay and silt size sediments.

Tubificids pump little water through their burrows, but the
creation of the pelletized layer enhances the diffusion of chloride
in this zone by a factor of two over deeper regions of the sediment.
The burrows of tubificids do increase the permeability of Lake Erie
sediment by a factor of two to four.

Sediment particle reworking by tubificids is highly directional
The worms feed at depth in the sediment and deposit material at the

sediment surface. Continued feeding results in an organized vertical circulation and mixing of the top 5 - 10 cm of sediment. The
depth of feeding and mixing may be density dependent. The rate of
sediment mixing by tubificids exceeds the sedimentation rate of new
material in the western and central basins of Lake Erie.

In the laboratory, tubificids can increase the oxygen demand
of lake sediments by a factor of two. Their feeding activity produces a thickened, porous, and oxidized pelletal zone at the sediment surface and causes of flux of oxygen demanding materials from
depth to the sediment surface. Fifty to seventy percent of the
enchanced oxygen demand may be associated with the flux of FeS from
depth to the surface. The mixing of sediment by tubificids inhibits
the release of phosphorous (PO_4^{-3}) from the sediment under anoxic
conditions by preventing the formation of an iron rich surface
layer of sediment and by decreasing the near surface phosphorous
concentration gradient in interstitial water.

INTRODUCTION

Tubificid oligochaetes have been used by applied ecologists
primarily as indicators of the trophic type and pollution levels
of freshwater lakes and rivers (see Brinkhurst, 1974). However,
the study of these oligochaetes is important for other reasons
as well. The U.S. Environmental Protection Agency has been interested for some time in the fate of hazardous materials in the
Great Lakes, and in Lake Erie in particular. They have of necessity also become interested in the movement of sediment particles,
because of many hazardous materials - pesticides, trace metal
and radionuclides - are adsorbed onto suspended sediment particles
upon their introduction into the lakes.

At some point in their passage through the lake, suspended
particles reach the bottom. But the bottom sediments are not a
permanent sink for hazardous materials. Materials can be remo-
bilized in at least two ways. Sediment particles may be resuspended under the influence of waves and currents, or particles
may be buried and eventually pass through different chemical
regimes where their coatings dissolve. Dissolved solutes may
then diffuse or be advected back toward the sediment-water
interface (SWI).

The interest in sediment research centers then on sediment
transport (the erosion and deposition of sediment particles),
pore water chemistry and the flux of dissolved materials across
the SWI. And this is where tubificid oligochaetes come into the
picture. It is our thesis that any study of the physical and
chemical properties of sediment-water systems that fails to take

into account the effects of benthic organisms on sediment properties
is likely to be in error by orders of magnitude.

Our work done to date supports this thesis. We have been
concerned with the effects of microflora and macrobenthos - primarily
bivalves, chironomids, and tubificids - on the physical properties
of the SWI, sediment transport, and the flux of dissolved materials
across the SWI. In this paper, we will summarize some of the field
observations and laboratory experiments that demonstrate the effects
of tubificids on sediment properties and processes. It is not our
intention to describe here the effects of the other benthic groups
(although in many cases they are just as important as the tubificids
in altering sediment properties) or to review the literature on
freshwater animal-sediment relations (for the most recent review
see Petr, 1977).

STUDY AREA

All the work described here was done with animals and sediments
from the western and central basins of Lake Erie and from the Ver-
million River, a stream which empties into the south shore of the
lake (Figure 1). Lake Erie is about 450 km long with its long
axis oriented parallel to the direction of the prevailing winds
and is about 80 km wide, but on the average the lake is only about
20 m deep. It is the shallowest of the Laurentian Great Lakes and
has a large surface to volume ratio. It is probably for these
reasons, plus the fact that the shore of the lake is densely populated and receives runoff rich in inorganic nutrients, that Lake
Erie is the most productive of the Great Lakes. The oligochaete
fauna of the lake has been surveyed several times in the last 40
years. The most recent survey was by Pliodzinskas (1978), who
described eighteen species of tubificids from the western and
central basins. The fauna is numerically dominated by *Limnodrilus
hoffmeisteri* and other limnodrilids, *Peloscolex ferox*, and in
nearshore and harbor regions by *Tubifex tubifex* and *Peloscolex
multisetosus*. We have found tubificid densities of 10^3-10^4
individuals/m^2 in the central basin, 10^4-10^5 individuals/m^2 in
the western basin, and 10^5-10^6 individuals/m^2 in Cleveland harbor.
Because the lake is shallow and the bottom sediments easily
resuspended by wind generated currents, and because the density
of oligochaetes is high, Lake Erie is an ideal environment to
begin a series of studies on the effects of freshwater tubificids
on sediment properties.

EFFECTS OF FEEDING ON GRAIN SIZE AND MASS PROPERTIES OF LAKE ERIE
SEDIMENTS

While many infaunal macrobenthos have the greatest effect on
sediment properties through burrowing or respiration (Rhoads, 1974;
Aller, 1977), our experiments suggest that it is through their

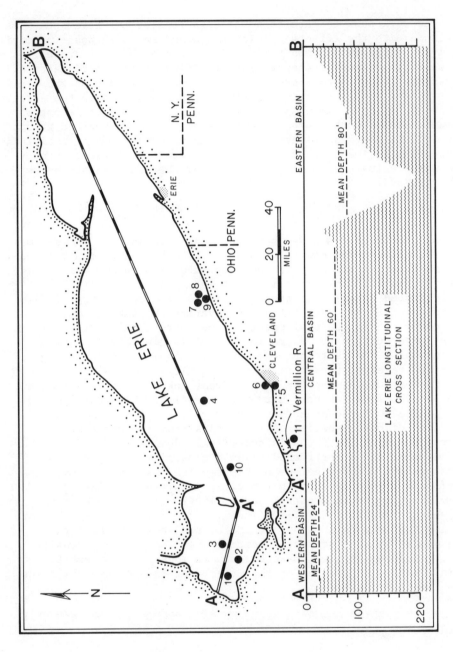

Figure 1. Map of Lake Erie and the Vermilion River showing sediment sampling sites.

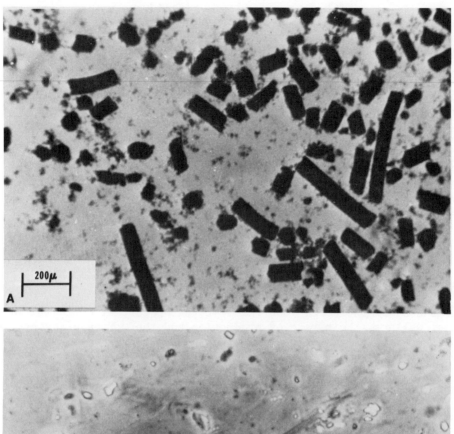

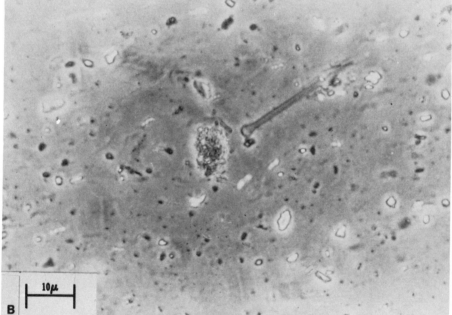

Figure 2.   a) 12x transmitted light photomicrograph of the top 1 mm
            of sediment from site 2 showing oligochaete fecal pellets.
            b) 400x transmitted light photomicrograph of sediment in
            (a) after blender and dispersant treatment.

deposit-feeding activities that the tubificids have the greatest
influence on Lake Erie sediments. Most of the bottom of the lake
is covered with high water content, post glacial, silt-clay size
sediment derived from river drainage and shore erosion (Thomas
et al., 1976). Most tubificids feed head down in muddy sediments,
consume sediments at depth, and egest undigested material at the
sediment water interface. This material is eventually buried and
falls back toward the feeding zone under the influence of gravity
and further feeding below. Such particle transfer can bring about
important changes in the physical properties of the topmost layer
of lake sediment, because egested sediment is packaged into mucus-
bound, sand size fecal pellets that maintain their integrity for
some time after being deposited at the SWI. If the rate of fecal
pellet production is high relative to the rate of deposition of new
sediment and the rate of pellet decay, then a pelletized layer of
sediment can form at the SWI. This layer will be comprised of
larger particles that have a higher settling velocity than the
underlying substratum. We might expect the pelletal layer to have
a different erodibility, porosity, and permeability than unpelletized sediment.

     We shall therefore look first at the properties of tubificid
fecal pellets, their rate of production and their rate of decay,
and then look at some field evidence for the existence of a
pelletized layer of sediment in Lake Erie.

Size and Settling Velocity

     Bulk samples of sediments and tubificids were collected from
the western basin of Lake Erie (site 2, Figure 1) by divers using
box cores. These samples were held at 20° C in laboratory aquaria.
Two weeks after collection, fresh fecal material was taken by
Pasteur pipet from the top 1 mm of the box cores, placed into a
petri dish, and photographed under a dissecting microscope (Figure
2a). Measurements made from the photographs showed a mean pellet
length of 280μm ± 66μm (N= 1500) and width of 86μm ± 27μm. The
same material was analyzed by standard geological techniques of
size grain analysis. Sediment was mixed in a Waring blender for
several minutes and put into a distilled water-dispersant solution.
Sediment particle settling velocity measurements (Figure 3) were
made by the pipet method (see Royse, 1970; Folk, 1965). Approximate equivalent spherical diameters were calculated assuming
Stokes' settling law. Figure 2b was taken at considerably higher
magnification than Figure 2a and shows particles subsequent to
the blender treatment. No pellets survived this treatment, and
constituent mineral grains were removed from their organic matrix.
The median particle size of the sediment calculated by this method
was 1.5μm. The effect of tubificid feeding then is to aggregate
clay sized grains into sand size particles.

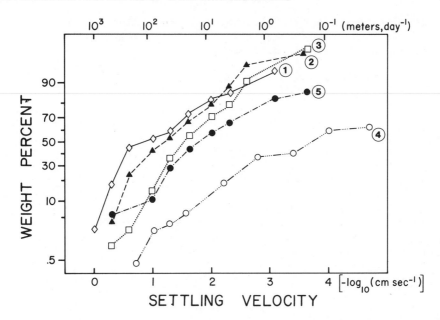

Figure 3.   Cumulative curve, probability ordinate showing settling
            velocity distribution of western Lake Erie sediment from
            site 2.   1 = top 1 mm of sediment settling in lake water.
            2 = top 10 mm of sediment in lake water.   3 = degraded
            pellets in lake water.   4 = top 10 mm of lake sediment
            with blender plus dispersant treatment in distilled
            water.   5 = top 10 mm of sediment in distilled water
            plus dispersant.

    The sand size pellets had a higher settling velocity than
their clay size constituents. The settling velocities of 50 fecal
pellets dropped one by one into glass settling cylinders containing
filtered lake water (22° C) were measured by timing their fall
from 10–30 cm below the water surface. The mean settling velocity
was $1.03 \pm .09$ cm sec^{-1}. The settling velocity distributions of
sediment collected from the box cores are shown in Figure 3. The
top 1 mm of deposit, which was the sediment most recently processed
by the tubificids, had a median settling velocity of .13 cm sec^{-1}
in lake water (22° C), while highly degraded particles had settling
velocity of .03 cm sec^{-1}. The top 1 cm of western basin sediment,
consisting of pellets in various states of decay, had an intermediate settling velocity of .06 cm sec^{-1}.

    When the standard analysis of the top 1 cm of sediment was
done, the median settling velocity was reduced to .0002 cm sec^{-1}.
Only a portion of the difference from the untreated case was due

to the differences in settling media. The median settling velocity
of the top 1 cm of sediment in distilled water and dispersant but
no blender treatment was 0.02 cm sec^{-1}.

Pellet Production Rate

Tubificids will have an important effect on sediment particle
size only if they can process sediment fast enough for a layer of
pellets to build up at the sediment surface. We measured the per
individual pellet production rate of the tubificid *Tubifex tubifex*
in the laboratory under the experimental conditions listed in
Table 1 in order to get an idea of the relative importance of
temperature, oxygen, worm density, and substratum on feeding and
egestion rate. Sediments were selected to correspond to different
'resource' levels. Cleveland Harbor sediments (site 5, Figure 1)
commonly contain natural population densities of 90,000-150,000
tubificids m^{-2} (and on occasion over 1,200,000 m^{-2}), while lake
sediments just outside the harbor (site 6) support densities of
10,000-50,000 tubificids m^{-2}. Sediment from these two sites was
passed through a 250μm mesh screen to remove worms from the sediment. Sieved sediment was added to glass vials (1.7 cm x 6.0 cm,
7.4 ml) to produce a column of sediment 4 cm tall. *T. tubifex*
were collected from Cleveland Harbor and acclimated to experimental temperatures in the laboratory. Worms were then added to the
vials and allowed to burrow and feed for one week prior to measurement of feeding rate. The experiment began when a white kaolin
clay-water mixture was pipetted onto the top of the experimental
sediments to create a visual marker horizon about 0.2 cm thick.
The worms put their tails through this marker bed to respire while
feeding. Tubificid egestion was measured as the amount of sediment deposited on top of the white marker bed in twenty days.
Feeding rate is reported in (ml of sediment)/(individual)/(hour).
Both sediment mixing and fecal pellet production rates are reported
on a per individual basis primarily for ease of comparison with
other work where biomass information is not available. These data
can be easily transformed into biomass units through multiplication
by the average dry weight of the worm population used in the experiments (.418 mg). Since the variance of the biomass of the experimental population was small by design (σ = .053 mg), and since
results were seldom based on groups of less than 50 worms, the
use of this mean biomass per individual figure is satisfactory.
Individual feeding rate was adjusted for worm mortality, which
was assumed to be linear over the course of the experiment. Percent mortality of worms was calculated as (initial density - final
density)/(initial density). Final density was determined by
sieving vial sediments at the end of the experiment.

No measurable feeding took place under low dissolved oxygen
conditions. Egestion rates measured under high dissolved oxygen
concentrations are shown in Table 2.

Table 1. Design of the factorial experiment used to study
          T. tubifex egestion rate in the kaolin clay marker
          horizon experiment.

FACTOR	LEVELS
Temperature:	$7.1 \pm 0.3^{\circ}C$; $15.3 \pm 0.1^{\circ}C$; $22.5 \pm 3^{\circ}C$
Dissolved Oxygen: (High; Low)	$(11.8 \pm 0.2$; 3.3 ± 1.5 ppm $[7^{\circ}C])$
	$(10.0 \pm 0.2$; 4.0 ± 1.0 ppm $[15^{\circ}C])$
	$(8.5 \pm 0.2$; 3.5 ± 1.2 ppm $[22^{\circ}C])$
Substratum:	Harbor sediment; Lake sediment
Initial tubificid population density:	1, 9, 18 *T. tubifex*/vial
	$(= 5000$; $50,000$; $100,000$ indiv./m^2)

TREATMENTS	
Factor Combinations:	3 temperature levels x 2 oxygen levels x
	2 substrata x 3 density levels = 36 factor combinations
Total treatments:	36 factor combinations x 5 replicates = 180 treatments (vials)

A linear model was fit to the data. This model may be
written as:

$$y_i = \beta_1 x_{1j} + \beta_2 x_{2i} + \cdots + \beta_{12} x_{12i} + e_i,$$

where y_i is the per individual egestion rate in the ith vial, the
x_{ij} are the values of variables x_j that define the experimental
conditions in the ith vial, β_j a fitted set of coefficients that
multiply the x_{ij} to reproduce y_i as closely as possible, and e
a variable that measures the degree to which the y_i cannot be
reproduced by the rest of the function. This residual error is
assumed to be normally distributed with zero mean and variance σ^2.
The significance of a particular variable x_j or group of variables
is assessed by comparing σ^2 under the full model with the residual
variance of the model with appropriate β_j set equal to zero. The
construction and interpretation of linear models has been fully
described by Seal (1964), Mendenhall (1968), and Finn (1974).

Table 2. Egestion rate (10^{-5} ml/individual/hour) of *T. tubifex* under experimental conditions.

Temperature		Lake Sediment			Harbor Sediment		
	Initial Population Density:	1	9	18	1	9	18
22° C		131.3	72.3	53.5	77.9	57.7	72.1
		160.4	75.2	92.9	75.8	82.9	79.8
		175.0	66.0	81.5	24.0	76.9	62.1
		118.8	84.2	100.8	65.4	86.3	69.0
		67.3	85.0	92.7	80.6	72.5	62.3
15° C		56.0	36.9	58.1	112.1	78.1	62.5
		7.5	42.9	53.6	22.5	52.3	71.0
		7.5	42.9	53.6	78.5	10.8	77.7
		7.5	29.8	37.7	63.5	69.0	47.9
		7.5	81.5	58.5	78.5	79.6	32.5
7° C		7.5	16.7	11.3	0	16.7	8.1
		7.5	16.5	8.5	0	10.8	7.5
		7.5	13.1	8.8	0	15.8	8.3
		7.5	19.0	12.9	0	19.4	10.8
		7.5	16.7	8.8	0	3.8	11.9

1×10^{-5} ml/individual/hour = 4.1×10^{-2} mg dry sediment/mg dry oligochaete/hour

= 2.4×10^{-8} cm/day/individual

The composition of the x_i, their interpretation, and significance are described in Tables 3 and 4. Particular variables or groups of variables are judged important if their associated β_i's are significantly different from zero. Overall, feeding rate increased linearly with temperature ($\beta_{3,6,7} = 0$; $\beta_{5,8,11,12} \neq 0$). Harbor and lake sediment, however, had significantly different temperature curvatures. A decrease in the temperature dependence of egestion rate took place at a lower temperature in lake than in harbor sediment ($\beta_5 < 0$). (Note that worm densities used in this analysis are adjusted for worm mortality and are thus labeled "effective" densities to distinguish them from "initial" densities). We will use these results later to account for variable depths of the tubificid feeding zone, but for now it is sufficient to conclude that the rate of production of pelletized sediment is most strongly dependent on water temperature and dissolved oxygen concentration.

Table 3.   Dependent variables used in analysis of the
egestion rate experiment.

x_0 = a vector of units, to represent the general mean

x_1 = -1 for 7° C treatments, 0 for 15° C treatments, +1 for 23° C treatments

x_2 = +1 for 7° C treatments, -2 for 15° C treatments, +1 for 23° C treatments

x_3 = -1 for lake sediment treatments, +1 for harbor sediment treatments

$x_4 = x_1 \cdot x_3$

$x_5 = x_2 \cdot x_3$

x_6 = effective density = (initial density + final density ÷ 2) - mean effective density

$x_7 = $ (effective density)^2 - mean (effective density)^2

$x_8 = x_1 \cdot x_6$

$x_9 = x_2 \cdot x_6$

$x_{10} = x_3 \cdot x_6$

$x_{11} = x_1 \cdot x_3 \cdot x_6$

$x_{12} = x_2 \cdot x_3 \cdot x_6$

Pellet Decay Rate

The rate of pellet decay was measured as a function of temperature (22° C, 15° C, 7° C) and mechanical agitation (at 22° C)
over a period of 21 days. Fresh pellets were collected with a
Pasteur pipet from the top 1 mm of box cores of western Lake Erie
sediment held in laboratory aquaria. Mechanical agitation was
provided by subjecting a sample of pellets to spinning in a 750 ml
beaker containing 400 ml of filtered Lake Erie water and a magnetic
stir bar spun at 50 rpm. At each of six sampling intervals, 250
pellets from each of the four treatments were photographed under
a Wild dissecting microscope (12x) and subsequently measured.
Data reported are the long diameters of discrete pellets as a
function of time. Pellet decay was not linear with time, but
rather followed most closely a power law relationship (Figure 4).
The slopes, α, of the lines were taken as a measure of decay rate.
The decay rate of stirred pellets was three times that of unstirred
pellets. Only the size of discrete pellets remaining during decay
was measured, since it was difficult to measure the size or volume

Table 4. Summary of analysis of egestion rate

Hypothesis	Effect	Sum of Squares	Mean of Squares	F	P
$\beta_1=\beta_2=0$	temperature slope and curvature	64212.55	32106.28	86.52	<.005
$\beta_3=0$	substratum difference	226.48	226.48	.61	NS*
$\beta_6=\beta_7=0$	density slope and curvature	661.34	330.67	.89	NS
$\beta_4=\beta_5=0$	temperature-substratum interaction	6109.19	3043.58	8.23	<.005
$\beta_8=\beta_9=0$	temperature-density slope interaction	4884.31	2442.16	6.58	<.005
$\beta_{10}=0$	density-substratum interaction	114.06	114.06	.31	NS
$\beta_{11}=\beta_{12}=0$	temperature-density substratum inter- action	7601.66	3805.33	10.25	<.005
	residual	26718.54	317.09		

$\beta_0 = 47.04$ $\beta_7 = -.07$ *not significant (p .05)

$\beta_1 = 35.45\pm6.73$** $\beta_8 = -1.64\pm1.32$** **$\beta_i \neq 0$

$\beta_2 = 1.30\pm3.63$ $\beta_9 = -.68\pm1.32$**

$\beta_3 = 1.62$ $\beta_{10} = .22$

$\beta_4 = 4.77\pm6.38$ $\beta_{11} = 1.26\pm1.22$**

$\beta_5 = -5.31\pm3.65$** $\beta_{12} = 1.24\pm.73$**

$\beta_6 = -.38$

of the amorphous mixture of sedimentary material and organic matrix that was the result of pellet decay. However, when the mean size of discrete pellets fell below 75-80μm, most of the pellets had completely decayed. Time to complete decay at 22° C was two days with stirring and (by extrapolation) sixty-six days with no stirring. Decay at 22° C was more than twice that at lower temperatures. Decay slowed greatly at 7° and 15° C and was not linear with temperature.

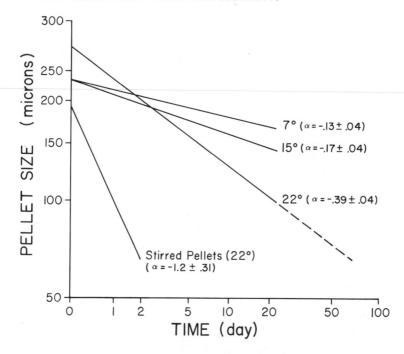

Figure 4.   Log linear relationship of pellet length and time, ln
            (pellet length) = α[ln (days + 1) − ln (days + 1)]
            + β for the four treatments shown. Coefficients of
            determination range from .89 − .96. Dashed line
            indicates extrapolation to complete decay.

Thickness of the Pelletal Layer

The sedimentation rate in the western basin of Lake Erie is
small (.1 − .6 cm year^{-1}; Kemp et al., 1977) relative to the
sediment reworking rate of tubificids (3.5 − 4 cm year^{-1}; Figure
21) and it was ignored in the calculation of the potential thickness of the pelletized zone in the lake. The production rate of
pellets is a complex function of temperature, substratum, worm
density, and oxygen content of the water, but at 22° C falls in
the range 1–2 x 10^{-6} cm day^{-1} individual^{-1} (Table 2). At in situ
densities (10^{4}m^{-2}) and 22° C, about .8 − 1.3 cm of pellets could
be produced in 66 days. Pellet decay drops dramatically at 7° C
and 15° C, and production rate drops to about .3−1 x 10^{-6}cm day^{-1}
individual^{-1}, from which we may estimate that perhaps another
.5−1 cm thick pellet zone ought to be produced during the rest
of the year. Mechanical agitation (storms, bioturbation) will

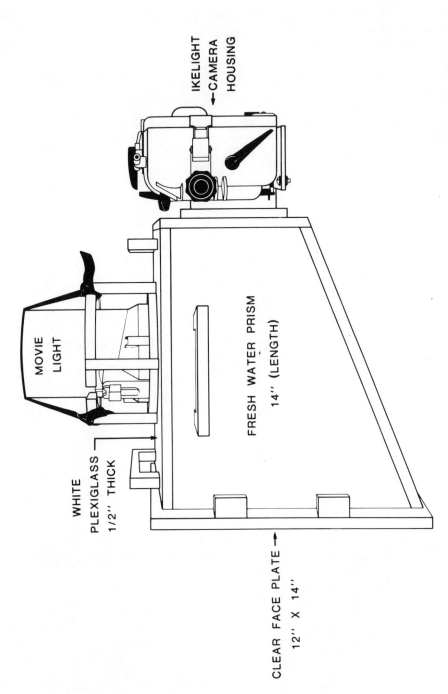

Figure 5. Sediment profile camera housing, a modified version of that described by Rhoads and Young (1970).

accelerate decay, but based on laboratory data it would not be surprising to find the top 1-2 cm of western basin bottom sediments pelletized.

Various locations in the western and central basins of Lake Erie were examined for evidence of pelletization of the surface of bottom sediments (see Figure 1). A sediment-water interface camera housing patterned after that of Rhoads and Young (1970) and containing a Minolta SRT-101 camera fitted with a 28mm wide angle lens was used to photographically examine the physical appearance of the substratum in vertical profile (Figure 5). An Ikelite Modular Movie Lite was used as a light source. The camera was diver positioned and actuated. A typical sediment is shown in Figure 6a. Pelletized layers of thickness 1 cm or more were evident at all locations examined during most of the year. The only exception was in the central basin (site 4) at the end of a period of anoxia in the hypolimnion in 1976 (Figure 6b). This part of the lake has a lower worm density ($\sim 10^3-10^4$ m^{-2}) than the western basin, and these worms probably slow or cease feeding during periods of anoxia.

THE EROSION OF MUDDY BOTTOMS

Tubificid oligochaetes egest sand-size fecal pellets on silt-clay bottoms, and a pelletized surface layer of sediment is present for most of the year in the central and western basins of Lake Erie. Even when pellets decay, mineral grains are still held in organic matrix. What is the effect of this pelletization on the erosion of bottom sediment?

The sedimentary properties that control the erosion of cohesive mud bottoms are only partly understood. The effects of sediment grain size distribution, porosity, and shear strength on erosion depend on the particular sediment studied (Sundborg, 1956; Postma, 1967; Southard, 1974), and for some cohesive sediments there is no relation at all (Parthenaides and Paaswell, 1968). Benthic organisms are known to alter sediment erodibility. Burrowing macrofauna and meiofauna can increase sediment water content (porosity) and thereby increase erodibility (Rhoads and Young,1970; Rhoads, 1970; Cullen, 1973). Tube dwelling macrofauna can increase sediment stability and decrease erodibility (Fager, 1964; Mills, 1969). Bacteria and algae secrete mucopolysaccharides that bind sediment particles together and decrease erodibility (Scoffin, 1970; Holland et al., 1974; Rhoads et al., 1978).

We will emphasize here the role that tubificids play in altering erodibility; the role of other organisms will be explored in a separate publication. We will suggest the inability of commonly measured physical sediment properties to account for the

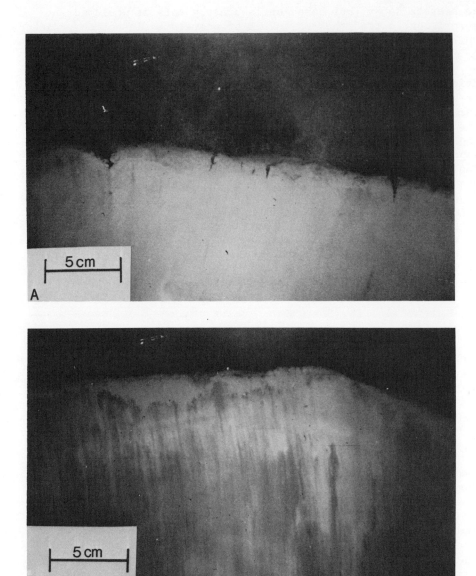

Figure 6. a) <u>in situ</u> SWI photograph taken at site 2 showing
        pelletized layer of sediment. b) <u>in situ</u> SWI photograph
        at site 4 after central basin anoxia. Irregularly
        developed pelletal zone (light color) overlies reduced
        sediments (dark color). Streaking results from inser-
        tion of camera into the bottom.

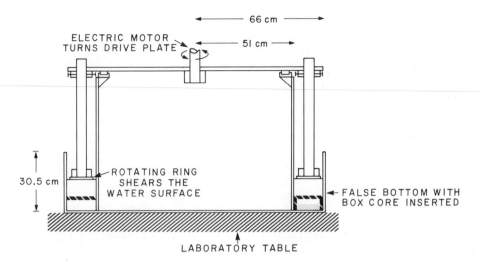

Figure 7.   Cross sectional view of annular flume.

critical erosion stress of Lake Erie sediment, and then show from
field data and a laboratory experiment how tubificids change the
erodibility of the lake bottom.

Sampling Procedure and the Annular Flume

        Bottom sediment was collected from nine locations in the
western and central basins of Lake Erie (sites 1-9, Figure 1) in
1976 and 1977.   SCUBA divers used plexiglas box cores (approximately 30 cm x 7.5 cm x 10.5 cm) shaped to fit into an annular
flume to collect intact a .03m^2 portion of the lake bottom.   In
the laboratory, the surface of the core sediment was brought flush
with the top edge of the core box by adjusting a leveling plate
that rested on top of the water tight base plate.   The box core
was then fitted into an opening in a plexiglas false bottom of a
132 cm diameter annular flume (Figure 7) so that the sediment
surface was even with the plastic bottom. The top plate of the
box core was then carefully removed, exposing the lake sediment
to dechlorinated tap water at 20° C in the flume.   The height of
the water column overlying the sediment was 7.6 cm.   A plexiglas
ring in contact with the water surface was rotated by a two horse-
power electric motor.   This rotation produced a water flow that
exerted a shear stress on the stationary bottom surface (sediment
and plexiglas false bottom).   Fukuda (1978) used a DISA boundary
layer hot film probe and constant temperature anemometer to
measure vertical velocity profiles in this flume.   He determined
bottom shear stress by fitting measured velocity profiles to the
two dimensional law of the wall and obtained the relationship
between ring rotation rate and average bottom shear stress that
was used in this study.   With the box core in place, the rotating

ring was accelerated by adjusting a rheostat connected to the electric motor. Ring rotation rate was measured with a stopwatch. The sediment surface of the box core was observed through a stereo-microscope (7x), and the rotation rate at which particles left the sediment surface and were entrained into the flow was noted and later translated into critical bottom stress, τ_e.

After these measurements were made, the box core was removed from the flume, and a sample of the top 1 cm of sediment was taken for measurement of sediment water content. Water content was taken to be the loss in weight on drying at 70° C for 48 hours as a percent of the total weight of sediment plus pore water. Other physical properties of sediment from the nine locations were measured on bulk samples collected with a grab sampler. Median grain size and percent clay were determined by the pipet method (Folk, 1965). Organic matter determinations were made by chromic acid digestion of sediments and a modified Walkley–Black titration (Gaudette et al., 1974).

Entrainment and Physical Properties

Entrainment of silt and clay size sediments is often related to sediment water content. High water content sediments have more pore space and fewer interparticle contacts; all other bottom properties being equal, they are more easily entrained than low water content sediments. This relationship is shown in Figure 8 for an illitic clay that has a mineralogic composition and grain size distribution similar to the Lake Erie sediments. This mate-rial is in fact decomposed Chagrin Shale, which outcrops along the southern lake shore and is a parent material for the lake sediments. The data were obtained by adding fluid mixtures of sediment and water to a series of box cores and then measuring critical entrainment stress as dewatering through self–compaction occurred. Most data showing the relation of sediment erosion to water content are obtained in this way (Sundborg, 1956; Postma, 1967; Southard, 1974; Fukuda, 1978). Here all other properties were equal. Measurements were obtained from the same sediment that differed (in time) only in water content. The correlation of water content and entrainment stress was high (r = .92, n = 26).

For lake sediments the situation is different. These sedi-ments had water contents that are not clearly associated with critical entrainment stress (r = .41, n = 30), although for the same water content all lake sediments were more difficult to erode than the illitic clay. In addition, the average τ_e for sites 2 and 9 were 100% and 75% greater, respectively, than the highest τ_e found by Fukuda (1978) for sediments that were collected from the same sites but that were sieved, stored in the laboratory for several weeks, and repeatedly suspended and redeposited in the flume. Some factor or factors act to aggregate and/or bind

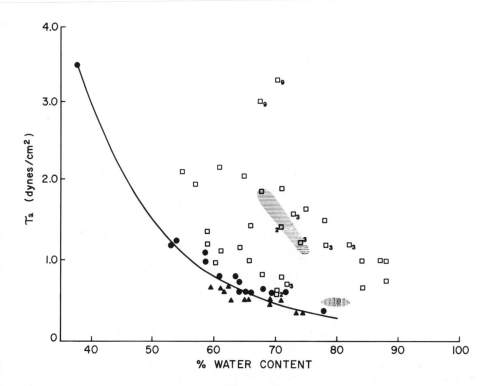

Figure 8.   Critical entrainment stress, τ_e, of oxidized box cores
            as a function of sediment water content. Closed circles
            represent box cores of shale based sediment run by Fuduka
            (1978). Closed triangles represent runs made with the
            entire flume covered with shale based sediment (also by
            Fukuda, 1978). Open boxes represent box cores collected
            from locations 1-9. Area shaded with horizontal bars
            is to be compared with box cores from location 9; area
            shaded with vertical bars is to be compared with box
            cores from locations 2 and 3.

particles together making the lake core sediment more difficult to entrain. The nine location averages of sediment median grain size, percent clay, percent organic matter, and percent water content (Table 5), all measured about their mean, were simultaneously regressed on the nine location averages of critical entrainment stress (measured about their mean) obtained under fully oxic conditions. The analysis of variance in Table 5 shows that knowledge of average sediment properties did not contribute significantly to the prediction of average critical entrainment stress ($\beta_{1,2,3,4}$ = 0).

Effects of Tubificids on τ_e

Average physical properties such as median grain size, clay content, and water content are not good predictors of critical entrainment stress of the fine silt and clay sediments in Lake Erie. Some of the observed variation of τ_e may be related to tubificid worm activity or to the activity of other biota. For instance, four of the box cores taken from the lake were covered with high densities of tubificid worm tubes (greater than 1000 tubes m^{-2}) projecting 3 mm or more above the SWI. We might expect dense fields of worm tubes to increase the critical entrainment stress, since they act to bind the sediment and prevent the transfer of stress to the SWI. Indeed the average τ_e of the high tube density box cores was 42% ($\pm$ 31%, n = 4) higher than comparable low tube density box cores.

A laboratory experiment illustrates another effect of tubificids. Sediment from site 2 was collected with grab samplers and sieved through a 250μm sieve to remove worms and other macrofauna from the sediment. A fluid mixture of sieved sediment and water was added to 18 core boxes and allowed to compact for one week. Six box cores received no further treatment. To six others were added enough worms (all belonging to the genus *Limnodrilus*) to equal a density of 10,000 individuals m^{-2}. Six other box cores had no worms added, and were treated with 10% formalin to stop biological activity. Box cores were incubated at 7° C and 24° C for twenty days and then placed in the flume for determination of critical entrainment stress.

None of the tubificids added to the box cores made tubes projecting above the sediment surface, so the effect of the worm additions was to pelletize the sediment surface and increase the median sediment grain size. The boxes with no worms contained microorganisms whose exudate could bind the sediment, and microfauna (ostracods, copepods) whose burrowing activity could loosen the sediment surface. Which process dominates is likely a function of organism density and temperature.

Tukey's T-test was used to make multiple comparisons of pairwise difference of the average critical entrainment stress between

Table 5. Average sediment properties for the nine sampling sites in Lake Erie and the relation to τ_e.

Location	Water Depth (meters)	$\tau_e = \beta_0 + \beta_1$ (dynes/cm²)	Median grain size+β_2 (μm)	Clay Content+β_3 (%<4μm)	Organic Matter+β_4 (%)	Water Content (%)
1	7	1.34	8	45	5.7	74
2	7	1.10	3	55	6.7	72
3	9	1.10	1	65	5.5	82
4	24	.90	1	64	6.9	88
5	8	1.07	4	50	9.0	60
6	10	1.28	19	37	7.6	61
7	16	1.46	7	42	6.0	68
8	14	2.16	10	38	5.8	64
9	13	3.15	21	21	6.5	68

Hypothesis	Variability on account of	df	Sum of Squares	Significance Level
$\beta_1 = 0$	grain size	1	.73	$p < .75$
$\beta_2 = 0$	clay content	1	1.49	$.90 < p < .75$
$\beta_3 = 0$	organic matter	1	.67	$p < .75$
$\beta_4 = 0$	water content	1	.89	$p < .75$
$\beta_1 = \beta_2 = \beta_3 = \beta_4 = 0$	all properties	4	4.03	$.90 < p < .75$
	Residual	4	.66	

a number of treatments (Scheffè, 1959; Belz, 1973). A 90% confidence level was used because it is a strong statement to say that the probability is .9 that all 15 possible pairwise differences with confidence intervals attached are simultaneously correct (Belz, 1973). The pairwise contrasts treatments of greater interest are underlined in Table 6.

Table 6. Critical entrainment stress of box cores incubated in the laboratory with worms (w), without worms (nw), and with 10% formalin (f).

	7°			24°		
	w	nw	f	w	nw	f
Replicates	.81	.72	.87	.87	.93	.81
	1.08	.61	.80	.80	.81	.84
	.89	.68	.81	.84	.85	.82
Mean	.93	.67	.83	.84	.86	.82

$7°w - 7°nw$				
$7°w - 7°f$	$7°nw - 7°f$			
$7°w - 24°w$	$7°nw - 24°w$	$7°f - 24°w$		
$7°w - 24°nw$	$7°nw - 24°nw$	$7°f - 24°nw$	$24°w - 24°nw$	
$7°w - 24°f$	$7°nw - 24°f$	$7°f - 24°f$	$24°w - 24°f$	$24°nw - 24°f$

=	.26*				
	.10	-.16			
	.09	-.17	-.01		
	.06	-.19	-.04	-.03	
	.10	-.15	-.00	.01	.04

*Contrasts whose absolute value is greater than .19 are significant at the 95% level; contrasts whose rate is greater than .17 are significant at the 90% level.

At 7° C the critical stress for boxes with worms was higher than for those without worms, indicating the importance of pelletization of the surface sediment at this temperature. Sediment in boxes with no worms was more difficult to entrain at 24° C than at 7° C, indicating that binding was more important at high temperatures. When box cores were treated with formalin and biologic activity stopped, this difference disappeared. At 24° C there was no difference in critical stress between box cores with and without worms; microbial processes were evidently as important as pelletization. This was also suggested by the equality of critical stresses in boxes with worms at both 7° C and 24° C. The remaining

significant difference was between box cores at 24° C with worms
and other biota and boxes at 7° C with no worms, indicating again
that pelletization and greater binding at 24° C increased critical
entrainment stress. Notice finally that box cores incubated at
7° C with no worms had a lower critical entrainment stress than
box cores incubated at 7° C and treated with formalin; the difference is nearly significant. This may be because the formalin
treated sediment was relatively more compact than untreated sediment and thus harder to erode. The untreated sediment was subjected
to bioturbation of the surface sediment by ostracods, copepods, and
other microfauna that perhaps loosened the sediment and made it
easier to entrain. Indeed, this treatment (7°C, no worms) in which
there is no pelletization and likely little binding but some
loosening of the surface sediment by meiofauna, had the lowest
τ_e of any treatment. Thus both activities of tubificids, tube
building by some species and fecal pellet egestion, act to make
the sediment surface more difficult to erode. Other biota may
reinforce or counteract these effects.

Fukuda (1978) was the first to measure not only the critical
entrainment stress, but also the entrainment rate of cohesive
sediments from freshwater environments. He found that at low
shear stresses, the entrainment rate of sediment from lake Erie
increased logarithmically with bottom shear stress. The significance of this finding is illustrated in Figure 9. If we assume
the same log linear relationship with stress found by Fukuda,
then we see that a 50-100% change in sediment τ_e (the region near
E = 0) effected by biological processes could mean a much larger
change -- 3-10 fold in this example -- in the entrainment rate of
sediment into the overlying water and a similar increase in the
sediment concentration in the water overlying the sediment.

ALTERATION OF PHYSICAL PROPERTIES OF SANDY SUBSTRATA:
THE VERMILION RIVER.

To this point we have presented results only from silt and
clay size sediments. We have concluded that tubificid oligochaetes
tend to aggregate sediments and make them more difficult to erode.
It is important to distinguish among sediment types, however, since
we have made just the opposite conclusion about the effects of
worms on the properties of sandy substrata (Tevesz et al., in
press).

Large populations (~3000 indiv. m^{-2}) of tubificids were found
all along on the fine sand bottom (median grain size 110μm, Figure
11d) of the Vermilion River, (Site 11, Figure 1) but their biodeposits were only found where current velocities were low (e.g. the
bottom of deep pools, the lowest reaches of the river, shallow
pools partially isolated from flowing water). This suggested that
the river biodeposits were easily eroded. A laboratory experiment

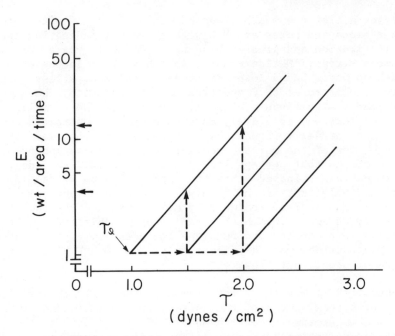

Figure 9.   Hypothetical entrainment rate of sediment, E, as a
            function of bottom shear stress, τ, showing the increase
            in E with 50% and 100% increases in critical entrainment
            stress, τ_e. While E is shown in arbitrary units, the
            slope of the line is the same as that found by Fukuda
            (1978) for sediment from the western basin of Lake Erie.

was done to test the effect of tubificids on the properties of the
near surface layer of the river sediment. Homogenized samples of
Vermilion River experiment were placed into six thin walled aquaria.
Three aquaria received no further treatment. To the other aquaria
oligochaetes were added to equal a density of 3000 worms m^{-2}.

     Eight weeks after the experiment began the feeding and defe-
cating activity of the worms had segregated the upper 1.5cm of the
aquaria with worms into four distinct layers (Figure 10). The top
layer consisted of an approximately 3-4mm thick accumulation of
fecal pellets. The second layer lay directly below and consisted
of an approximately 5mm thick layer of compacted pellets and
residue from pellet decay. The maximum size of mineral particles
in these layers was limited to the maximum size of particles that
were ingested by the oligochaetes. The third layer, about 4-5mm
thick, was a sandy concentrate which corresponded to the location
of the worm feeding zone. Many of the mineral grains comprising
it were too large to be ingested by the worms and thus remained
behind instead of being transported to the SWI. A thicker layer,

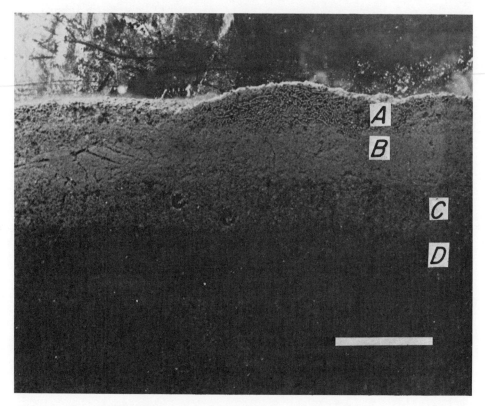

Figure 10.   Layering of Vermilion River sediments produced by
             tubificid feeding.   Scale = 1 cm.

comprising the rest of the sediment column in the tanks lay under
the sandy concentrate. This more poorly sorted layer appeared to
be unaltered by the oligochaetes. The aquaria containing no worms
were not layered.

Analysis of physical properties of these layers showed that
the pelletal layers A and B (Figure 10) had a higher water content,
higher organic matter content, and smaller median grain size than
sediment from the same depth in aquaria with no worms (Table 7).
This change in properties comes about because the worms feed
selectively on the smaller size particles in the sediment. These
particles have greater surface area for the attachment of a greater
number of digestible bacteria and have associated with them more
organic matter than larger ($<4\phi$ = 62.5µm) particles (Newell, 1965).
Thus when the grain size of sediment from aquaria with worms was
analyzed by wet sieving, layers A and B were found to be enriched
in silt and clay size particles ($<4\phi$) relative to unaltered layer

Table 7. Physical properties of sediment layers in laboratory
          aquaria.

GRAIN SIZE (μm)

Layer	Mean with 95 % confidence	n
A	47±26	3
B	46±16	3
C	133±43	3
D	111±14	3

WATER CONTENT (% by weight)

Layer	Treatment	Mean with 95 % confidence	n
0-1.5 cm (A & B)	no worms	27.42 ± 3.71	3
	worms	48.47 ± 3.58	3
1.5-3.0 cm (C & D)	no worms	20.75 ± 4.85	3
	worms	25.24 ± 4.20	3

ORGANIC MATTER (% by weight)

Layer	Treatment	Mean with 95% confidence	n
0-1.5 cm (A & B)	no worms	.75 ± 0.50	3
	worms	1.59 ± 0.20	2
1.5-3.0 cm (C & D)	no worms	.53 ± 0.27	3
	worms	.80 ± 0.47	3

D, while layer C was relatively depleted of the same sizes
(Figure 11). The 62.5-2.5μm particle size fraction was further
analyzed on a Coulter counter (Figure 12). The only statistically
significant difference in grain size distribution was between
layers B and C. The feeding layer C had relatively fewer clay

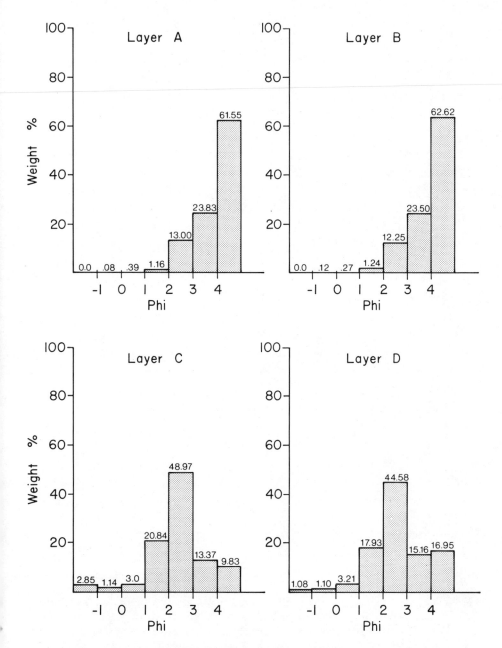

Figure 11. Grain size distribution of wet-sieved sediment from laboratory aquaria containing tubificids. Letters refer to layers in Figure 10. Phi size scale equals $\log_2$ (grain size in mm).

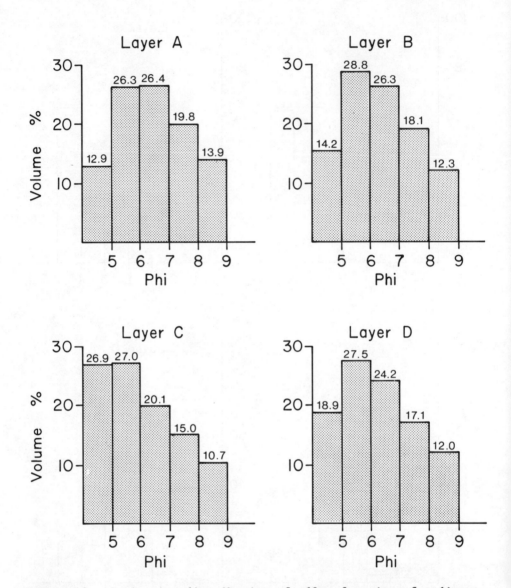

Figure 12. Grain size distribution of <62μm fraction of sediments
            from aquaria containing tubificids determined on the
            Coulter counter. Letters refer to layers in Figure 10.
            Phi size scale equals - log$_2$ (grain size in mm).

size (<8φ) particles than the fecal layer B. Apparently the
oligochaetes feed selectively but not exclusively on the smaller
clay size particles.

The effect of the size-selective feeding is to increase the
downstream transport of fine-grain river sediments. The fecal
layers consist of an open fabric, high water content mixture of
particles whose "ultimate" grain size is small. In general, high
water-content bottoms are more easily resuspended than low water-
content bottoms (Postma, 1967; Rhoads, 1970; Fukuda, 1978). The
silt and clay-size particles are packaged into sand-sized fecal
pellets. But these pellets are loosely bound together and contain
a relatively high concentration of low-density organic matter, so
their densities are undoubtedly lower than equivalent size quartz
sand grains.

EFFECTS OF TUBIFICIDS ON SEDIMENT DIAGENESIS

Tubificids generate numerous burrows in the upper 10 cm of
the sediment. These provide a direct link between two distinctly
different biogeochemical regimes, one oxidized and one reduced.
Various workers have found that tubificids alter vertical profiles
of the master chemical variables, Eh and pH (Schumacher, 1963;
Hargrave, 1972; Davis, 1974b), and affect the flux of materials
(Rhodamine B dye, Hg, P, N) across the SWI (Austin, 1970; Jernelöv,
1970; Byrnes, 1971; Wood, 1975; Davis et al., 1975; Chatarpaul
et al., 1978).

The purpose of the work presented below was to investigate
some of the mechanisms by which tubificids alter sediment chemistry,
and to examine the effects of tubificids on two processes of great
limnological interest, sediment oxygen demand and phosphorous
release. A series of controlled laboratory experiments were
conducted to quantitatively investigate the effects of tubificid
feeding and burrowing activity on: (1) conservative (e.g. non-
reactive) solute flux across the SWI, (2) sediment permeability,
(3) sediment mixing, (4) sediment oxygen demand, and (5) phosphorous release under anoxia.

Tubifex tubifex was chosen as the experimental animal for all
our experiments because this species is widespread and abundant
in Lake Erie (Pliodzinskas, 1978), could readily be obtained in
large numbers from Cleveland Harbor, and can easily be differentiated from other tubificid species found in Cleveland Harbor.
Sediments used in the experiments were collected from Lake Erie
sites 2 and 10 with a Ponar grab sampler, and macrofauna were
removed by passing the bulk material through a 250μm mesh containers.
Just prior to use in the experiments, the sediments were homogenized by stirring.

ALTERATION OF DIFFUSIVE FLUX

     The purpose of this experiment was to determine the effect of
tubificid burrowing and feeding on the flux of a non-reactive
(i.e. conservative) substance across the SWI. Tubificid feeding
creates a surface pelletal layer which appears to be more porous
than unpelletized sediment, and tubificid burrows may act as
conduits for the flux of dissolved materials.

Conservative Solute Flux Experiment

     Sediment from site 2 was mixed with water and labeled with
Cl^- by adding sufficient NaCl to achieve a Cl^- concentration of
$1.41 \times 10^{-2}M$ in the interstitial water. In side experiments, this
Cl^- concentration was found to have no effect on T. $tubifex$
behavior or survivorship. Microcosms used in the experiment were
sections of cellulose acetate butyrate tubing having an internal
cross sectional area of $17.8 \ cm^2$. Enough sediment was added to
achieve a total sediment column thickness of 14 to 17 cm after two
weeks of settling. Water overlying the sediment was aerated and
stirred with a bubbler, and the microcosms were maintained at
$20 \pm 0.5^\circ$ C in a waterbath.

     After two weeks of settling, 178 T. $tubifex$ were added to
four of the microcosms to simulate a population density of 10^5
indiv. m^{-2}. The remaining four microcosms were maintained without
worms. This conditioning period allowed the worms to establish a
pelletal zone of steady-state thickness ($\sim$ 1 cm) at the SWI and a
network of burrows extending to 8 to 10 cm depth. During this
four week preparation period, except for deionized water added to
make up for evaporative loss, all overlying water was derived from
the compacting sediment. Thus, Cl^- concentration gradients were
prevented from forming in the sediment column.

     To initiate the experiment, the overlying water was removed
and replaced by water having a low (8.5×10^{-4} M) Cl^- concentration.
The overlying water Cl^- concentration of each microcosm was monitored regularly for 293 hours. At the termination of the experiment, the sediment columns were sectioned into 1 cm intervals,
and the water content and Cl^- concentration of each section was
determined. A mass balance for Cl^- was calculated for each microcosm, and 99 - 100% of the Cl^- was accounted for. During the
experiment, 20 ml of water was withdrawn from each microcosm.
This constituted a volume reduction due to sampling of $6.3 \pm 0.2\%$.
The measured overlying water concentrations were adjusted for evaporation and initial Cl^- concentration after water exchange. The
concentration each microcosm would achieve at equilibrium (C_{eq})
was calculated, and the adjusted concentrations ($C_w(t)$) were
expressed as fractions of C_{eq} ($C_w(t)/C_{eq}$).

The Time History of Chloride Diffusion

If it is assumed that Cl^- is transported only by diffusion, the appropriate equation describing the time history of Cl^- concentration is:

$$\frac{\partial C}{\partial t} = \partial(D\partial C/\partial Z)/\partial Z \qquad (1)$$

where C is the concentration (mg cm^{-2}) of Cl^- in the interstitial water, D is a molecular diffusion coefficient (cm^2 sec^{-1}), Z is depth in the sediment (cm) measured positively downward from the SWI, and t is time (sec). The boundary conditions at the SWI were: 1) the Cl^- concentration in the overlying water, $C_w(t)$, was equal to the Cl^- concentration of the interstitial water at the SWI, $C(0,t)$, and 2) changes in Cl^- concentration in the overlying water were solely due to cl^- flux from the sediment, i.e.,

$$C_w(t) = C(0,t) \qquad (2)$$

$$h(dC_w/dt) = -\phi(\partial C(0,t)/\partial Z) \qquad (3)$$

where h is the depth of the overlying water and ϕ is the porosity, or the volumetric ratio of water to sediment. For the cases presented here, porosity was approximately 0.84, while h was 13.98 cm when no oligochaetes were present and 13.55 cm when oligochaetes were present.

A solution was first obtained for the case in which no oligochaetes were present. In this case, it was sufficient to assume that D was constant with depth in the sediment. In Figure 13a, the results of the integration of Equation (1) for a value of $D = 5.5 \times 10^{-6}$ cm^2 sec^{-1} are shown and compared with the experimental data. The agreement between theory and observation is excellent.

The results from the case in which $T.$ $tubifex$ were present is shown in Figure 13b. Also shown are the theoretical results where a constant diffusion coefficient of 5.5×10^{-6} cm^2 sec^{-1} has been assumed.

The differences between the theoretical curve and the experimental observations are small but significant. The deviation of the experimental values from the constant diffusion calculation is small at the start of the experiment, increases with time for a short period, and then remains approximately constant as time increases.

The Role of Tubificids

This behavior cannot be explained by assuming an "effective" diffusion coefficient which is greater than the molecular diffusion

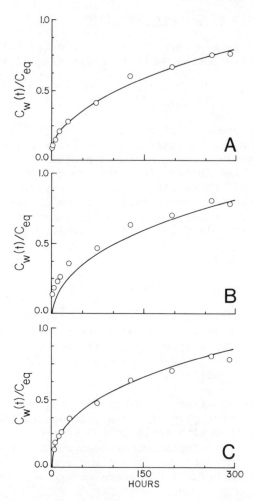

Figure 13.   (A) Non-dimensionalized Cl$^-$ concentration in the
             overlying water versus time. Oligochaetes absent.
             Circles = experimental observations. Solid line =
             model results. (Eq. (1), D = 5.5 x 10^{-6} cm^2/sec).
             (B) Non-dimensionalized Cl$^-$ concentration in the
             overlying water versus time. Oligochaetes present.
             Circles = experimental observations. Solid line =
             model results. (Eq. (1), D = 5.5 x 10^{16} cm/sec).
             (C) Non-dimensionalized Cl$^-$ concentration in the
             overlying water versus time. Oligochaetes present.
             Circles = experimental observations. Solid line =
             model results. (Eq. (1), variable D: D = 11 x 10^{-6}
             cm^2/sec at the sediment-water interface, D = 5.5 x
             10^{-6} cm^2/sec at 0.75 cm). 95% C.I. for observed
             concentrations < diameter of circles.

coefficient in the absence of oligochaetes over the entire tubificid life zone (8 - 10 cm) due to worm burrowing. Such an assumption would cause the predicted results to deviate increasingly with
time (for the time period considered here) from those predicted in
the constant diffusion case. For the same reason, the observed
deviation cannot be due to intake of interstitial water at depth
by the tubificids, and subsequent pumping of this water to the SWI.

Another possibility is that the oligochaete feeding and
defecating activities are continually subducting interstitial
water. It was found that a proper order of magnitude agreement
between the calculated curve and the experimental observations
could be obtained only by assuming that tubificid particle
reworking activities subduct near surface interstitial water at
a velocity of .04 cm/day, or approximately one third the rate of
sediment particle subduction measured in another experiment (see
Mode and Rate of Sediment Mixing, below). In addition, an interstitial water subduction mechanism would cause the calculated curve
to deviate increasingly with time (for the period considered here)
from the constant diffusion case.

A more reasonable explanation for the results of the two
cases is the following: Defecation at the sediment-water interface by the tubificids produces a surficial layer of fecal pellets.
These fecal pellets are relatively large, cylindrical (86μm dia.,
250-300μm long) particles comprised of the fine grain material
which the oligochaetes ingest. As the pellets are subducted by
continued oligochaete feeding activity, pore space between pellets
is decreased due to compaction, and the pellets are degraded to
finer material by mechanical action and bacterial attack. Eventually, they become indistinguishable from the host sediment. The
thickness of the porous fecal pellet layer is approximately 1 cm
in the experiments considered here. Due to the large effective
grain size of the fecal pellets, their loose packing at the SWI,
and stirring action in the overlying water, it is reasonable to
assume diffusion within the pelletal zone is enhanced.

To analyze this case, Equation (1) was used, but it was
assumed that D varied linearly from a value of 11×10^{-6} cm^2 sec^{-1}
at the SWI to a value of 5.5×10^{-6} cm^2 sec^{-1} at a depth of
approximately 1 cm, and was constant with depth thereafter. It
was found that best fit could be obtained if D reached a value of
5.5×10^{-6} cm^2 sec^{-1} at a depth of 0.75 cm. The results of this
calculation are shown and compared with the experimental data in
Figure 13c. It can be seen that there is excellent agreement
between this theory and the experimental observations.

Interstitial water profiles of Cl⁻ concentration at the
termination of the experiment (+293 hours) for both with and without worm cases are shown in Figure 14. The data presented here

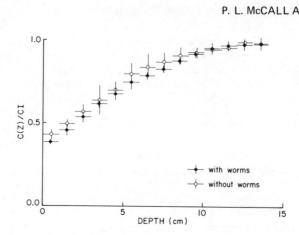

Figure 14. Interstitial water profiles of Cl⁻ concentration after
            295 hours. Circles = oligochaetes absent. Diamonds =
            oligochaetes present. Horizontal lines are 95% C.I.

are the observed interstitial water concentration C(Z), non-
dimensionalized as fractions of the initial concentration $(C(Z)/(C_i)$.
The profile observed for the with worm case is indistinguishable
from that of the without worm case. This result is not surprising
in view of the analysis of the overlying water data, and provides
further evidence that tubificids do not pump a significant amount
of interstitial water.

The Importance of Enhanced Diffusion

Even though the activities of tubificids appear to enhance
diffusion only in the upper 1 cm or so of their substratum, the
biogeochemical consequences of this zone of enhanced diffusion
may be large, since this is the most biologically active zone of
sediments. For example, such enhanced diffusion could greatly
modify sediment oxygen demand, speed the mineralization of organic
matter entering sediments, and facilitate silica release. But it
is equally important to note that such enhanced diffusion will
not significantly affect long term diffusive fluxes of solutes
into or out of sediments. Long term diffusive fluxes will be
controlled by the value of D extant in the bulk of the sediment
column.

The importance of tubificid enhanced diffusion in natural
environments will depend largely upon environmental temperature
and tubificid abundance, since the thickness of the pelletal zone
and the rate of sediment particle convection by tubificids is
strongly dependent upon these two parameters. In Lake Erie, for
example, tubificid enhanced solute movement is likely to be
important in the western basin, river mouths, harbour areas, and
perhaps portion of the central basin.

The diffusion coefficient of 5.5×10^{-6} cm^2 sec^{-1} for the sediment below the pelletal zone is a reasonable value for silty clay material (Manheim, 1970. The enhanced diffusion coefficient of 11×10^{-6} cm^2 sec^{-1} determined for the top of the pelletal zone is probably a reflection of non-molecular processes due to stirring of the overlying water.

EFFECT OF TUBIFICIDS ON PERMEABILITY

We have seen that the presence of tubificid burrows does not appear to have a significant effect on diffusive flux across the SWI. However, their presence may significantly affect the movement of interstitial water and its dissolved constituents if lateral pressure gradients caused by waves or currents are present. Consequently, we conducted some laboratory experiments to determine the effect of oligochaete burrows on sediment permeability by measuring the coefficient of saturated permeability, the rate at which a column of water passes through a thickness of sediment, of sediments with and without tubificids.

Determination of Permeability

Falling head permeameters of the type described by Gillot (1968) were made of 1.7 cm I.D. glass tubes 30 cm in length. In one experiment to test the effect of worm density on permeability, sixteen permeameters were randomly allocated to four treatment groups. Each treatment group of four permeameters had a different simulated tubificid population density (0, 4396, 48,350, and 101,116 indiv. m^{-2}).

Sediment from site 2 was added to the permeameter tubes and allowed to settle to a stable height. The sediment height ranged between 3.6 and 5.3 cm ($\bar{x}$ = 4.71 cm, s = .49 cm). Sediment loss was prevented by inserting a cottong plug at the base of the permeameter. Worms were then added, and except for daily water addition, the permeameters were left undisturbed for two weeks. The permeability of the sediment in each permeameter was then determined by measuring the change in the height of the water overlying the sediment column over a 24 hour period. Thirteen successive 24 hour determinations were made for each permeameter. The initial height of the water column overlying the sediment, H_o, was the same (20 cm) for each permeameter. Head heights were measured to the nearest 0.01 cm using a mechanical depth gauge accurate to .001 cm. The permeability coefficient of the sediment was determined using the relationship

$$k = \frac{L}{t} \ln \frac{H_o}{H_i} \ cm \ sec^{-1}$$

where t = time (in seconds), H_i is the height of the water column
after t seconds, and L is the length of the sediment sample.

In another experiment, 8 permeameters were prepared with
sediment from site 5. The mean height of the sediment columns
was 4.5 cm (s = .24 cm). To four of the permeameters enough worms
were added to simulate densities of 101,000 worms m^{-2}. No worms
were added to the other four permeameters. Permeability coefficients of the two treatment groups were then measured each day for
five days after the addition of the worms.

Enhancement of Permeability by Tubificids

In the experimental permeameters, no significant difference
(p<0.05) in water content was found among the four treatment groups
(tubificid population densities) using one-way analysis of variance.
The mean sediment water content was 73.19% (95% C.I. = ± 0.12%).
This water content corresponds to a porosity of 87.61% (95% C.I. =
± 0.26%).

One-way anova was also used to test for significant differences
in permeability among the four treatment groups. Scheffé type
confidence intervals (α = 0.05)(Scheffe, 1953) were calculated to
allow multiple comparisons among the treatment groups. No significant difference in permeability was found between the control
treatment group and the treatment group having 4,396 indiv. m^{-2} or
between the 48,350 indiv. m^{-2} and the 101,116 indiv. m^{-2} treatment
group. The permeability of sediments containing high population
densities of tubificids was about twice as great as both the
control sediment and sediment with low tubificid population
density (Figure 15). The permeability of sediments from site 5
was increased by a factor of four in three days by the addition of
10^5 worms m^{-2} (Figure 16). The difference in the increase in
the two experiments and the different absolute values of k was
probably due to the different sediments used and to slightly
different preparation techniques.

The increase in permeability of sediments was probably due to
the construction by the worms of burrows that decrease the tortuos-
ity of the sediment. However, until the advection of water through
lake sediments by pressure gradients can be determined in situ,
the importance of this penomenon for increasing sediment-water
chemical exchange will remain unknown.

THE MIXING OF SEDIMENTS BY TUBIFICIDS

Tubificid worms not only alter the way in which dissolved
materials move through near surface sediment, but they also move
particulate matter through different chemical regimes as they
feed on material at depth and deposit it on the sediment surface.

The rate of sediment particle movement and the depth of mixing
under the influence of tubificid feeding was studied by following
the circulation of a marker horizon of cesium-137 labeled illite
clay through the sediment.

Experimental Methods

Each of the sixteen microcosms used in the experiment was a
narrow rectangular polycarbonate container (25.4 cm high x 5.08 cm
wide b 1.27 cm thick), having a cross sectional area of 6.45 cm^2,
and a total contained volume of 164 cm^3. The container walls of
largest dimension were 0.16 cm thick. During the experiment, these
microcosms were maintained at 20° C in a 40 1 polycarbonate
aquarium. Each microcosm was fitted with a bubbler to maintain
rapid water exchange with the main aquarium.

The sixteen microcosms were divided into four equal treatment groups. Each group of four microcosm had a different initial
depth of sediment labeled with cesium-137 (0, 2, 4, 8 cm). Just
prior to the addition of tubificids, each microcosm was scanned
to establish the initial distribution of cesium-137.

The experiment was initiated by adding 65 $T.$ $tubifex$ to each
microcosm. This number of oligochaetes simulated a population
density of 100,000 individuals m^{-2}. The distribution of cesium
-137 activity within each microcosm was followed at regular
intervals for 120 days. A detailed description of the 2 x 2 NaI
crystal scanning system and the method of labeling illitic clay
with cesium-137 is given in Robbins et al., (1979) and Fisher
(1979).

The Mode of Rate of Sediment Mixing

The progressive descent of the cesium-137 activity maximum
from the sediment-water interface of three microcosms is shown
in Figure 17. The activity maximum moved downward at a constant
rate until it reached a depth of approximately 7 cm. Beyond this
point, the rate decreased markedly and approached zero at a depth
of approximately 8.5 cm. Labeled material was continuously
returned to the sediment-water interface from 6 - 8 cm, while
less labeled material was transported to surface sediments from
lesser depths. This pattern, illustrated in Figure 18, indicates
that the bulk of $T.$ $tubifex$ feeding activity in these three
microcosms was confined to an interval between approximately 6
and 9 cm deep in the sediment.

A per individual particle reworking rate was calculated from
the time history plot of the position of the cesium-137 activity
maximum relative to the sediment-water interface during the first
60 days of the experiment (Figure 17). Ten other microcosms begun

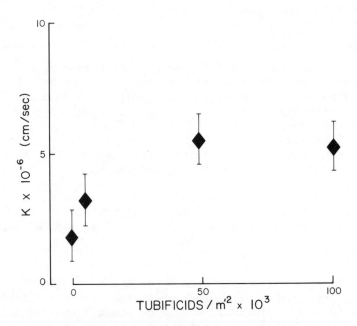

Figure 15.   Coefficient of permeability, k, as a function of
             tubificid population density.   Confidence intervals
             shown are 95% Scheffé type intervals.

at the same time as these showed no redistribution of cesium–137,
and indications of massive mortality, such as the lack of a thick
pelletized sediment layer and the migration of the black FeS zone
to just below the sediment surface, developed prior to the end of
the first thirty days of the experiment.   Since in the remaining
six microcosms reported on here the rates of descent of the
cesium–137 maxima were constant over the first thirty days of the
experiment, it is unlikely that significant mortality occurred
in these microcosms during this period.   Therefore, it was assumed
in this experiment that all tubificid mortality (measured as
initial density minus density after 120 days) occurred early in
the experiment before any substantial redistribution of cesium–137.
The most likely cause of the observed mortality was mechanical
damage and thermal shock sustained by the worms during sorting
and transport to the radioisotope laboratory (Robbins et al.,
1979) where this experiment was conducted.   The calculated per

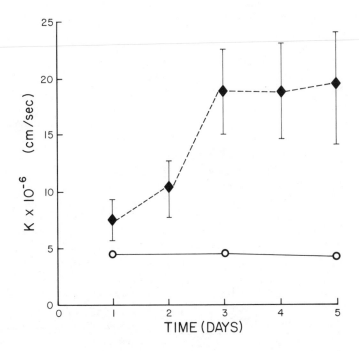

Figure 16. Coefficient of permeability, k, as a function of time. Diamonds represent observations from permeameters with worms; Circles, permeameters with no worms. Confidence intervals for values in aquaria with no worms equal the diameter of the circles.

individual particle reworking rate was $59.7 \pm 10.1 \times 10^{-5}$ cm^3 $hour^{-1}$ at 20° C. With a population density of 100,000 individuals $/m^2$, the downward velocity of the marker horizon due to tubificid feeding would be $0.12 \pm .02$ cm day^{-1}.

The distribution of cesium-137 activity in microcosm C at successive time intervals is shown in Figure 19. Initially, all of the cesium-137 activity was near the SWI. The small downward displacement of the peak was due to the presence of an overlying white Kaolin clay layer which served as a visual marker horizon. The broadness of the peak is due to detector optics (see Robbins

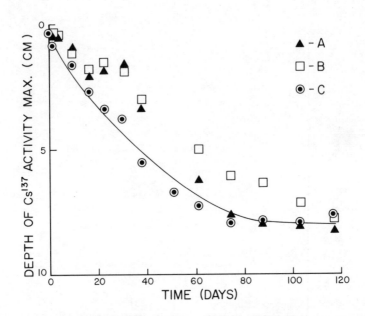

Figure 17.   Position of cesium-137 activity maximum in microcosms
A, B, and C.   Solid line shows the calculated position
of the cesium-137 activity maximum for microcosm C
using the mixing model of Fisher (1979).

et al., 1979).   Within a few minutes after being introduced into
the microcosm, the tubificids penetrated the marker horizon, and
some individuals burrowed to the base of the sediment column
within a few hours.   Even so, after 4 days, no downward smearing
of the cesium-137 label was detected.   After 38 days of exposure
to tubificid reworking activity, the cesium-137 activity maximum
had descended to 4.5 cm.   It is not clear, however, whether or
not any cesium-137 labeled material had been returned to the
surface due to the tail of cesium-137 activity extending upward
from the cesium-137 activity maximum.   Although most of the labeled
material descended as a discrete unit, heterogeneities in the worm
induced subduction caused some of the material to descend at a
slower velocity.   Results from microcosms labeled at 2 and 4 cm
(see Figure 18) showed that a small amount of material was transferred from these depths to the sediment-water interface.   No
substantial movement of labeled material ahead of the activity

maximum was observed. At the termination of the experiment (120 days), the activity maximum had moved to a depth of 7.7 cm, and some transfer of labeled material to surface sediments was indicated by regions of excess activity superimposed on the upward extending cesium-137 activity tail. By this time, a small amount of label was smeared downward from the activity maximum. This could be due to a drop off in numbers of tubificids feeding at depths greater than 7.7 cm or to movement of labeled material due to sediment slumping down tubificid burrows. Sediment reworking by tubificids in this experiment was successfully analyzed mathematically by treating the sediment as a continuum and including terms describing both the convection and "diffusion" of sediment particles. Details of the tubificid reworking model can be found in Fisher (1979).

The Depth of Mixing

The depth distribution of tubificid feeding and burrowing determines the depth of biogenic sediment mixing. Numerous field and laboratory investigations have been conducted to determine the depth distribution of these activities (Cole, 1953; Poddubnaya and Sorokin, 1961; Poddubnaya, 1961; Brinkhurst and Kennedy, 1965; Jernelöv, 1970; Stockner and Lund, 1970; Milbrink, 1973; Davis, 1974a,b; Fisher and Beeton, 1975; Wood, 1975; Kirchner, 1975; Krezoski, 1976). Most workers have inferred that the depth of maximum tubificid feeding activity corresponds with the depth of peak tubificid abundance. The results obtained by Appleby and Brinkhurst (1970), however, suggest that it would be more reasonable to conclude that the depth of peak feeding should correspond to the depth of peak feeding biomass.

The depth distribution of tubificid feeding activity can be derived from the depth distribution of tubificid abundance assuming 1) the depth distribution of tubificid abundance is the expression of the length distribution of the tubificid population, 2) a direct proportionality exists between worm length and biomass, and 3) a direct proportionality exists between biomass and feeding rate. Using data compiled from available literature by Davis (1974b) on the depth distribution of tubificid abundance, the distribution of tubificid feeding activity was calculated (Figure 20). Although 70% of the tubificid population is confined to the upper 3 cm of deposit, the calculated zone of maximum feeding occurs somewhere between 3 and 6 cm.

The depth of maximum feeding may be expected to vary with the distribution of length of the tubificids examined (Poddubnaya and Sorokin, 1961), the species composition of the population (Milbrink, 1973), and method of activity determination. Sorokin (1966) determined the depth distribution of feeding activity using ^{14}C labeled horizons of bacteria and algae. However, since the

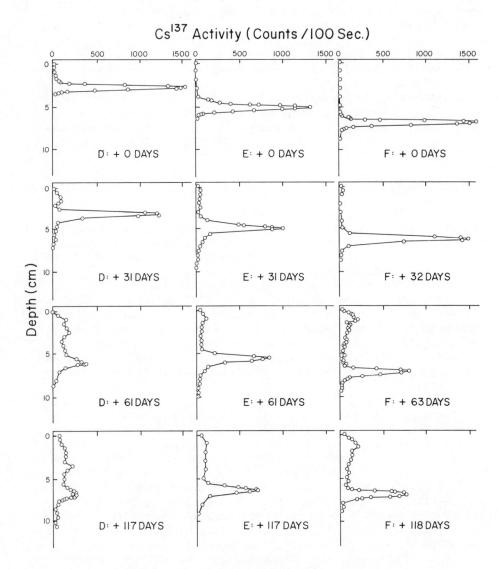

Figure 18.   Cesium–137 activity profiles at selected times in
             microcosms D, E, and F.   Initial depth of cesium–137
             label = 2, 4, and 8 cm respectively.

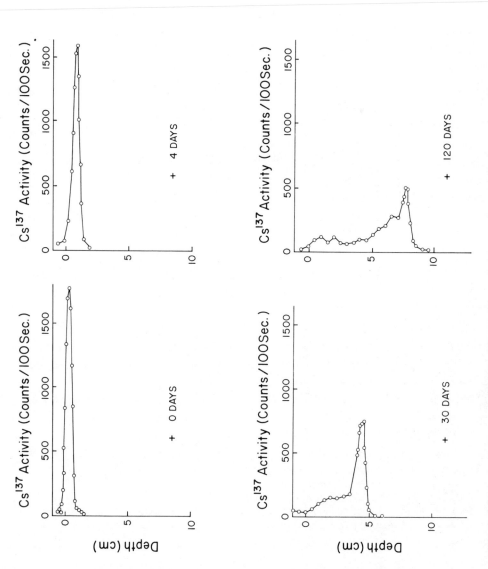

Figure 19. Cesium–137 activity profiles at selected times in microcosm C.

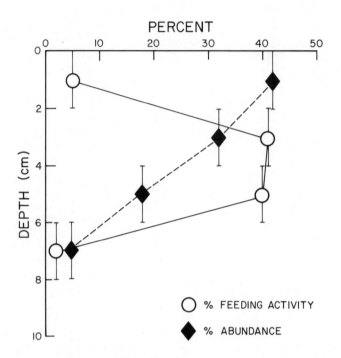

Figure 20. Tubificid abundance and feeding activity (both percent)
            as a function of depth below the SWI.

experiment entailed determining the ^{14}C activity of the tubificids
after a short exposure (3 days) to the labeled material, it is
more likely that the depth distribution of tubificid growth and
assimilation was determined. While a reliable sediment particle
tracer was used in our experiment, the results also are biased
by the use of mostly adult worms (fully extended lenths 3 to 8 cm)
belonging to a single species.

The depth of maximum feeding (and, therefore, the mixing
depth) appears also to be dependent on worm density. In experiments
performed by us, the deepest maxima were associated with the highest
tubificid densities. Microcosm C had an initial population density
of 100,000 m^{-2} and experienced 35% initial mortality; its feeding
maximum was at 7.7 cm. Microcosm D had the same initial density

and experienced 52% initial mortality; its feeding maximum was at
5.5 cm.   Robbins et al., (1979) began other mixing experiments
with $T.$ $tubifex$ densities of 50,000 m^{-2}.   While the depth of
maximum feeding at this density was not determined, a massive flux
of radioactive label was measured from a depth 3 cm, shallower than
any noted in subsequent experiments at higher initial densities.
We think that this density dependence of feeding depth is related
to the food supply of tubificids.   Tubificids feed at depth on
bacteria attached to sediment grains and deposit fecal pellets of
low food content on the sediment surface (Brinkhurst et al., 1972;
Wavre and Brinkhurst, 1971).   As these pellets fall below the
sediment surface under the influence of further feeding activity,
they will be recolonized by bacterial food species.   Sediment
circulates faster when driven by more worms and perhaps must
descend deeper below the sediment surface before food is replenished.

The maximum feeding depth in the egestion rate experiment
described earlier was fixed at 4 cm by the experimental apparatus.
The significant two and three way interactions in the egestion rate
experiment (Table 4) support the idea that resource limitation
occurred at high densities (100,000 m^{-2}) and high temperatures
(22° C) in 'low food' substrata.   Per individual feeding rate
slowed under these conditions, perhaps because of interference
competition, longer search time for suitable food, or a slow down
in the passage of food through the gut to increase digestion
efficiency.   The results of Brinkhursts et al.   (1972) also support
our idea of resource limitation.   Holding total density of worms
constant, Brinkhurst et al. (1972) found that $T.$ $tubifex$ exhibited
better growth in multispecific cultures (2 or 3 species) than in
monospecific cultures.   If we make the reasonable assumption that
intraspecific competition for food is more severe than interspecific competition, then $T.$ $tubifex$ were feeding and growing at
effectively higher densities in pure than in mixed cultures.

Although tubificids have been observed at considerable depth
in sediments (Davis, 1974b), there is a consensus that feeding by
most tubificids occurs only to a depth of about 10 cm (Sorokin,
1966; Davis, 1974b; Robbins et al., 1979).   At this depth, tubificids can still respire in the aerobic zone of the sediments and
overlying water while they feed.   Excursions beyond the maximum
depth of feeding should not greatly contribute to the redistribution of sediment particles.   If tubificids match their feeding
depth to available food resources, then within this 10 cm zone the
actual mixing depth will be a complex and varying function of
biotic (worm length, population density, and bacterial regeneration
rate) and abiotic (temperature, sediment grain size, and sediment
chemistry) factors.   Our experiments, for instance, were conducted
with population densities (100,000 m^{-2}) at the high end of the
range of densities found in Lake Erie, and our observed mixing
depths were somewhat deeper than those calculated or observed for

most field populations. Recent investigations of sedimentation in
the Great Lakes and other bodies of water using both natural (lead-
210) and anthropogenic (cesium-137) radionuclides have found a
homogenous zone of varying thickness (2-6 cm) to occur in the
presence of macrobenthos (Robbins and Edgington, 1975, 1976).
Additionally, for two locations in Lake Huron, it has been shown
that the depth of the homogenized zone corresponds well with the
depth distribution of tubificids and that the macrobenthic populations present at these two locations are capable of producing
the homogenous zone (Krezoski, 1976; Robbins et al., in press).

Sediment Mixing in Lake Erie

     In order to evaluate the relative importance of sediment
mixing by tubificids in various parts of Lake Erie, the results of
the feeding rate experiment were combined with available literature
data on Lake Erie tubificid abundance, water temperature, and
dissolved oxygen (Veal and Osmond, 1968; Burns and Ross, 1972;
Burns, 1976; Pliodzinskas, 1978) to calculate approximately the
annual amount of sediment that tubificids process. While there
exist statistically significant substratum and population density
interaction effects on per individual feeding rates, these are
small relative to the effects of temperature and oxygen (Table 4)
and may be safely ignored in this approximate calculation. Additionally, available literature information on tubificid abundance
in Lake Erie (Veal and Osmond, 1968; Pliodzinskas, 1978) probably
underestimates the actual population density of tubificids, since
both studies used relatively large Ponar grabs (.05 m^2) and large
mesh sieves (650μm and 420μm, respectively). Our recent sampling
of several stations of Pliodzinskas (1978) using a diver operated
multiple coring device and a 250μm mesh sieve revealed resident
tubificid populations that were consistently an order of magnitude
greater than those reported by Pliodzinskas (1978). Therefore,
mixing rates calculated from his abundance data were multiplied
by ten. The calculated particle mixing rates were expressed as
the mean downward velocity of the fecal layer in units of cm day
$^{-1}(\bar{V}_t)$. The calculated tubificid particle mixing rates ($\bar{V}_t$) were
compared with sedimentation rates ($\bar{V}_s$) determined by Kemp et al.
(1977). The calculated values of $\bar{V}_t$ are in excess of sedimentation
in the central and western basins and at several stations in the
eastern basin (Figure 21). This indicates that even with all the
errors inherent in the extrapolation, it is likely that the particle reworking activities of tubificids alone result in significant
sediment mixing to a depth of several centimeters throughout much
of Lake Erie.

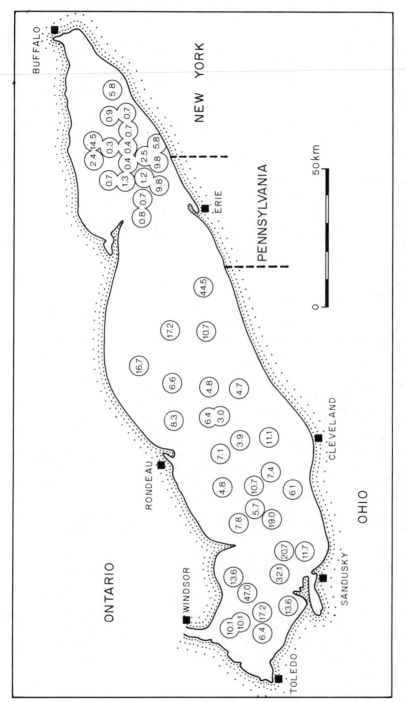

Figure 21. The ratio of tubificid induced subduction velocities ($\bar{V}_s$) to the mean sedimentation velocities ($\bar{V}_t$) ($\bar{V}_t\bar{V}_s$) in Lake Erie.

THE CHEMICAL EFFECTS OF SEDIMENT MIXING: OXYGEN DEMAND AND
PHOSPHOROUS RELEASE

The muddy sediment of lakes are chemically layered. The
surface sediments in aerobic waters receive a supply of oxygen
from overlying water and are commonly oxidized. In shallow lakes,
the sediments are an important site of detrital decomposition and
consequently, of oxygen consumption. At some depth in the sediment,
oxygen consumption exceeds oxygen supply, the sediments become
anaerobic, and oxidized compounds are microbially reduced. Important chemical species - toxic trace metals and inorganic phosphate,
a major algal nutrient, for instance - are redox sensitive, and are
adsorbed onto particles in the oxidized sediment layer and dissolved
in the pore water in the reduced layer. Both the mixing of sediments and the transfer of reduced sediments to the oxidized zone
have important chemical effects. We will illustrate this with the
results of laboratory experiments on the effects of tubificids on
sediment oxygen demand (SOD) and on phosphorus release from the
sediments.

The Sediment Oxygen Demand Experiment

In highly productive lakes, the sediment oxygen demand (SOD)
engendered by the decay of sedimenting organic matter may be
sufficient to cause hypolimnetic anoxia. In the central basin of
Lake Erie, SOD accounts for 81% of the oxygen consumed (Burns and
Ross, 1972). As a result of this demand, the hypolimnion of the
central basin is subject to complete oxygen depletion during late
summer (Dobson and Gilbertson, 1972).

SOD is a combination of the oxygen demands for respiration
by the resident benthic community and for the oxidation of reduced
chemical species reaching the oxidized zone. Macrobenthos can
alter surficial sediment properties in such a way that more oxygen
is consumed by the deposit than can be accounted for by organismal
respiration and molecular diffusion of reduced substances alone.
Edwards (1958) and Neame (1975) studied sediments inhabited by
insect larvae and concluded that their sediment stirring and water
pumping activity could account for this discrepancy, and that molecular diffusion alone could not account for SOD when insect larvae
were reasonably abundant.

Tubificid oligochaetes feed within the anaerobic zone of the
sediment and egest fecal pellets which contain oxygen demanding
materials, FeS for example, at the SWI. Concomitantly, oxidized
material at the SWI is subducted to depth in the sediment where
it is reduced.

We set up some laboratory respirometers containing sediment
and tubificids to examine the effect of feeding on SOD. Oxygen

consumption determinations were made for the following cases: 1)
kaolin sediment alone, 2) kaolin sediment + *T. tubifex,* 3) lake
sediment alone, and 4) lake sediment + *T. tubifex.* Oxygen consumption measured in each of these cases was interpreted as shown in
Table 8.

Table 8.   Interpretation of oxygen consumption measurements.

CASE	CONDITION	OXYGEN CONSUMPTION COMPONENTS
1	Kaolin sediment	S
2	Kaolin sediment + *T. tubifex*	S + T
3	lake sediment	S + M + C
4	lake sediment + *T. tubifex*	S + M + C + T

S = measurements system oxygen demand
T = *T. tubifex* oxygen demand
M = microbial oxygen demand
C = chemical oxygen demand

The respirometers used in the experiment were constructed
of polycarbonate cylinders 45 cm in height having an internal
diameter of 7.62 cm, and a wall thickness of 0.635 cm. The bottom
of each respirometer was permanently sealed with a polycarbonate
base. During experimental runs, the respirometers were sealed
with a gas tight silicon rubber gasketed cap through which oxygen
(4S1 Model 5739) and temperature (4S1 Type 401) probes were fitted.
Water within the respirometers was constantly stirred using a
slide mounted magnetic stirrer. Respirometer temperature was
maintained at $20^{\circ} \pm 0.5^{\circ}$ C by immersion in a temperature controlled
water bath. The oxygen probe used to measure consumption was air
calibrated before and after each experimental run. At no time was
calibration drift greater than $6.24 \mu M$ O_2 1^{-1} $(0.2$ mg $1^{-1})$.

An attempt was made to partition the chemical and microbial
components of oxygen consumption by adding ·a mixture of erythro-
mycin (200 mg 1^{-1}) and minocycline (10 mg 1^{-1}) to the respirometers. Oxygen consumption measured after antibiotic treatment

was always greater than before treatment and varied erratically.
This phenomenon has been reported elsewhere (Viner, 1975), and may
be due to the growth of fungi after the removal of bacteria (Bradshaw, 1961). Treatment with other antimicrobial agents was not
attempted.

For each of the two sediments used, four respirometers were
prepared. A thick suspension of sediment was added and permitted
to compact. The resulting sediment column was adjusted to a
stable height of 20 cm. An air bubbler was inserted in each
respirometer, and stirring was initiated. The respirometers were
left undisturbed for fourteen days at 20° C. At the end of this
time, the experiments were begun by sealing each respirometer in
turn, and monitoring the decrease in dissolved oxygen until a
concentration of $\leq 30 \mu M$ O_2 1^{-1} was observed. Two oxygen consumption
determinations were made for each respirometer. After these
measurements were completed, 456 *T. tubifex* were added to each
respirometer. This simulated a resident tubificid population
density of 100,000 individuals m^{-2}. The tubificids were thoroughly
cleaned of adhering detritus and allowed to void their gut contents
prior to introduction. The worms were permitted fourteen days to
establish burrows within the sediment and a steady-state thickness
pelletal zone at the sediment-water interface. Then, a further
series of two oxygen consumption determinations were made for each
respirometer. Observed tubificid mortality was 3 - 5% in the
kaolin sediment and 2 - 4% in the lake sediment.

The respirometers were cleaned with detergent and 15% H_2O_2
and thoroughly rinsed prior to sediment addition. Except for
brief periods, all respirometers were kept in darkness after
sediment addition, and all oxygen consumption determinations were
made in darkness.

Enhancement of SOD by Tubificids

Results obtained in the experiments are shown in Figure 22.
The curves describing O_2 concentration in the overlying water
through time were approximately linear to the first 20 hours,
therefore, the decrease in oxygen concentration during this period
was approximated at a zero order reaction, i.e.

$$dC/dt = -k_o ;$$

integrating,

$$C_t = C_o - k_o t,$$

where C_t = concentration at time t, C_o = concentration at time
t = 0, and k_o in the zero order rate constant. Regression lines
were fitted to these data, and the calculated slopes were

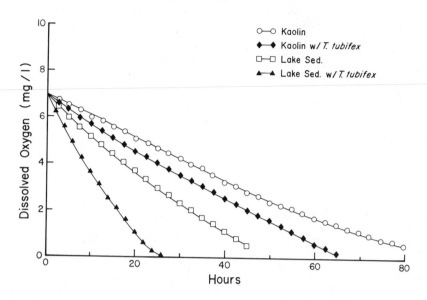

Figure 22. Dissolved oxygen concentrations in the respirometers versus time for the cases examined.

interpreted as the respective $-k_o$'s. The results of the regression analysis are given in Figure 23. As can be readily observed in Figure 22, oxygen loss from the overlying water observed in the experiments does not follow a zero order rate equation. The use of a zero order reaction approximation assumes that all oxygen is consumed at the sediment-water interface before it can diffuse into the sediments. Further, this approximation assumes that oxygen loss is not concentration dependent. These assumptions are not strictly correct. However, the first 20 hours of data are reasonably linear, and the implicit or explicit assumption of zero order kinetics for sediment oxygen consumption has been made in a number of studies (e.g., Hayes and MacAuly, 1959; Carey, 1967; Lucas and Thomas, 1972; Wyeth, 1977). Additionally, oxygen consumption by both microflora and *T. tubifex* is known to be independent of oxygen concentration while oxygen is present in excess of about 30µM O_2 1^{-1} (Zobell and Stadler, 1940; Palmer, 1968).

The oxygen consumption of the measurement system (blank effect) was subtracted from cases 2, 3 and 4 (Table 8). The blank adjusted results of the experiments are given in Table 9. If no interaction between sediments and tubificids occurred to modify sediment oxygen consumption, the oxygen consumption of the tubificid inhabited lake sediment should simply be the sum of tubificid respiration and the oxygen demand of the lake sediment without tubificids. Clearly, this was not the case. The oxygen consumption of tubificid-inhabited sediment was approximately a factor of two greater than the simple sum of tubificid respiration and the oxygen consumption of lake sediment without a tubificid population.

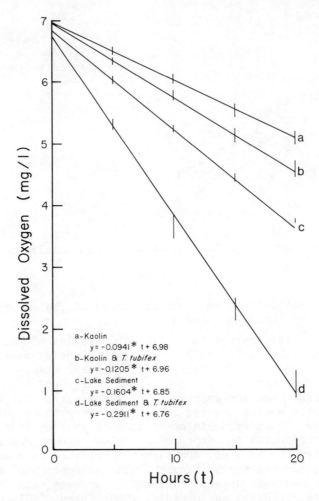

Figure 23.   Dissolved oxygen decrease for the first 20 hours for
the cases examined and linear regression models
for the four cases.

As stated earlier, attempts to partition chemical from microbial oxygen consumption by destroying the aerobic bacteria were
unsuccessful. In order to gain some understanding of the relative
contribution of direct chemical and microbial oxygen demand to
the total oxygen demand of tubificid inhabited sediments, a calculation of the oxygen demand expressed by iron sulfide fluxed to
the sediment-water interface was made. The FeS concentration
profile of a sediment column prepared in the same manner as those
used in the respirometers was determined by measuring acid vola-

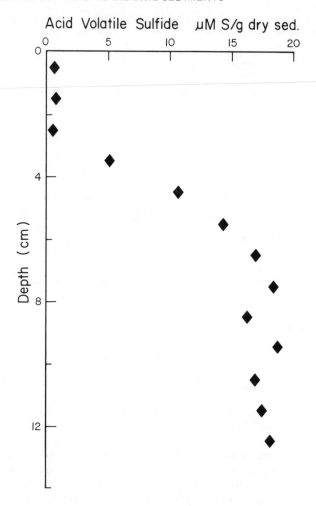

Figure 24. Depth profile of acid volatile sulfide in a sediment
            column prepared in the same way as those used in
            the respirometers.

tile sulfide at 1 cm intervals using the method of Aller (1977).
The results of these measurements are given in Figure 24. Within
the zone of maximum tubificid feeding (6 - 9 cm), acid volatile
sulfide was present at a concentration of 15 to 20μM FeS g^{-1} dry
sediment. Assuming a sediment porosity of ~90%, and using our
tubificid particle transport rate data, a FeS flux of 19 - 26μM
day^{-1} was calculated for the daily flux of sediment from the
depth to the SWI.

Table 9. Blank Adjusted Oxygen Consumption Results.

CASE	OXYGEN CONSUMPTION COMPONENTS	k_o (μ moles O_2/m^2)
2	T	$0.90 \pm .02$
3	M + C	$2.26 \pm .03$
4	T + M + C	$6.72 \pm .03$

T = *T. tubifex* oxygen demand
M = Microbial oxygen demand
C = Chemical oxygen demand

Doyle (1967) showed that FeS oxidizes via a two step reaction

$$FeS + 3/4\ O_2 + 1/2\ H_2O = FeOOH + S^o$$

$$S^o + 3/2\ O_2 + H_2O = H_2SO_4$$

Consequently, for every mole of FeS oxidized, 2.25 moles of O_2 are consumed. The complete oxidation of FeS fluxed to the sediment-water interface in these experiments by tubificid particle reworking activities would account for ~50 to 70% of the observed enhanced oxygen demand. Since the second reaction requires the action of thiobacillaceae (Doyle, 1967; Kuznetsov, 1970), this demand is not purely chemical. Admittedly, this is a crude calculation. However, it does show that the flux of reduced material from depth to the sediment-water interface caused by tubificid particle reworking activities could account for a substantial portion of the enhanced oxygen demand observed. The cause of the remaining enhanced oxygen demand observed for tubificid inhabited lake sediment must be related to some aspect of tubificid interaction with sediments, but its source is not obvious. Enhanced diffusion in the upper 1 cm of the sediment may play a role in increasing SOD. Increased diffusivity increases the depth to which oxygen is present (from 3-4 mm to 10-15 mm below the SWI) and expands the habitat available to aerotic bacteria. If the biomass of aerobic bacteria were increased, SOD would be further enhanced. In fact, there is an embarrassment of choices to account for increased sediment oxygen demand when worms are present, and there is an obvious need for more research to determine the exact partitioning of this increased demand.

It is interesting to note here that the development of an anoxic hypolimnion may be a self-accelerating process not only through the feedback mechanism of cyclic self fertilization (Mortimer, 1941; 1942; 1971), but also because of the changes in benthic community structure which accompany reduced levels of dissolved oxygen (e.g. Jonasson, 1969; Brinkhurst, 1974). The macróbenthic organisms best adapted to low oxygen tensions, such as certain tubificid and chironomic species, appear to enhance SOD (Edwards, 1958; Edwards and Rolley, 1965; Neame, 1975).

The Phosphorous Release Experiment

Of the many profound changes produced by hypolimnetic anoxia, one of singular interest is the release of algal nutrients such as phosphorous from their sediment store (Mortimer, 1941, 1942, 1971). The effects of tubificids on the release of phosphorous from sediments during periods of overlying water anoxia were also studied in laboratory microcosms.

Each of the twenty microcosms used in the experiment was comprised of an acrylic tube 7.6 cm in diameter and 50.8 cm long containing homogenized lake sediment from site 10 and an overlying water reservoir (5 liters) corresponding to a 1 meter high column of overlying water. The overlying water was constantly stirred, and the microcosms were maintained at $16 \pm 1^{\circ}$ C in a large water bath.

Tubificids (456 T. *tubifex*/tube) were added to ten of the microcosms to simulate a population density of 100,000 individuals m^{-2}. The remaining ten microcosms were maintained without worms. The tubificids were allowed to rework the sediments for thirty days before all microcosms were sealed and purged with helium. The concentration of chloride, ammonia, bicarbonate, silicate, phosphate, calcium ferrous iron, manganese, and hydrogen ion in the overlying water of each microcosm was monitored regularly. At intervals, two microcosms (one with worms and one without) were sacrified to determine pore water gradients of dissolved materials.

Inhibition of Inorganic Phosphorous Release by Tubificids

The flux of phosphorous observed in one microcosm with worms and one without under anoxia is shown in Figure 25. The pattern· and extent of phosphorous release for bioturbated sediments is strikingly different from that of non-bioturbated sediments. In the non-bioturbated sediment an initial pulse of phosphorous is observed after anoxia. After this initial pulse, phosphorous flux decreases, becomes negative, goes positive then decreases again. In the bioturbated sediment no initial large pulse of phosphorous is observed. Again, the initial positive flux of phosphorous is

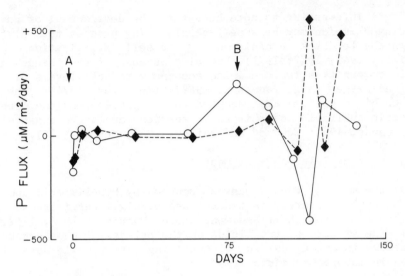

Figure 25.   Phosphorous flux observed in bioturbated (diamonds) and
              non-bioturbated (circles) sediments.   (A) Overlying
              water reservoirs filled.   (B) Both microcosms anoxic.

followed by a negative flux period, and phosphorous flux oscillates
about zero.

     A tentative explanation of the differences in the initial
phosphorous flux between bioturbated and non-bioturbated sediment
is that in non-bioturbated sediment a ferric hydroxide rich surface
layer forms due to upward diffusion of Fe^{+2} and its oxidation at
the interface. This iron hydroxide rich layer is the site for the
precipitation of upward diffusing phosphate. Phosphorous entering
this surface layer forms a ferric hydroxide-orthophosphate complex
$(Fe_x(OH)_{3(x-y)}(PO_4)_{(y)} \cdot 2\ H_2))$ (Stam and Kohlschutter, 1965;
Williams et al., 1976a,b). Phosphorous thus bound will be readily
available for release to the overlying water when the ferric hydroxide dissolves after anoxia. In sediments bioturbated by tubificids
such an iron rich surface layer is prevented from forming because
the worms constantly subduct material from the SWI. Thus the
actions of the tubificids during oxic conditions prevent a high
initial flux of phosphorous immediately after anoxia (later changes
in flux are more complicated and are probably not related to worm
activity). Further, examination of the interstitial water profiles
of phosphorous concentration observed in the two microcosms at the
termination of the experiment (Figure 26) shows that the worms
prevented a strong phosphorous concentration gradient from forming
at the sediment-water interface. Indeed, quite a long period of
worm inactivity would be required before enough phosphorous could
diffuse to establish a phosphorous gradient in the bioturbated

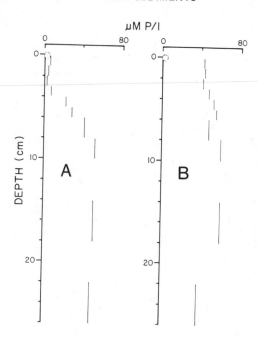

Figure 26. Interstitial water profiles of phosphorous concentration observed in bioturbated (A) and non-bioturbated (B) sediment.

sediment equivalent to that in the non-bioturbated sediment. Thus, the actions of tubificids during oxic periods prevent an initial large pulse of phosphorous from entering the overlying water after anoxia and decrease diffusive flux of phosphorous across the SWI for a reasonably long time after the anoxia.

BIOGENIC MIXING AND SEDIMENT DIAGENESIS:
A CONCEPTUAL MODEL

Sediment particle subduction and reflux from depth of tubificids coupled with solution, precipitation, and absorption reactions of materials associated with sediment particles may determine whether a material is stripped from or retained by sediments and/or rate at which material stripping or accumulation occurs.

Schematic outlines of possible material pathways in sediments containing tubificids are given in Figure 27. In Case I, "chemically indifferent" materials (e.g. detrital quartz, pollen grains, etc). fall onto the surface. This material is subducted, and, if ingested by tubificids, is conveyed to the SWI. Alternatively, the material is rejected by the tubificids. Although tubificids

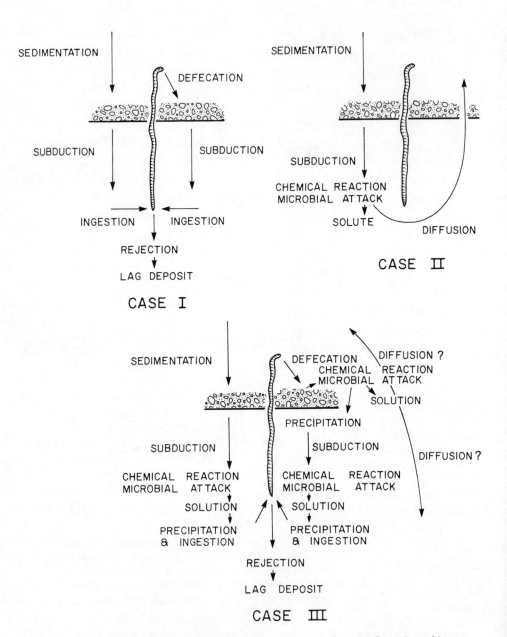

Figure 27.   Conceptual model of the effect of tubificid sediment mixing on diagenesis.

are capable of ingesting particles of relatively large size (Figure 11), they select silt size and smaller particles ($\leq$62.5µm) (Figure 10, 11 and Davis, 1974a). In addition, particle selection on some basis other than size may take place. In Case II, a material associated with sediment particles is subducted and released to solution by chemical or biochemical action at some depth in the sediment. The material undergoes no further transformation and diffuses from the deposit (e.g. ammonia). If Case II is modified so that the solution species released at depth interacts with other materials present to form a solid phase (Case III), the precipitated material can be ingested and conveyed to the surface. If the solid phase is stable under oxidizing conditions, it will behave as in Case I. If the material is released to solution under oxidizing conditions, however, the resulting solute can diffuse into the sediment or overlying water, or become associated with sediment particles (e.g. phosphorous in association with ferric hydroxide). In the latter case, the material is subducted again. If small precipitates act as nucleation sites, continuation of this cycle may result in the production of large grain precipitates that will not be conveyed to the SWI, but rather lagged at the base of the tubificid feeding zone.

The latter process may be important, for instance, in increasing the transfer of phosphorous from temporary storage at the SWI to permanent storage as vivianite at depth. In Lake Erie, the dominant non-detrital form of phosphorous present is vivianite (Dell, 1973; Nriagu and Dell, 1975; Williams, 1976a; Williams et al., 1976b). In interesting accordance with this conceptual model, Nragu and Dell (1975) in their study of authigenic vivianite in Lake Erie found that vivianite nodules present averaged ~60µm in diameter. This "worm reflux" mechanism for the movement of phosphorous from the SWI to depth should be more rapid than phosphorous removal due to sedimentation alone, provided that the downward velocity of the sediment surface engendered by tubificid particle reworking activities is greater than its upward velocity due to sedimentation. In all of the western basin, most of the central basin, and parts of the eastern basin of Lake Erie, this condition is met.

CONCLUSION

It should perhaps not be surprising that tubificid oligochaetes have a marked effect on physical and chemical properties of lake sediment - they are so intimately associated with the sediments by virtue of being infaunal burrowing deposit feeders. Moreover, their feeding and egestion activities connect two otherwise quite separate biochemical regions, the oxidized and reduced layers of the sediment. Indeed, virtually all sediment properties and processes examined - sediment grain size, settling velocity, erodibility, porosity, permeability, vertical gradients of particulate

and dissolved sedimentary materials, diffusive flux across the SWI
(chloride, phosphorous), and SOD - may at times be altered by
factors of at least two and sometimes ten or more by the presence
of tubificids.

While the laboratory demonstration of these effects is certainly instructive, the results can be used to infer in situ
effects only with caution. We have succumbed here to a few such
inferences about effects in the field, but we regard these extra-
polations as only approximate. The effects of tubificids on sediment properties depend very much on the substratum they are in and
on a variety of hydrographic conditions. The effects of tubificids
may be reinforced or counteracted by other organisms. Tubificids
are the most abundant macrofauna in Lake Erie, but the role of
other important organisms cannot be neglected. Chironomids are
second in abundance, and they are burrowing filter feeders and
surface deposit feeders. Filter-feeding unionid clams are the
largest invertebrates in the lake and account for most of the
secondary production in the western basin. They plow the top ten
centimeters of sediment as they burrow, and they pump large quantities of water through the sediment (McCall et al., in press).
Sediment bacteria probably exert an important effect on the distribution of macrobenthos. In addition, macrobial exudates bind
sediments and make them more difficult to erode; the burrowing
activities of other organisms have the opposite effect. The
effects of tubificids and other organisms are large, but the net
biogenic alteration of sediment properties is still largely
undetermined.

REFERENCES

Aller, R., 1977, The influence of deposit feeding benthos on
    chemical diagenesis of marine sediments, Ph.D. Thesis, Yale
    University. 600 p.
Appleby, A.D., and R.O. Brinkhurst, 1970, Defecation rate of
    three tubificid oligochaetes found in the sediment of Toronto
    Harbour, Ontario, J. Fish. Res. Bd. Can., 27:1971.
Austin, E.R., 1970, The relese of nitrogenous compounds from lake
    sediments, M.S. Thesis, Univ. of Wisconsin. 78 p.
Belz, M., 1973, "Statistical Methods in the Process Industries,"
    J. Wiley, New York. 706 p.
Bradshaw, J.S., 1961, Laboratory experiments on the ecology of
    foraminifera, Contrib. Cushman Found. Foraminiferal Res.,
    12:87-105.
Brinkhurst, R.O., 1974, "The Benthos of Lakes," St. Martin's
    Press, New York. 190 p.
Brinkhurst, R.O., and C.R. Kennedy, 1965, Studies on the biology
    of the Tubificidae (Annelida, Oligochaeta) in a polluted
    stream. J. Anim. Ecol., 34:429-443.

Brinkhurst, R.O., K.E. Chua, and N.K. Kaushik, 1972, Interspecific
    interactions and selective feeding of tubificid oligochaetes,
    Limnol. Oceanogr., 17:122-123.
Burns, N.M., 1976, Temperature, oxygen, and nutrient distribution pat-
    terns in Lake Erie, 1970, J. Fish Res. Bd. Can., 33:485-511.
Burns, N.M., and C. Ross., p972, Oxygen - nutrient relationships
    in the Central Basin of Lake Erie, In: "Project Hypo," N.M.
    Burns and C. Ross, eds., Can. Cent. Inland Waters. Pap.6, 180 pp.
Byrnes, B.H., 1971, Ammonium-N released from lake sediments to
    water, M.S. Thesis, University of Wisconsin. 93 p.
Carey, A.G., 1967, Energetics of the benthos of Long Island Sound.
    I. oxygen utilization by sediment. Bull. Bingham Oceanogr.
    Coll., 19:136-144.
Chatarpaul, L., J.B. Robinson, and N.K. Kaushik, 1978, Role of
    oligochaete worms in nitrogen transformations in stream sedi-
    ments, Abstracts, 26th Annual Meeting, N. Am. Benthological
    Soc., Winnipeg, Manitoba.
Cole, G.A., 1953, Notes on the vertical distribution of organisms
    in the profundal sediment of Douglas Lake, Michigan, Am. Midl.
    Nat., 49:252-256.
Cullen, D.J., 1973, Bioturbation of superficial marine sediments
    by interstitial meiobenthos, Nature, 242:323-324.
Davis, R.B., 1974a, Stratigraphic effects of tubificids in pro-
    fundal lake sediments, Limnol. Oceanogr., 19:466-488.
Davis, R.B., 1974b, Tubificids alter profiles of redox potential and
    pH in profundal lake sediment, Limnol. Oceanogr., 19:342-346.
Davis, R.B., D.L. Thurlow, and R.E. Brewster, 1975, Effects of
    burrowing tubificids on the exchange of phosphorous between
    lake sediment and water, Verh. Int. Verein. theor. angen.
    Limnol., 19:382-394.
Dell, C.I., 1973, Vivianite: An authigenic phosphate mineral in
    Great Lakes Sediments, Proc. 16th Conf. Great Lakes Res., 1027-28.
Dobson, H.H. and M. Gilbertson, 1972, oxygen depletion in the
    hypolimnion of the central basin of Lake Erie, 1929 to 1970,
    In: "Project Hypo," M.N. Burns and C. Ross, eds., Canad. Cent.
    Inland Waters, Hamilton, Ontario. Pap. 6. 180 p.
Doyle, R.W.S., 1967, Eh and thermodynamic equilibrium in environ-
    ments containing dissolved ferrous iron, Ph.D. Thesis, Yale
    University. 100 p.
Edwards, R.W., 1958, The effect of larvae of Chironomus riparius
Meigen on the redox potentials of settled activated sludge, Ann.
    Appl. Biol., 46:457-464.
Edwards, R.W., and H.L.J. Rolley, 1965, Oxygen consumption of
    river muds, J. Ecol., 53:1- 19.
Fager, E.W., 1964, Marine sediments: Effects of a tube-building
    polychaete, Science, 135:356-359.
Finn, J.D., 1974, "A General Model for Multivariate Analysis,"
    Holt, Rinehart and Winston, New York. 423 p.

Fisher, J.B., 1979, Effects of tubificid oligochaetes on sediment
    movement and the movement of materials across the sediment–water
    interface, Ph.D. Thesis, Case Western Reserve University, 132 pp.
Fisher, J.A. and A.M. Beeton, 1975, The effect of dissolved oxygen
    on the burrowing behavior of *Limnodrilus hoffmeisteri* (oligo-
    chaeta), Hydrobiologia, 47:273–290.
Folk, R.L., 1965, "Petrology of Sedimentary Rocks," University of
    Texas, Austin. 159 p.
Fukuda, M.K., 1978, The entrainment of cohesive sediments in fresh-
    water, Ph.D. Thesis, Case Western Reserve University, 210 pp.
Gaudette, H.E., W.R. Flight, L. Toner, and D.W. Folger, 1974, An
    inexpensive titration method for the determination of organic
    carbon in recent sediments, J. Sedim. Petrol, 44:249–255.
Gillot, J.K., 1968, "Clay in Engineering Geology," Elsevier,
    Amsterdam. 296 p.
Hargrave, B.T., 1972, Oxidation–reduction potentials, oxygen
    concentration and oxygen uptake of profundal sediments in a
    eutrophic lake, Oikos, 23:167–177.
Hayes, F.R., and M.A. MacAulay, 1959, Lake water and sediment. V.
    Oxygen consumed in water over sediment cores, Limnol.
    Oceanogr., 4:291–298.
Holland, E.F., R.G. Zingmark, and M.J. Dean, 1974, Quantitative
    evidence concerning the stabilization of sediments by marine
    benthic diatoms, Mar. Biol., 27:191–196.
Jernelov, A., 1970, Release of methyl mercury from sediments with
    layers containing inorganic mercury at different levels,
    Limnol. Oceanogr., 15:958–960.
Jonasson, P.M., 1969, Bottom fauna and eutrophication, In:
    "Eutrophication: Causes, Consequences, Connectives", Nat.
    Acad. Sci., Washington, D.C. 661 p.
Kemp, A.L.W., R.L. Thomas, C.I.,and J.M. Jaquet, 1978, Cultural
    impacts on the geochemistry of sediments in Lake Erie,
    J. Fish. Res. Bd. Can., 33:440–462.
Kemp, A.L.W., G.A. MacInnis, and N.S. Harper, 1977, Sedimentation
    rates and a revised sediment budget for Lake Erie, J. Great
    Lakes Res., 3:221–233.
Kirchner, W.B., 1975, The effect of oxidized material on the
    vertical distribution of freshwater benthic fauna, Freshwat.
    Biol., 5:423 – 429.
Krezoski, J.R., 1976, Macrobenthos and mixing processes in Lake
    Huron sediments, M.S. Thesis, University of Michigan, Ann
    Arbor. 34 p.
Kuznetsov, S.I., 1970, "The Microflora of Lakes and its Geoche-
    mical Activity," University of Texas Press, Austin. 503 p.
Lucas, A.M. and W.A. Thomas, 1972, Sediment oxygen demand in Lake
    Erie's Central Basin, 1970, In: "Project Hypo," N.M. Burns and
    C. Ross, eds., Can. Cent. Inland Waters Pap. 6, 180 pp.
Manheim, F.T., 1970, The diffusion of ions in unconsolidated
    sediments, Earth. Planet, Sci. Lett., 9:307–309.

McCall, P.L., M.J. Tevesz, and S. Schwelgien, in press, Sediment mixing by *Lampsilis radiata siliquoidea* (Mollusca) from western Lake Erie, J. Great Lakes Res.

Mendenhall, W., 1968, "Introduction to Linear Models and the Design and Analysis of Experiments," Wadsworth, Belmont, California. 465 p.

Milbrink, G., 1973, On the vertical distribution of oligochaetes in lake sediments, Rep. Inst. Freshwater Res. Drottingholm, 53:34-50.

Mills, E., 1969, The community concept in marine zoology, with comments on continua and instability in some marine communities: a review, J. Fish. Res. Bd. Can., 26:1415-1428.

Mortimer, C.H., 1941, The exchange of dissolved substances between mud and water in lakes, J. Ecol., 29:280-329.

Mortimer, C.H., 1942, The exchange of dissolved substances between mud and water in lakes, J. Ecol., 30:147-201.

Mortimer, C.H., 1971, Chemical exchanges between sediments and water in the Great Lakes - Speculations on probable regulatory mechanisms, Limnol. Oceanogr., 16:387-404.

Neame, P.A., 1975, Benthic oxygen and phosphorous dynamics in Castle Lake, California, Ph.D. Thesis, University of California, Davis. 234 p.

Newell, R.C., 1965, The role of detritus in the nutrition of two marine deposit feeders, the prosobranch *Hydrobia ulvae* and the bivalve *Macoma balthica,* Proc. Zool. Soc. Lond., 144:25-45.

Nriagu, J.O. and C.I. Dell, 1974, Diagenetic formation of iron phosphates in recent lake sediments, Am. Mineralogist, 59:934-946.

Palmer, M.F., 1968, Aspects of the respiratory physiology of *Tubifex tubifex* in relation to its ecology, J. Zool. (Lond.), 154:463 - 473.

Parthenaides, E.,and R.E. Passwell, 1968, Erosion of cohesive soil and channel stabilization, Civil Engineering Rept. 19., State Univ. of New York, Buffalo. 51 p.

Petr, T., 1977, Bioturbation and exchange of chemicals in the mud-water interface, In: "Interactions between Sediments and Fresh Water," H.L. Golterman, ed., Junk, The Hague. 473 p.

Pliodzinskas, A.J., 1978, Aquatic oligochaetes in the open water sediments of western and central Lake Erie, Ph.D. Thesis, Ohio State University. 160 p.

Poddubnaya, T.L., 1961, Concerning the feeding of the high density species of tubificids in the Rybinsk Reservoir, Tr. Inst. Biol. Vodokhran. Akad. Nauk. SSS, 4:219-231.

Poddubnaya, T.L., and V.S. Sorokin, 1961, The thickness of the nutrient layer in connection with movements of the Tubificidae in the ground, Byull. Inst. biol. Vodokhr., 10:14-20.

Postma, H., 1967, Sediment transport and sedimentation in the estuarine environment, In: "Estuaries," G.H. Lauff, ed., AAAS, Washington, D.C. 757 p.

Rhoads, D.C., 1970, Mass properties, stability, and ecology of marine muds related to burrowing activity, In: "Trace Fossils," J.P. Crimes and J.C. Harper, eds., Geol. Jour. Spec. Issue, 3:391-406.

Rhoads, D.C., 1974, Organism - sediment relations on the muddy sea floor, Oceanogr. Mar. Biol. Ann Rev., 12:263-300.

Rhoads, D.C. and D.K. Young, 1970, The influence of deposit-feeding organisms on sediment stability and community trophic structure, J. Mar. Res., 28:150-178.

Rhoads, D.C., J.Y. Yingst, and W.J. Ullman, 1978, Seafloor stability in Central Long Island Sound: Part I. Temporal changes in erodibility of fine-grained sediment, In: Estuarine Interactions," M.L. Wiley, ed., Academic Press, New York. 603 pp.

Robbins, J.A., and D.N. Edgington, 1975, Determination of recent sedimentation rates in Lake Michigan using Pb-210 and Cs-137, Geochim. Cosmochim. Acta., 39:285-304.

Robbins, J.A., and D.N. Edgington, 1976, Depositional processes and the determination of recent sedimentation rates in Lake Michigan. Proc. Second Federal Conf. on the Great Lakes.

Robbins, J.A., J.R. Krezoski, and S.C. Mozley, 1977, Radioactivity in sediments of the Great Lakes: Post-depositional redistribution by deposit feeding organisms, Earth Planet. Sci. Sett., 36:325-333.

Robbins, J.A., P.L. McCall, J.B. Fisher, and J.R. Krezoski, 1979, Effect of deposit feeders on migration of cesium-137 in lake sediments, Earth Planet Sci. Lett., 42:277-287.

Royse, C.F., 1970, "An Introduction to Sediment Analysis," Arizona State University, Tempe. 180 p.

Seal, H.L., 1964, "Multivariate Statistical Analysis for Biologists", Meuthen and Co., London. 209 p.

Scheffé, H., 1953, A method for judging all contrasts in the analysis of variance, Biometrika, 40:87-104.

Scheffé, H., 1959, "The Analysis of Variance," J. Wiley, New York. 477 p.

Schumacher, A., 1963, Quantitative Aspekte der Beziehung Zwischen Starke der Tubificidenbesiedlung und Schichtdicke der Oxydationzone in den Susswasserwatten der unterelbe, Arch. Fischwiss., 14:48-51.

Scoffin, T.P., 1970, The trapping and binding of subtidal carbonate sediments by marine vegetation in Bimini lagoon, Bahamas, J. Sedim. Petrol., 40:249-273.

Sorokin, Yi. I., 1966, Carbon-14 method in the study of the nutrition of aquatic animals. Int. Rev. ges. Hydrobiol., 51:209-224.

Southard, J.B., 1974, Erodibility of fine abyssal sediments, In: "Deep Sea Sediments," A.L. Inderbitzen, ed., Plenum Press, New York. 432 p.

Stamm, H.H.,and H.W. Kohlschutter, 1965, Die Sorption von Phosphationene an Eisen (III) - Hydroxide, J. Inorg. Nucl. Chem., 27:2103 -2108.

Stockner, J.G., and J.W.G. Lund, 1970, Live algae in postglacial lake sediments, Limnol. Oceanogr., 15:41-58.

Sundborg, A., 1956, The River Klaralven: A study of fluvial processes, Geogrof. Ann., 38:127-136.

Tevesz, M.J.S., F.M. Soster, and P.L. McCall, in press, The effects of size-selective feeding by oligochaetes on the physical properties of river sediments. J. Sedim. Petrol.

Thomas, R.L., J.M. Jaquet, A.L.W. Kemp, and C.F.M. Lewis, 1976, Surficial sediments of Lake Erie, J. Fish Res. Bd. Can., 33:385-403.

Veal, D.M., and D.S. Osmond, 1968, Bottom fauna of the western basin and near-shore Canadian waters of Lake Erie, Proc. 11th Conf. Great Lakes Res.:151-160.

Viner, A.B., 1975, The sediments of Lake George (Uganda) I: Redox potentials, oxygen consumption, and carbon dioxide output, Arch. Hydrobiol., 76:181-197.

Wavre, M., and R.O. Brinkhurst, 1971, Interactions between some tubificid oligochaetes found in the sediments of Toronto Harbour, Ontario, Jour. Fish. Res. Bd. Can., 28:335-341.

Williams, J.D.H., J.M. Jaquet, and R.L. Thomas, 1976a, Forms of phosphorous in the surficial sediments of Lake Erie, J. Fish. Res. Bd. Can., 33:413-429.

Williams, J.D.H., T.P. Murphy, and T. Mayer, 1976b, Rates of accumulation of phosphorous forms in Lake Erie sediments, J. Fish. Res. Bd. Can., 33:430-439.

Wood, L.W., 1975, Role of oligochaetes in the circulation of water and solutes across the mud-water interface. Verh. Internat. Verein. Limnol., 19:1530-1533.

Wyeth, R.K., 1977, Changes in sediment oxygen demand rates as a function of deposition of dredged materials, Abstracts 20th Conference on Great Lakes Research, Ann Arbor, Mich.

Zobell, C.E., and J. Stadler, 1940, The effect of oxygen tension on the oxygen uptake of lake bacteria, J. Bacteriol.,39:307-322.

ECOLOGY OF TIDAL FRESHWATER AND

ESTUARINE TUBIFICIDAE (OLIGOCHAETA)

Robert J. Diaz

Virginia Institute of Marine Science
and the School of Marine Science of the
College of William and Mary
Gloucester Point, Virginia 23062
U.S.A.

ABSTRACT

There is an abrupt shift in the community composition of
Tubificidae as one proceeds from tidal freshwater to estuarine
habitats. Not only does the species composition change but also
their relative trophic importance and, to a more variable degree,
their importance to the community. Communities from tidal freshwater areas tend to resemble the fauna of large eutrophic lakes
while the fauna of estuarine areas is comprised mainly of estuarine
endemic and euryhaline marine species. Tidal freshwater forms
tend to be larger than estuarine forms and are the major sediment
burrowers and bioturbators. Estuarine forms do not appear to be
important bioturbators. Competition or interaction with the large
and diverse polychaete fauna most likely lessens the importance of
oligochaetes in estuarine areas. Where pollution or other factors
result in extreme environmental conditions or reduced habitat
diversity, estuarine tubificids tend to become more important,
while in tidal freshwater areas these conditions result in little
change in community composition. The common bond between the
faunas inhabiting tidal freshwater and estuarine areas is their
opportunistic nature.

319

INTRODUCTION

Recently, interest in oligochaetes has increased. They have
been found by many workers to be numerically dominant in tidal
river-estuarine systems (Diaz, 1977; Chapman, 1979; Hunter and
Arthur, 1978), but their ecological role in such systems is nebu-
lous. While most studies have focused on oligochaetes as indicators
of polluted conditions (Peeters and Wolff, 1973; Brinkhurst and
Cook, 1974; Howmiller and Scott, 1977) there are an increasing
number of studies, exemplified by Hunter and Arthur (1978) and
Rofritz (1977), that indicate the value of oligochaetes to ecosystems. In this paper, I will attempt to focus on the role of
tidal freshwater and estuarine tubificids as community members in
terms of their resource value, predator prey interactions, cycling
of nutrients, reworking of sediments, and competitive interactions.

Tidal freshwater and estuarine areas are ecologically inseparable. They are part of a single system, providing nursery grounds
for many commercially important anadromous and catadromous fish, and
are sites of large concentrations of organic matter originating from
rivers, and production by phytoplankton and wetlands within the
system.

Tidal freshwater is defined as the area from the inland limit
of tidal influence, which in many cases is the fall line, seaward
to where salt intrudes. The estuarine segment is the area from the
edge of salt intrusion to the mouth of a river or bay system.

COMMUNITY COMPOSITION

Oligochaete communities that inhabit tidal freshwater areas
are composed of fewer species than communities in lotic waters of
similar sedimentary character. In the tidal freshwater James River,
Virginia, Diaz and Boesch (1977a) found the tubificid fauna com-
prised of eight species, Crumb (1977) found 10 species in the
Delaware River. Delaware and Pfannkuche et al. (1975) reported
seven species from a freshwater tidal mudflat in the Elbe Estuary,
Germany. These authors found a total of 15 species of tubificids
with 3 species common to all the systems (Table 1). For comparison,
Veal and Osmond (1968) reported 16 species of tubificids from Lake
Erie, Johnson and Brinkhurst (1971) collected 16 from parts of
Lake Ontario, and Howmiller (1977) 18 from Wisconsin lakes. These
lake studies represent a total of 25 species with 8 species in
common. The particular choice of lake studies is intended to show
that more tubificids can occur in lake environments and is not
intended to imply that all lakes have such a high diversity of
species.

Part of the reason for the apparent difference in number of
species between lakes and tidal freshwater may be due to a lack of

Table 1. Tubificids found in tidal freshwater areas.

Species	James[1]	Delaware[2]	Elbe[3]
Branchiura sowerbyi	+[4]	+	
Ilyodrilus templetoni	+++	+	
Limnodrilus angustipenis		+	
L. cervix	++	+	
L. claparedeianus			+++
L. hoffmeisteri	+++	+++	+++
L. profundicola	+	+	++
L. udekemianus	+	+	+
Peloscolex ferox		+	
P. freyi	+		
P. multisetosus	++	+	
Potamothrix moldaviensis			+
Psammoryctides barbatus			+
P. curvisetosus		+	
Tubifex tubifex	+		+++

[1]Diaz and Boesch, 1977a   [4]+++ dominant
[2]Crumb, 1977   ++ common
[3]Pfannkuche et al., 1975   + present or no indica-
                                         tion of abundance

investigation in the latter, but the primary differences can be traced to the physical environment. Tidal freshwater areas can be characterized by a lack of diverse habitats. The bulk of alluvial sediments received from limnic waters tends to deposit when tides are encountered, reducing available habitats to those consisting mostly of mud, with isolated sandy substrates where wind and wave energy keep fine sediment from accumulating.

The tidal freshwater fauna is generally most similar to that of large lakes, such as the United States Great Lakes system (Johnson and Brinkhurst, 1971), the profundal zone of smaller lakes (Brinkhurst 1964; Howmiller, 1977), or polluted harbors (Brinkhurst, 1967, 1970; Johnson and Matheson, 1968) where sediments usually consist of silt, clay, and organic-rich mud. The dominant species in all these habitats is invariably *Limnodrilus hoffmeisteri*, which suggests a similarity of physical environments. *Limnodrilus hoffmeisteri* is the most eutopic and tolerant oligochaete, and occurs in large numbers in most freshwater environments that consist of fine grained sediments containing a high organic content of either natural or anthropogenic origin.

As tidal freshwater areas grade to estuarines there is a
reduction in number of oligochaete species, and also a general
reduction of species from all faunal groups, that is related to
the presence of salt. Salinities in this transition zone are in
the oligohaline range (0.5-5 ^{o}/oo) and fluctuate greatly with the
tides, making this zone very inhospitable for all organisms (Remane
and Schlieper, 1971) of both freshwater and marine origin. Within
the transition zone between fresh and salt waters the fauna is
composed of euryhaline freshwater, euryhaline marine, and estuarine
endemic species.

In the Chesapeake Bay three species of tubificids have been
found in this transition zone: one of freshwater origin, *Limnodrilus
hoffmeisteri*; an estuarine endemic, *Peloscolex heterochaetus*; and
P. gabriellae, a euryhaline marine form. In the Fraser River
estuary, British Columbia, Chapman (1979) found only two tubificids.
L. hoffmeisteri and *P. gabriellae*. In contrast, oligohaline areas
of the Baltic Sea, which do not fluctuate in salinity as do estuaries, have a richer oligochaete fauna represented by seven species
of freshwater and marine origin (Bagge and Ilus, 1973) (Table 2).
The majority of oligochaetes that inhabit the low transition zone
salinities are of freshwater origin (six of nine) with only three
of marine origin. Densities reached by these species even in the
stable oligohaline Baltic Sea are much reduced from densities
occurring in their favored habitats. Bagge and Ilus (1973) found
total oligochaete densities less than 500/m^2 in 313 of 335 stations
collected in the oligohaline salinities of the Baltic Sea. The
only exception is the estuarine endemic, *Peloscolex heterochaetus*,
which reaches highest densities at lowered salinities of the transition zone (1700/m^2 at 1.7 ^{o}/oo and 2120/m^2 at 4.8 ^{o}/oo - Diaz
and Boesch 1977b) and is rarely found at salinities less than 0.1
^{o}/oo or higher than 8 ^{o}/oo.

Seaward from the transition zone the number of annelid species
increases greatly, but the increase is due mainly to polychaetes.
The number of oligochaete taxa does not greatly increase in estuarine waters. The limited species increase may be artificial,
resulting from a confused taxonomic status of the two dominant
forms, *Peloscolex gabriellae* and *P. benedeni*. At present, the
definition of both species is very plastic and each probably
represents an agglomeration of two or more species.

It is important to keep in mind that estuaries are semi-
enclosed bodies of water where sea water is diluted with freshwater
runoff. This differentiates them from coastal waters where oligochaete diversity is considerably higher. For example, from the
east coast of North America, Cook and Brinkhurst (1973) list 21
marine tubificids, nine euryhaline marine species, and 12 probable
stenohaline marine species. Recently the reported number of stenohaline marine species has increased with the addition of about 10
newly described species (Erseus, personal communication).

Table 2.   Transition or oligohaline zone tubificids.

Species	Type	Salinity %₀ Chesapeake Bay	Baltic Sea[1]
Limnodrilus hoffmeisteri	F[2]	0.01-5.2	0.05-4.31
L. Profundicola	F	--	2.32-4.31
L. udekemianus	F	--	0.17-3.00
Peloscolex ferox	F	--	0.05-2.76
P. heterochaetus	E	0.1-8.0	--
P. gabriellae	M	3.8-up	--
Potamothrix hammoniensis	F	--	0.05-6.20
Psammoryctides barbatus	F	--	0.17-5.15
Tubifex costatus	M	--	4.25-up

[1]Bagge and Ilus, 1973
[2]F= euryhaline freshwater; E= estuarine endemic;
M= euryhaline marine

Typically, estuarine areas are dominated by one or two species. While there are a number of oligochaetes reported from estuaries, most, as in tidal freshwater areas, are not dominant (Table 3).   In Europe, *Tubifex costatus* and *Peloscolex benedeni* are the most important estuarine tubificids (van den Broek, 1978; Brinkhurst, 1964; Muss, 1967; Hunter and Arthur, 1978) while in North America *Peloscolex gabriellae* is the most dominant (Brinkhurst and Simmons, 1968; Diaz, 1977; Burke, 1976; Dauer et al., 1979; Chapman, 1979). These are all euryhaline taxa that form part of a much more multi-phyletic community, in contrast to the overwhelming importance of oligochaetes in tidal freshwater areas.   While estuarine tubificids can numerically dominate a community their importance to the community has generally been discounted.

In summary, there is a rather abrupt shift in the community composition of oligochaetes proceeding from tidal freshwater to estuarine habitats.   The shift is related mainly to salinity which changes the community composition from oligochaete to polychaete dominance.   While estuarine oligochaetes are not as obvious as their freshwater counterparts, recent work by Hunter and Arthur (1978) and van den Broek (1978) indicates that they are important to the estuarine-marine ecosystem in transferring bacterial energy to higher trophic levels.

The greatest similarity between the tubificid species inhabiting tidal freshwater and estuarine areas is their opportunistic nature.   Species in both areas respond to physical disturbance and organic pollution in a similar manner by increasing population size (Diaz 1977).   The general eurytopy of this tubificid fauna explains their success in these relatively stressful environments.

Table 3. Tubificids found in estuaries.

Species	James[1]	San Francisco Bay[2]	Thames[3]	Fraser[4]
Peloscolex apectinatus		x		
P. benedeni			x	x
P. gabriellae	x	x		
P. heterochaetus	x			
P. nerthoides		x		
T. costatus			x	

[1]Diaz, 1977   [3]Hunter and Arthur, 1978
[2]Brinkhurst and Simmons, 1968   [4]Chapman, 1979

RELATIVE TROPHIC IMPORTANCE

Not only does tubificid species composition change from tidal
freshwater to estuarine habitats but their relative trophic importance also changes. Freshwater oligochaetes are important food
sources for many fish (Brinkhurst 1974) and probably birds (Rofritz
1977). Since many species, and consequently their life habits, are
the same from lakes to tidal freshwater, it can be inferred that
tidal freshwater oligochaetes have the same potential value as a
food resource as to lake oligochaetes. Feeding habit studies of
fish from the tidal freshwater James River confirm this (Diaz et
al. 1978), despite the fact that oligochaete importance in fish
diets was probably underestimated because of rapid digestion.

There is less direct evidence for the trophic importance of
estuarine oligochaetes. While they are certainly consumed by fish
(Giere 1975) and at times may form a major component in a fish's
diet (van den Broek 1978) there are several factors that make
estuarine oligochaetes trophically less important relative to tidal
freshwater species. Estuarine oligochaetes are generally smaller
than tidal freshwater species. The mean live weights of the three
most common estuarine tubificids and a mixture of mature and immature tidal freshwater *Limnodrilus* (mainly *hoffmeisteri*) and *Branchiura sowerbyi* are presented in Table 4. *Peloscolex benedeni* and
Limnodrilus spp. seem to be approximately the same size, while
mature *Tubifex costatus* is slightly smaller and *P. gabriellae* much
smaller. Had the *Limnodrilus* spp. been separated into mature and
immature individuals the size differences would have been much more
pronounced. *Branchiura* is certainly the largest, with individuals
weighing over 50 mg.

Table 4. Live weight of some estuarine and tidal freshwater
          oligochaetes.

Species	Mean Weight (mg)	Reference
Tubifex costatus (immature)	0.70	Brinkhurst, 1964
T. costatus (mature)	1.77	Brinkhurst, 1964
Peloscolex benedeni	3.72	Giere, 1975
P. gabriellae	0.18	Diaz and Boesch, 1977b
Limnodrilus spp. (mainly *hoffmeisteri*)	3.43	Diaz et al. 1978
Branchiura sowerbyi	38.93	Diaz et al. 1978

Susceptibility to predation is another factor determining
resource value. Many of the *Limnodrilus* spp. undulate their
caudal ends above the sediment surface, possibly making them easier
for a predator to detect. However, the life habits of at least two
estuarine species *Peloscolex benedeni* (Erseus, personal communication), and *P. gabriellae*, are different from *Limnodrilus* spp. Both
benedeni and *gabriellae* hold their caudal ends above the sediment
surface but do not move them or rotate them slowly while feeding,
possibly making them less detectable. Kajak and Wisniewski (1966)
found a high percentage of predator cropped tails among shallow
water lake tubificids. Although cropped tails were found in only
1 to 2% of the tidal freshwater James River *Limnodrilus* spp. they
do indicate a certain susceptibility to predation that has never
been observed in estuarine tubificids possibly because the estuarine
species are smaller and when preyed upon would tend to be taken
whole.

From an energetic standpoint oligochaetes are a very nutritious
food source. Caloric measurements compiled by Cummins and Wuycheck
(1971) demonstrate that oligochaetes are very high in caloric
content and as a group are only surpassed by the Copepoda (Table 5).
Although all oligochaete caloric measures are from freshwater
species it is likely that estuarine species will have similar values.

Size, availability, and density are important in comparing
resource value. In tidal freshwater areas where oligochaetes are
the dominant community members they also have the potential for
higher resource value than estuarine oligochaetes. While estuarine
oligochaetes may numerically dominate a community their importance
is lessened by competitive interactions with a large and diverse
polychaete fauna. Polychaetes are the major community dominant and
despite a much lower caloric content than oligochaetes (Table 5),

Table 5. Caloric value for oligochaetes and other animal
          groups (from Cummins and Waycheck 1971).

Group	cal/g dry weight
Copepoda	5765
Oligochaeta	5575
Ephemeroptera	5469
Hirudinea	5443
Chironomidae	5424
Cladocera	5241
Odonata	5117
Bivalves	4530
Sphaeridae	4321
Gammaridae	4050
Decapoda	3944
Polychaetes	3503
Gastropoda	2024

provide the majority of the annelid resource to higher trophic
levels. However, in areas where pollution or other physical factors
result in extreme environmental conditions or reduced habitat
diversity, estuarine oligochaetes tend to become more important
through an increase in density and a reduction in polychaete
species that are able to tolerate the extreme conditions.

The most important trophic role of tidal freshwater and
estuarine tubificids is the direct transfer of bacterial energy
to top carnivores (mainly fish) or to primary carnivores which
make the energy available to top carnivores. Many studies have
demonstrated that animals other than fish seem to depend upon
oligochaetes (Thorhauge, 1976; Loden, 1974; Brinkhurst, 1974;
Brinkhurst and Kennedy, 1965). These animals are in turn very
important as a food source for fish. But the actual portion of
oligochaete production utilized by other trophic levels and
remineralized is unknown. Oligochaetes from freshwater can be
quantitatively important in the trophic transfer of energy,
(Brinkhurst, 1974) but so little is known about the interaction
of estuarine oligochaetes that there can only be speculations as
to their importance. Giere (1975), generalizing on the ecological
role of marine oligochaetes, provides evidence that oligochaetes
are utilized by primary carnivores but concludes they are of only
minor trophic importance. Giere's conclusion, which seems based
upon few adequate studies, may not hold for estuarine areas.
Van den Broek (1978) in the Medway estuary, England, found
Peloscolex benedeni to be a major component in the diets of
flounder and plaice.

FUNCTIONAL ECOLOGY

Oligochaetes are the primary sediment burrowers and bioturbators in freshwater environments because of the virtual exclusion
of polychaetes. When large densities of oligochaetes occur they
have a significant effect upon sedimentary structure mainly through
their subsurface ingestion of sediments and surface egestion.
Davis (1974a) found that feeding and subsequent movement of sediments to the surface occurred mainly at 3-4 cm sediment depth,
but small amounts of deposits from as deep as 8-9 cm could also be
transported to the surface. Oligochaetes are found at depths
greater than 10 cm, and Davis (1974b) reports tubificids from
over 35 cm depth, but it is doubtful that they do anything other
than burrow at those depths; if they did feed the sediments would
certainly not be egested at the surface but could be moved nearer
the sediment surface.

At least one estuarine form, *Peloscolex gabriellae*, has the
ability to burrow deep into the sediment. Dauer et al. (1979)
found specimens as deep as 17 cm in muddy habitats, but most were
in the top two to three cm. In sandy habitats they found *P.
gabriellae* to burrow only 4 to 6 cm, probably because the worms
are too small to move the sand grains and too large to go between
them.

The amount of sediment that can be moved by oligochaetes in
freshwater environments is quite large. Appleby and Brinkhurst
(1970) noted that the top 4 to 6 cm of sediment in various areas
of Toronto Harbor was turned over by tubificids up to 12 times a
year. Associated with turnover of sediments is the regeneration
of nutrients from the sediments to the water column and vice versa.
Davis et al. (1975), for example, found that tubificids
enhanced the movement of phosphorus into the sediments from the
water column, but did not effect release of phosphorus to the
water column from sediments. When tubificids are active they
can also greatly lower the redox potential discontinuity within
the sediments (Davis 1974b), especially controlling all the
geochemical cycling in approximately the top 4 cm of sediments.
McCall (in this volume) explains in detail the mechanisms involved
in tubificid control of geochemical cycling in lake environments.

Despite life habits similar to tidal freshwater worms,
estuarine tubificids do not appear to be important bioturbators.
Estuarine species are smaller than tidal freshwater oligochaetes
and even smaller when compared to most estuarine polychaetes.
Assume that an average estuarine tubificid which is one-half the
size of a tidal freshwater form can rework about 30 ml of sediment
per worm each year or one-half the highest rate reported by Davis
(1974a) for limnetic species. Then comparing tubificids with
Clymenella torquata, a polychaete with a similar feeding mode to

tubificids and common in areas where tubificids are found, can
rework from 96 to 274 ml of sediment per worm each year (Rhoads
1974), or three to nine times as much sediment as the hypothetical
tubificid. Considering that *Clymenella* is but one of hundreds of
estuarine polychaetes, many of which are larger, the role of estu-
rine tubificids as sediment processors becomes even more obscure.
Again, in areas where tubificids dominate greatly over polychaetes
they may take on a more important role in bioturbation.

ACKNOWLEDGEMENTS

I sincerely appreciate the assistance given by Mike Barbour
and Peter Chapman in critically reviewing the manuscript and
Christer Erseus for sharing thoughts on estuarine tubificids.
Acknowledgement is also due to the U.S. Army Corps of Engineers
who supported much of the work in the Chesapeake Bay that I relied
upon for information. Drafts and final copies of this report
were prepared by the VIMS Report Center.

REFERENCES

Appleby, A.G. and R.O. Brinkhurst, 1970, Defecation rate of three
     tubificid oligochaetes found in the sediments of Toronto
     Harbour, Ontario. J.Fish Res. Bd. Canada, 27:1971-1982.
Bagge, P. and E. Ilus, 1973, Distribution of benthic tubificids
     in Finnish coastal waters in relation to hydrography and
     pollution. Oikos Suppl., 15:214-225.
Brinkhurst, R.O., 1964, Observations on the biology of the marine
     oligochaete *Tubifex costatus*. J. Mar. Biol. Assoc. U.K.,
     44:11-16.
Brinkhurst, R.O., 1967, The distribution of aquatic oligochaetes
     in Saginaw Bay, Lake Huron. Limnol. Oceanogr. 12:137-143.
Brinkhurst, R.O., 1970, Distribution and abundance of tubificid
     (Oligochaeta) species in Toronto Harbour, Lake Ontario. J.
     Fish. Res. Bd. Canada, 27:1961-1969.
Brinkhurst, R.O., 1974, The Benthos of Lakes. St. Martin's Press,
     New York, p. 190.
Brinkhurst, R.O. and D.G. Cook, 1974, Aquatic earthworms (Annelida:
     Oligochaeta). pp.143-156. In: C.W. Hart, Jr., and S.L.H.
     Fuller (eds.). Pollution Ecology of Freshwater Invertebrates.
     Academic Press, New York.
Brinkhurst, R.O. and C.R. Kennedy, 1965, Studies on the biology
     of the Tubificidae (Annelida, Oligochaeta) in a polluted
     stream. J.Anim. Ecol., 34:429-443.
Brinkhurst, R.O. and M.L. Simmons, 1968, The aquatic oligochaetes
     of the San Francisco Bay system. California Fish Game, 54:
     180-194.
Burke, W.W. III., 1976, Vertical and horizontal distribution of
     macroinvertebrates on the cord grass, *Spartina alterniflora*
     in a Louisiana salt marsh. M.S. Thesis, Louisiana State U.

Chapman, P., 1979, Seasonal movements of subtidal benthic commu-
    nities in a salt-wedge estuary as related to interstitial
    salinities. Ph.D. Thesis. University of Victoria. 222 pp.
Cook, D.G. and R.O. Brinkhurst, 1973, Marine flora and fauna of
    the Northeastern United States. Annelida: Oligochaeta. NOAA
    Tech. Rept. NMFS CIRC-374, 23 pp.
Crumb, S.E., 1977, Macrobenthos of the tidal Delaware River between
    Trenton and Burlington, New Jersey, Chesapeake Sci., 18: 253-265.
Cummins, K.W. and J.C. Wuycheck, 1971, Caloric equivalents for
    investigations in ecological energetics. Int. Assoc. Theor.
    Appl. Limnol., Comm. No. 18, 158 pp.
Dauer, D.M., W.W. Robinson, C.P. Seymour and A.T. Leggett, Jr.,
    1979, Effects of non-profit pollution on benthic invertebrates
    in Lynnhaven River system. Va. Water Resour. Res. Cent. Bull.,
    117, 112 pp.
Davis, R.B., 1974a, Stratigraphic effects of tubificids in
    profundal lake sediments. Limnol. Oceanogr., 19:466-488.
Davis, R.B., 1974b, Tubificids altered profiles of redox potential
    and ph in profundal lake sediment. Limnol. Oceanogr., 19:
    342-346.
Davis, R.B., D.L. Thurlow, and F.E. Brewster, 1975, Effects of
    burrowing tubificid worms on the exchange of phosphorus
    between lake sediments and overlying water. Verh. Internat.
    Verein. Limnol., 19:382-394.
Diaz, R.J., 1977, The effects of pollution on benthic communities
    of the tidal James River, Virginia. Dissertation of the U.
    of Virginia, Charlottesville, 148 pp.
Diaz, R.J. and D.F. Boesch, 1977a, Habitat development field
    investigations Windmill Point marsh development site James
    River, Virginia. Appendix C: Environmental impacts of marsh
    development with dredged material: Acute impacts on the
    macrobenthic community. U.S. Corps Engin., Vicksburg, MS.
    Tech. Rept. D-77-23, 122 pp.
Diaz, R.J. and D.F. Boesch, 1977b, Environmental effects of the
    James River sewage treatment plant outfall construction: soft
    bottom macrobenthos. Rept. to Hampton Roads Sanitation Dist.,
    Virginia Beach, Va., pp.59-92.
Diaz, R.J., D.F. Boesch, J.L. Hauer, C.A. Stone and K. Munson, 1978,
    Habitat development field investigations, Windmill Point marsh
    development site, James River, Virginia. Appendix D: Envi-
    ronmental impacts of marsh development with dredged material:
    Botany, soils, aquatic biology, and wildlife. U.S. Corps
    Engin., Vicksburg, MS. Tech. Rept. D-77-23, 18-54 pp.
Giere, O., 1975, Population structure, food relations and ecolo-
    gical role of marine oligochaetes, with special reference to
    meiobenthic species. Mar. Biol., 31:139-156.
Howmiller, R.P., 1977, On abundance of tubificidae (Annelids:
    Oligochaeta) in the profundal benthos of some Wisconsin Lakes.
    Amer. Midl. Nat., 97:211-215.

Howmiller, R.P. and M.A. Scott, 1977, An environmental index based
    on relative abundance of oligochaete species. J. Water Poll.
    Cont. Fed., 49:809-815.
Hunter, J. and D.R. Arthur, 1978, Some aspects of the ecology of
    Peloscolex benedeni Udekem (Oligochaeta: Tubificidae) in the
    Thames Estuary. Estuar. Coast. Mar. Sci., 6:197-208.
Johnson, M.G. and R.O. Brinkhurst, 1971, Associations and species
    diversity in benthic macroinvertebrates of Bay of Quinte and
    Lake Ontario. J. Fish. Res. Bd. Canada,28:1683-1697.
Johnson, M.G. and D.H. Matheson, 1968, Macroinvertebrate communi-
    ties of the sediments of Hamilton Bay and adjacent Lake
    Ontario. Limnol. Oceanogr.,13:99-111.
Kajak, Z. and R.J. Wisniewski, 1966, Proba oceny intensywnosci
    syzerania Tubificidae pazez drapiezce. Ekologia Polska
    Seria B,12:181-184.
Loden, M.S., 1974, Predation by chironomid (Diptera) larvae on
    oligochaetes. Limnol. Oceanogr.,19:156-159.
Muus, B.J., 1967, The fauna of Danish estuaries and lagoons.
    Distribution and ecology of dominating species in the shallow
    reaches of the mesohaline zone. Meddr. Danm. Fisk. og
    Havunders, 5:1-316.
Peeters, J.C.H. and W.J. Wolff, 1973, Macrobenthos and fishes
    of the rivers Meuse and Rhine, the Netherlands. Hydrobiolo-
    gical Bull.,7:121-126.
Pfannkuche, O., Jelinek and E. Hartwig, 1975, Zur fauna eines
    susswasserwattes im Elbe-Aestaur. Arch. Hydrobiol.,76:4750-
    498.
Remane, A. and C. Schlieper, 1971, Biology of brackish water.
    Die Binnengewasser., 25, 372 pp.
Rhodes, D.C., 1974, Organism-sediment relations on the muddy sea
    floor. Oceanogr. Mar. Biol. Ann. Rev., 12:263-300.
Rofritz, D.J., 1977, Oligochaetes as a winter food source for the
    Old Squaw. J. Wildlife Manage, 41:590-591.
Thorhauge, F., 1976, Growth and life cycles of (Tubificidae,
    Oligochaeta) in the profuncal of eutrophic Lake Esrom. A
    field and laboratory study. Arch. Hydrobiol.,78:71-85.
van den Broek, W.L.F., 1978, Dietary habits of fish populations
    in the Lower Medway Estuary. J. Fish. Biol.,13:645-654.
Veal, D.M. and D.S. Osmond, 1968, Bottom fauna of the western
    basin and near-shore Canadian Waters of Lake Erie. 151-160.
    In: Proc. 11th Conf. Great Lakes Res. Internat. Assoc.
    Great Lakes Res.

THE ECOLOGY OF TUBIFICIDS IN THE THAMES ESTUARY

WITH PARTICULAR REFERENCE TO *TUBIFEX COSTATUS*

(CLAPARÈDE)

I.K. Birtwell[1]
D.R. Arthur

Department of Zoology
King's College
University of London
Strand, London WC2R 2LS
England

[1]Present address:
Habitat Protection Division
Department of Fisheries and Oceans
1090 West Pender Street
Vancouver, British Columbia
V6E 2P1, Canada

ABSTRACT

    Tubifex tubifex, Limnodrilus hoffmeisteri, Tubifex costatus,
and *Peloscolex benedeni* were sequentially the numerically dominant
tubificids along 65 km of the intertidal zone in the Thames estuary
seaward of London Bridge. Laboratory experiments on their salinity
tolerance, thermal tolerance, anaerobic tolerance, particle size
preference and respiration assisted in explaining the location of
each species in the estuary. Salinities, and to a lesser extent,
dissolved oxygen levels were considered to be primary factors
limiting distribution, while particle size had an effect on
abundance. Population studies were carried out in a relatively

polluted section of the Thames estuary. At this location the
intertidal macrofauna was almost exclusively of *Tubifex costatus*.
Population densities, which attained a mean maximum of $601.10^3 m^{-2}$,
were studied in relation to environmental factors. Spatial dispersion (aggregation and vertical distribution) and specific details
of the life cycle stages of *T. costatus* were examined in relation
to biological and physical factors. The tubificid populations
were contagious and the negative bionomial distribution was
applicable (k=8.046). Populations of *T. costatus* moved seasonally
within the beach substrate and were proximal to the surface during
the summer and early autumn. Evidence is presented from an examination of the growth of size classes within the population of *T.
costatus* in association with stages in the life history, to suggest
that this tubificid has a two year life cycle and that temperature
influences the timing of reproduction.

INTRODUCTION

    The river Thames is 236 km in length and originates from a
drainage area of about 9,800 km^2. Upstream tidal influence is
eliminated by Teddington Weir seaward of which the coastal plain,
slightly stratified estuary, extends 70 km or more, depending
upon freshwater flows. An important feature of this estuary is
its long retention time which is related to a short tidal excursion
(12.9 km to 14.5 km) and freshwater flows which are maintained
above a statutory minimum $9m^3 s^{-1}$. Even with an 18 fold increase
in freshwater flow, the residence time of material discharged at
the head of the estuary will only be reduced by a factor of 3,
and at times of low freshwater flows, may remain within the
estuarine system for 2 to 3 months (HMSO, 1964).

    Increasing population and industrial developments have been
primarily responsible for the discharge of large quantities of
wastes into the Thames estuary. Most of these wastes originate
within the Greater London Council area, resulting in a daily
transfer of effluent greater than $2.5.10^9 1$ from a population
larger than 8 million. Approximately 79% of the effluent volume
entering the Thames estuary is from sewage treatment plants, 12%
from industrial discharges, 7% from freshwaters and 2% from storm
waters. The net effect of these effluent loadings has been to
reduce estuarine water quality.

    Historically, sewage effluent and heated cooling water from
electrical power generating plants have, respectively, been
primarily responsible for the occurrence of a significant dissolved
oxygen "sag curve" and a 4.5°C overall elevation in water temperature within the estuary (Arthur, 1972). The position of the
dissolved oxygen sag curve varied with tidal action, but the nadir
was usually located in the vicinity of the major sewage treatment
plants at Bekton and Crossness. Between 1940 and 1960, anaerobic

conditions often occurred in the estuary and on one occasion
extended for 48 km. These adverse water quality conditions,
simply illustrated with reference to dissolved oxygen, were of
extreme concern and it was only after a major government research
effort and modification of administrative control of the Thames
estuary that efforts were made to improve waste disposal practices
and receiving water quality (Train and White, 1971). Within the
last decade, better waste management and treatment practices have
led to an increase in water quality such that serious consideration
has been given to the introduction of salmon stocks. This consideration exemplifies the progress that has been made to improve
estuarine water quality; it was in 1830 that the last salmon was
captured upstream of London Bridge. (Thames Migratory Fish
Committee, 1977).

    Research was initiated at London University, under the direc-
tion of Professor D.R. Arthur, to describe the transient and
resident biological communities in the Thames estuary. Ecological
and physiological research was effected on planktonic populations
(Lumkin, 1971), fish and crustaceans (Huddart, 1971; Sedgwick,
1978) and benthic organisms (Huddart, 1971; Birtwell, 1972; Hunter,
1977), thus providing detailed information in relation to the
pollution of the estuary.

    Prior to the initiation of these research projects, there had
been few reports and studies on the ecology of tubificids in the
Thames estuary (e.g. Stephenson, 1930; Palmer, 1964, 1968). Preliminary ecological studies in the oligohaline regions were reported
by Palmer (1968), who recorded large populations (up to $5 \cdot 7.10^6 m^{-2}$)
of *Tubifex tubifex* and *Limnodrilus hoffmeisteri* close to the heated
effluent discharge from an electrical generating plant upstream of
London Bridge. However, it was the work of Huddart (1971) that
influenced the direction of subsequent detailed research on tubificids in the Thames estuary. Huddart (1971) sampled the intertidal benthos at 8 sites along the estuary (Figure 1), and revealed
the existence of large populations of tubificids, the numerically
dominant species of which changed with progression seaward to All
Hallows, 64.7 km below London Bridge. Not only were large populations of tubificids recorded at most sampling sites but they were
frequently the dominant faunal component of the intertidal benthos.
This unique situation facilitated research on the life histories,
physiological processes, preferences and tolerances of *Tubifex
tubifex, T. costatus, Limnodrilus hoffmeisteri* and *Peloscolex
benedeni*. The results of this research are documented by Huddart
(1971), Huddart and Arthur (1971a, b), Hunter and Arthur (1978).

    This paper summarizes some of the research undertaken on the
factors which may affect the distribution and abundance of tubificids within the Thames estuary. Experiments were carried out on
the tolerance of *T. tubifex, L. hoffmeisteri, T. costatus* and

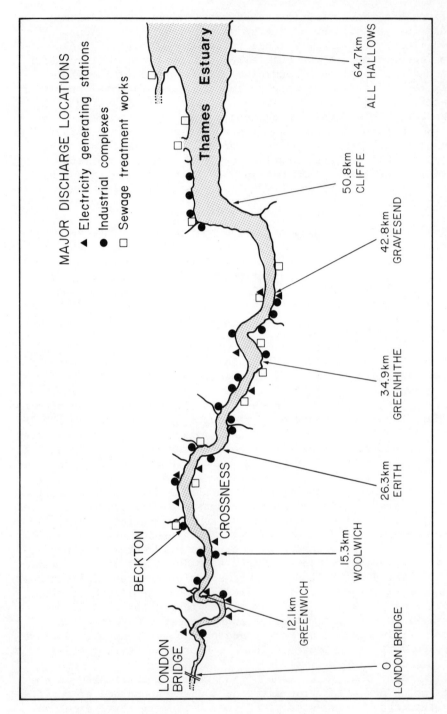

Figure 1. A diagram of the estuary of the River Thames and the location of sampling sites used in benthic studies.

P. benedeni to salinity, temperature, anaerobic conditions, and
their preference for particular sediment fractions. Determinations
of the dependence of their respiration on external oxygen tension
and the development of an oxygen debt after exposure to anaerobic
conditions were examined. This paper also documents research on
the ecology and life history of *T. costatus*.

MATERIALS AND METHODS

Limiting and Controlling Factors

     The results of research on some of the factors which may limit
or control the distribution and abundance of *T. tubifex, L. hoffmeisteri, T. costatus* and *P. benedeni* along the Thames estuary is
the subject of another paper (Birtwell and Arthur in preparation).
Accordingly, only brief mention will be given to the methodology
associated with these studies; specific information has been documented by Birtwell (1972).

Collection and Laboratory Treatment

     Tubificids were collected from intertidal locations in the
Thames estuary. *T. tubifex* was collected from London Bridge;
L. hoffmeisteri from Woolwich; *T. costatus* from Erith and *P. benedeni* from Cliffe (refer to Figure 1 for site location). Each
species was maintained in substrate from the collection site.
Aerated dilutions of seawater covered the stocks. *T. tubifex* was
maintained in water with a salinity of 3.4 ^{o}/oo; *L. hoffmeisteri*
at 5.1 ^{o}/oo; *T. costatus* at 17.0 ^{o}/oo and *P. benedeni* at 24.0 ^{o}/oo.
All the tubificids were progressively acclimated to the required
temperature and remained under laboratory conditions for 1 to 7
weeks during which time samples were taken for experimentation.
Whenever possible adult tubificids were used.

Experimentation

     All the preference and tolerance experiments were carried out
in darkness. The criterion of death during the tolerance experiments was the lack of movement of the individual on physical
stimulation.

     Salinity tolerance experiments were carried out in 0.5 1 glass
dishes which were ground to provide a perfect fit with lids through
which passed a glass sinter that permitted diffused air flow into
the vessel and test solutions (aged and filtered seawater dilutions). Within each glass dish, a 20 compartment plexiglass
"Repli dish" was used to separate and hold 20 tubificids. The
temperature was maintained at ± 0.1^o C of the experimental temperature. Salinity tolerance experiments were undertaken for 168h,
a time chosen for response stabilization based upon initial

experiments of 13 days duration with *L. hoffmeisteri*. Five
replicate experiments were used for each salinity tested, and
the immersion water was changed daily.

Thermal tolerance experiments employed the same vessels and
similar techniques to those used in salinity tolerance experiments,
except that the duration of experimentation was 96h, and temperature was varied from the 20° C acclimation temperature.

From the results of the final mortalities in the salinity and
thermal tolerance experiments the respective 168h LC50 (^{0}/oo) and
96h LC50 (°C) were calculated together with 95% confidence limits
(Finney, 1971).

The tolerance of anaerobic conditions by the tubificids
necessitated the use of deoxygenated dilutions of seawater. High
purity nitrogen was used to deoxygenate the solutions and was
continuously admitted to the 0.5 l glass test vessels which were
surrounded with polyethelene covers thus maintaining an atmosphere
of nitrogen. Frequent observations were made to determine the
mortality of the tubificids at the three test temperatures (20° C,
25° C, and 30° C) allowing determination of the LT50 at each
temperature. The tubificids were acclimated to a temperature of
20° C.

Substrate preference experiments were used to determine the
preference of tubificids to "estuarine" and "sterile" substrates
with particle sizes of 1.0mm to 0.5mm; 0.5mm to 0.25mm; 0.25mm to
0.063mm; <0.063mm. Equal quantities of each substrate fraction
were placed into quadrants of 0.5 l glass dishes. The area covered
by each fraction was approximately 25 cm^2 and the depth 2 cm. The
substrates were immersed in aerated water of the required salinity
before 15 tubificids of a particular species were placed on each
substrate fraction, a lid with sinter placed over the dish and
aeration commenced. The experimental duration was 48h after which
the number of tubificids in each substrate fraction was determined.

Respiration studies using mono-specific groups of tubificids
within a continuous flow apparatus with polarographic oxygen determination (Birtwell, 1972) were used to study the effect of a number
of variables: e.g. dissolved oxygen level (100% to 2.5% air saturation) temperature, and salinity on oxygen uptake.

Population Studies

An area of 10m^2 was delineated at mid-tide level at Erith,
26.3 km seaward of London Bridge. Between 21 September 1970 and
20 September 1971, ten samples, 15 cm deep, 15.2 cm^2 surface area,
were taken with a plexiglass corer at 14 ± 1 day intervals.
Tubificids were extracted from the sediment samples by wet sieving

through a 0.25 mm screen. Subsequently, the samples were enumerated and identified using the keys of Brinkhurst (1963a, b, 1966).
Subsamples of the tubificids were retained for length and weight
determinations and examination of the state of maturity.

For length measurements, 100 tubificids were narcotized in
5% magnesium sulphate, killed with 70% alcohol and rapidly placed
onto a grooved plexiglass block and covered with a saline solution
and a cover slip. Measurements of all tubificids longer than 4 mm
were made directly with a graduated scale, and smaller specimens
were measured by the use of a microscope. The length was recorded
in mm.

Weight determinations were made on groups of 100 randomly
collected tubificids, after allowing 24h for the displacement of
gut contents. A micro balance was used to determine their wet
weight followed by their dry weight after incubation for 7 days at
60° C.

In order to determine the life cycle stages through which
T. costatus passes, several criteria were used to assist in the
description. Examination of random samples of the population
were made after the tubificids had been mounted in Ammans lactophenol. Those *T. costatus* in which there was no gonadial development were referred to as being immature. Maturing *T. costatus*
show gonadial growth but lack penis sheaths; this probably represents a most imprecise grouping because the earliest appearance
of the gonads may be obscured. Mature *T. costatus* possess chiti-
nous penis sheaths but lack spermatophores in their spermathecae,
in contrast to breeding worms which do. Post-breeding *T. costatus*
lack spermatophores although retaining penis sheaths and are
characterized by gonadial regression.

The vertical distribution of tubificids in the sediment of
Erith beach was determined from 24 August 1970 to 20 August 1971.
Five cores (15 cm deep, 15.2 cm^2 surface area) were taken on each
sampling occasion, extruded and sectioned at 3 cm intervals. The
tubificids within each section were subsequently enumerated and
expressed as a percentage of the total number in the core samples.

On each sampling occasion at Erith beach, sediment temperature,
salinity, particle size and the depth of the black reduced (ferrous
sulphide) layer were determined. The temperature of the air, and
that of the sediment was recorded at depths of 1 cm, 3 cm and 6 cm
from the surface using mercury thermometers.

The salinity of the sediment was determined by a method similar
to that used by Smith (1956) and Huddart (1971). Five sediment samples were used on each sampling occasion (6 cm depth, 15.2 cm^2
surface area). They were weighed wet (A), dry (B), and subsequently

mixed with a known quantity of distilled water (V). The salinity
(S) was recorded by a calibrated conductivity bridge and the salinity
of the original sample calculated $(V \cdot S)/(A-B)$ ^{o}/oo.

Twenty core samples were also taken on each sampling occasion
and the depth of the black reduced layer from the substrate surface
was measured at the time of collection.

Particle size analysis was carried out on 5 cores (6 cm depth,
15.2 cm^2 surface area) after each sampling occasion using the wet
sieving method prescribed by Morgans (1956) which used the Wentworth
scale (Wentworth, 1922) to classify sediment fractions.

In a laboratory study the effects of *T. costatus* on the depth
of the black reduced layer in sediments was investigated using two
100 cm^2, 20 cm deep cores from Erith beach. The surface layer to
5 cm depth was removed, water of 17 ^{o}/oo added and deoxygenation
commenced using high purity nitrogen. The temperature varied
between 20° C and 23° C, and deoxygenation continued for 6 days.
There was no evidence of surviving tubificids in either sample
after this time. The compliment of *T. costatus* previously screened
from one sample was readmitted to one core only and aeration of
the water overlying both cores commenced. Measurements of the
depth of the black reduced layer in each core were made over a
period of 42 days.

RESULTS AND DISCUSSION

Distribution and Abundance

The distribution of the four numerically dominant species of
tubificids (*T. tubifex, L. hoffmeisteri, T. costatus* and *P. benedeni*) remained relatively constant in the intertidal zone along
the length of the Thames estuary during the period of study from
1968 to 1972, but their abundance fluctuated particularly in
relation to the nature of the substrate, proximity to heated
cooling water discharges, and elevation on the beaches.

Most changes in tubificid abundance that occurred at any
particular site were attributed primarily to recruitment within
the population following breeding, and to a much lesser extent
immigration of individuals from other locations (Huddart, 1971,
Birtwell, 1972, Hunter, 1977, Hunter and Arthur, 1978). *T. tubifex*
was recorded at London Bridge in mixed populations with *L. hoffmeisteri*, and densities (approximately $300 \cdot 10^3 m^{-2}$) tended to
increase down the beach. In the finer sediments at this location
greater densities of tubificids occurred, and just upstream of
London Bridge, Palmer (1968) recorded $5.7 \cdot 10^6 m^{-2}$. The seaward
distribution of *T. tubifex* was variable, and this species was a
transient member of the tubificid fauna at Woolwich, 15.3 km below
London Bridge.

L. hoffmeisteri occurred in large numbers ($500 \cdot 10^3 m^{-2}$) 12.1 km and 15.3 km seaward of London Bridge at Greenwich and Woolwich respectively. In contrast to the situation at London Bridge, densities of *L. hoffmeisteri* decreased towards the low water level (Huddart, 1971; Hunter, unpublished data). This species of tubificid was a transient component of the benthic community at Erith 26.3 km seaward of London Bridge.

T. costatus attained a maximum population size of $600 \cdot 10^3 m^{-2}$ at Erith, and, as at Woolwich and Greenwich, the densities of tubificids decreased towards the low water level. Further seaward from London Bridge (34.9 km) the tubificid populations at Green-hithe were of *T. costatus* and *P. benedeni*, and densities rarely exceeded $40 \cdot 10^3 m^{-2}$. Further penetration of *T. costatus* into the estuary was recorded at Woolwich, but populations were not sustained in this location.

At Cliffe, 50.8 km from London Bridge, maximum populations of *P. benedeni* occurred in densities of $142 \cdot 10^3 m^{-2}$, and were admixed with *T. costatus* at densities of $2.4 \cdot 10^3 m^{-2}$. (Hunter, 1977; Hunter and Arthur, 1978). The penetration of *P. benedeni* into the Thames estuary was limited to Erith where it was a transient component of the benthos, however, persistent populations occurred further seaward at Greenhithe, 34.9 km below London Bridge.

The occasional recordings of *T. tubifex* and *L. hoffmeisteri* further seaward of their general positions in the estuary occurred at times of high freshwater flows. At such times lower salinities were prevalent in the middle reaches of the estuary and it is presumed that the transport of *T. tubifex* and *L. hoffmeisteri* into these areas had been aided by bed load transport and drift. Conversely, the occasional recordings of *T. costatus* and *P. bene-deni* further into the estuary occurred at times of low freshwater flows when salinity incursion was enhanced.

Limiting and Controlling Factors

The distribution of some of the four dominant species of tubificids along the Thames estuary seemed to be surprising when consideration was given to the field observations of other workers in different estuaries and brackish water areas, and in view of the recorded water quality variation within the Thames estuary. While recognizing the limitations of extrapolating the results of laboratory experiments to assist explanation of observations made in the field we considered that such an approach was justifiable, particularly since very little research had been undertaken on the factors which may limit or otherwise affect *T. tubifex, L. hoff-meisteri, T. costatus* and *P. benedeni* in estuaries. Some of the experimental results we obtained revealed that certain factors could affect the location of *T. tubifex, L. hoffmeisteri,*

T. costatus and *P. benedeni* within the estuary and a summary of the experimental results is shown in Table 1. Pertinent average water quality conditions at particular sites in the estuary are presented in Table 2.

Salinity tolerance experiments with *T. tubifex* inferred that this species could extend into the Thames estuary as far as Erith where average salinity values were just less than the 168h LC50 level. However, it is more likely than the maximum salinities at any one location are more effective in limiting the seaward distribution of *T. tubifex*. Maximum salinites of 22 °/oo were recorded at Erith while further upstream 13.5 °/oo occurred at Woolwich and 8 °/oo at Greenwich. It may be expected that salinities of 8 °/oo would be stressful to some individuals of *T. tubifex* populations. Hence there was some agreement between the field observations on the distribution of *T. tubifex* and lethal salinity levels recorded in the laboratory.

The tolerance of *T. tubifex* to temperature revealed a 96h LC50 value of 33.9° C a temperature which would not be encountered within the main water body of the estuary but possibly close to discharges of heated cooling water from electrical generating plants.

Tubifex tubifex abstracted oxygen from the overlying water at a relatively constant rate down to about 7.5% to 10% air saturation and exhibited a relatively low tolérance of anaerobic conditions, after which it appeared to incur an oxygen debt. Based on the dissolved oxygen data presented in Table 2, the average dissolved oxygen levels would not be expected to limit populations, however, dissolved oxygen levels frequently below 5% air saturation were a regular occurrence in some more seaward locations. It is likely that the low dissolved oxygen levels were not conducive to the establishment of populations of *T. tubifex* in these more seaward locations.

In the substrate preference experiments, *T. tubifex* selected the finer fractions (<0.25 mm). While such fractions predominate in many regions of the Thames estuary it was concluded that factors other than sediment particle size may be limiting the seaward distribution of *T. tubifex*. However, in the more oligohaline areas with higher dissolved oxygen levels relatively high populations of *T. tubifex* were recorded in fine sediments (Palmer, 1968).

It was concluded that salinity and to a lesser extent minimum dissolved oxygen levels were probably responsible for limiting the seaward distribution of *T. tubifex* within the Thames estuary while sediment particle size may have affected the abundance of the species.

Table 1. A summary of results from experiments on the respiration, anaerobic, thermal and salinity tolerance and substrate preference of four dominant tubificids in the Thames estuary.

Experiment	n	Tubifex tubifex	Limnodrilus hoffmeisteri	Tubifex costatus	Peloscolex benedeni
Salinity tolerance 5°C	5	FW to 9.8±0.2	FW to 10.7±0.1	FW to >34.0	2.8±0.3 to >34.0
168h LC50 (°/oo) 20°C	5	FW to 9.0±0.1	FW to 14.7±0.1	0.45±0.1 to>34.0	5.9±0.03 to>34.0
Temperature tolerance					
96h LC50 (°C)	5	33.9±0.8	37.5±0.2	32.4±0.9	28.5±0.2
Anaerobic tolerance					
LT50 (h) 20°C	12	28.1±1.0	52.0±0.2[a]	31.8±1.0[b]	58.8±1.6[c]
25°C		19.8±0.7[d]	25.0±0.7[d]	17.1±0.6	26.6±0.8
30°C		12.5±0.6[e]	18.1±0.8	7.5±0.3	17.8±0.8
Demonstration of Oxygen Debt (after anaerobic conditions)	4	+	-f	+	-f
(Tc - % oxygen)*	5	1.5 to 2.0	1.0 to 3.0[g]	3.0 to 5.0	5.0 to 6.0[g]
Particle size Preference					
(mm) "sterile"	10	<0.063	<0.063	1.0 to 0.25[h]	<0.063[g]
"natural"	5	0.25 to 0.063	<0.063	<0.063[i]	<0.063

FW-Freshwater: a-24, b-13, c-8, d-10, e-11, f-3, g-6, h-15, i-23. *Critical oxygen level.

Table 2. Average temperature, salinity and dissolved oxygen values derived from weekly determinations between December 1968 and September 1971 (data courtesy of the Great London Council, Port of London Authority).

Location	Distance from London Bridge (km)	Temperature (°C)	Salinity (°/oo)	Dissolved Oxygen (% Air Sat.)	Dominant Species of Tubificid
London Bridge	0	15.2	0.7	44.6	*Tubifex tubifex*
Woolwich	15.3	15.3	3.9	16.4	*Limnodrilus hoffmeisteri*
Erith	26.3	15.2	8.7	12.7	*Tubifex costatus*
Cliffe	50.8	13.6	20.6	48.6	*Peloscolex benedeni*

Limnodrilus hoffmeisteri was found to be relatively more tolerant of salinity and the other variables examined. Its ability to withstand higher salinites was reflected in its location further seaward in the estuary at Woolwich where maximum salinities were typically 13.5 °/oo, just below the 168h LC50 value of 14.7 °/oo (20° C).

The thermal tolerance of L. hoffmeisteri was greater than that of the other 3 tubificids examined, thus conferring upon this species the possibility of exploiting habitats adjacent to thermal discharges (Palmer, 1968; Aston, 1973; Hunter, 1977).

Relatively large populations of L. hoffmeisteri occurred at Greenwich and Woolwich where dissolved oxygen levels frequently dropped to below 5% air saturation. The change from independent to dependent respiration (Tc value) was between 5% and 15% air saturation. It was expected that this species would have occurred in regions where dissolved oxygen levels exceeded the Tc level. This was obviously not the case and the low metabolic rate coupled with a relatively better ability to survive anaerobic conditions possibly without incurring an oxygen debt suited this tubificid to its downstream locations of Greenwich and Woolwich.

The findings of the substrate preference experiments were in agreement with field observations on the distribution of L. hoffmeisteri. At Greenwich and Woolwich where large populations of L. hoffmeisteri occurred there was a substantial deposition of fine sediment (<0.063 mm).

In summary, increasing salinity with seaward progression into the Thames estuary was the factor most likely to restrict L. hoffmeisteri from penetration much further seaward of Woolwich. The ability of this tubificid to dwell in such regions of low dissolved oxygen is probably related to its relatively low metabolic rate and tolerance of anaerobic conditions.

Tubifex costatus was determined to be euryhaline and only in experiments carried out at 20° C was there mortality at low salinities. The penetration of this tubificid into the Thames estuary was to a region just landward of Erith where annual fluctuations in salinity were from about 1 °/oo to 19 °/oo. The absence of T. costatus in the lower salinity regions suggests that some other factor(s) may have prevented occupation of these regions.

Thermal tolerance data revealed a 96h LC50 value of 32.4° C, a temperature which would not generally be encountered within the main estuarine water body, but unlike T. tubifex and L. hoffmeisteri, T. costatus would not be expected to occur close to electrical generating plant outfalls where water in excess of 32° was often discharged.

T. costatus, like *T. tubifex*, had a relatively high metabolic
rate, appeared to incur an oxygen debt and was relatively intolerant
of anaerobic conditions. Hence it was expected that *T. costatus*
would be found in locations where the dissolved oxygen levels were
above the Tc value (25% to 30% air saturation), but this was not
so. They occurred in large populations at Erith where dissolved
oxygen minima were frequently less than 5% air saturation. Their
occurrence at Erith could be explained if oxygen uptake occurred
upon tidal emersion. During late summer when high temperatures
and low dissolved oxygen levels prevailed in the estuary, these
tubificids were almost entirely within a layer 3 cm from the
substrate surface and at low tide were often observed on the
surface. It is possible that oxygen uptake was facilitated at
such times thus compensating for reduced dissolved oxygen levels
which would have been experienced on tidal immersion. Further
upstream of Erith at Woolwich the more fluid and mobile nature of
the beach sediment may not have been conducive to the burrowing of
T. costatus and access to the substrate surface may not have been
possible. Hence oxygen uptake at the substrate surface may not
have been facilitated on tidal emersion. These opinions are not
supported entirely by the substrate choice experiments for the
finer (<0.063 mm) sediments were chosen; and the finer fractions
predominate at Woolwich.

Most probably, a combination of low dissolved oxygen levels,
and the prevalence of lower salinities landward of Erith may have
influenced the upstream penetration of *T. costatus* into the
Thames estuary.

Populations of *Peloscolex benedeni* (approximately $40 \cdot 10^3 m^{-2}$)
penetrate the Thames estuary to Greenhithe approximately 34.9 km
below London Bridge. Although annual salinity fluctuations are
large at this location, minimum salinities are often 4 ^{o}/oo during
the high freshwater flow periods which occur during the winter
months. *P. benedeni* populations would be expected to be limited
by salinities lower than about 2.8 ^{o}/oo in winter, a level just
lower than typical minimum salinity values recorded at Greenhithe.
Even if salinity levels decreased to very low levels, the ability
of *P. benedeni* to tolerate anaerobic conditions, apparently without incurring an oxygen debt, and its relatively low metabolic
rate would make retreat into the more stable environment of the
substrate an alternative that may favour its survival.

Temperature tolerance was relatively low, but it is unlikely
that temperature limited the penetration of *P. benedeni* into the
Thames estuary.

The critical dissolved oxygen level (Tc) infers that *P. benedeni* would preferentially occur in waters with relatively high
dissolved oxygen levels. At Cliffe, where *P. benedeni* reaches

its maximum densities along the south shore of the Thames estuary, dissolved oxygen levels are generally greater than 10% air saturation and averaged 20.6% air saturation between December 1968 and September, 1971.

P. benedeni preferred fine substrate fractions, and such a preference is reflected by observations made in the estuary. Maximum densities occurred in locations such as Cliffe where the <0.063 mm fractions comprised 80% of the substrate. This species of tubificid was absent in the same location in substrates of sand and gravel.

It was concluded that salinity was the factor most likely to limit the penetration of *P. benedeni* from marine waters into the Thames estuary.

Overall, salinity seemed to be the major factor affecting the distribution of the tubificids, followed by or in combination with the level of dissolved oxygen. Particle size probably had more influence on the abundance of the tubificids. The importance of temperature is difficult to assess because of its indirect effect on many metabolic processess. In *T. costatus* and *P. benedeni* low salinities are tolerated at low temperatures, and this would be an aid to survival when low salinities prevail, during the winter period. However, higher than ambient temperatures as a result of thermal pollution could influence the distribution at such times. *T. costatus* and *P. benedeni* breed in the spring and summer and early autumn at a time of low freshwater flows. Tidal action could serve to disperse the tubificids to lower salinity regions which could be tolerated because of the high temperatures that prevail at such times. Progressive acclimation to reduced salinities may lead to enhanced survival during the winter (low salinity) period and in this way enable these tubificids to penetrate further up the estuary. However, the effect of low dissolved oxygen may also have influenced the upstream distribution of *T. costatus* and, to a lesser extent, *P. benedeni.* *T. tubifex* and *L. hoffmeisteri* breed in the winter and early spring respectively. The recruited tubificids in spring and early summer would thus be subjected to low salinities and relatively high fresh water flows. The high flows would tend to disperse the tubificids down the estuary and thus increase their distribution. Progressive salinity increases during the late spring and summer periods could then result in acclimation to promote a higher salinity tolerance; and this would seem to be the case for *L. hoffmeisteri*. However, *T. tubifex* remains much less tolerant of salinity. The effect of temperature on the salinity tolerance of *T. tubifex* and *L. hoffmeisteri* is minimal and therefore does not appear to have any adaptive significance. Low dissolved oxygen would seem to be a factor in the distribution of *T. tubifex* but not of *L. hoffmeisteri*, which is very tolerant of a wide range of environmental factors.

Studies on the Ecology of Tubificids at Erith

A temporal study was initiated at Erith, 26.3 km seaward from London Bridge, to provide baseline biological data in relation to the prevailing environmental conditions. It was anticipated that such studies would provide a basis for future studies concerned with the environmental condition of the Thames estuary. Erith beach was chosen because of its proximity to the outfalls from the major sewage treatment plants at Beckton and Crossness, its location within a region of the Thames estuary frequently encompassed by a major part of the "oxygen sag curve", and the presence of large numbers of tubificids which were primarily mono-specific (*T. costatus*).

While the ecology and life history of tubificids in freshwaters have received some attention (e.g. Brinkhurst, 1964a, Brinkhurst and Kennedy, 1965, Aston, 1973, Ladle, 1971) there has been fewer similar studies in estuarine and marine waters (e.g. Dahl, 1960, Brinkhurst, 1964b, Hunter and Arthur, 1978). The objective of the present contribution is to consider the ecology and life history of *Tubifex costatus* (Claparède, 1863) in an estuarine habitat exemplified by the estuary of the River Thames.

a)   Environmental Factors

Certain environmental factors were studied during the 12 month sampling period to assist in interpretation of the biological data. Additional water quality data were supplied by the Port of London Authority and Greater London Council. For standardization purposes these data were corrected to the half-tide situation within the estuary.

Figure 2 shows the range of air, water and sediment temperature recorded at Erith beach from which it is evident that water temperature drops more or less uniformly from about 21° C in late August to a minimum of 10° C in mid-January after which it rises similarly to mid-June. Except that on 14 June there was a marked atypical decline due to a significant increase in rainfall, after which it rose to a maximum in mid-July (22.5° C) before declining to about 18° C. Air temperatures, though following a similar overall pattern, were subject to greater oscillations and, except for one instance in mid-May and another in January, were always lower. The temperatures within the sediments tended to follow air temperature oscillations (Figure 2) and normally lay between air and water temperatures except for a mid-May sampling when sediment temperatures (at 1 cm and 3 cm) exceeded both air and water temperatures. The depth of the sediments also influences the temperature gradient, which tends to be more stable at depth, whereas in the surface layers temperature changes are more rapid. These buffering effects of the sediments are reflected in the

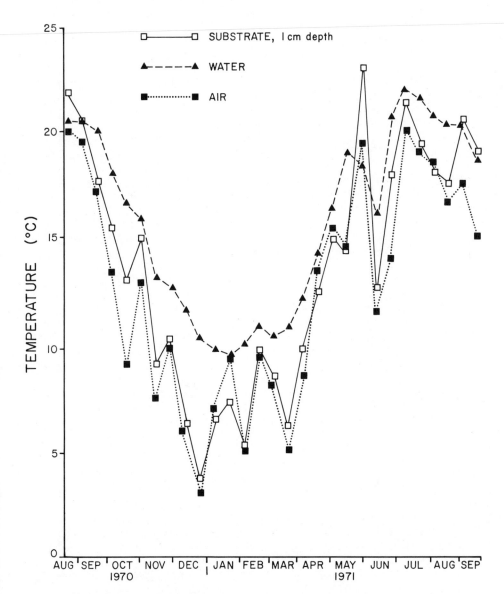

Figure 2. Temporal changes in the temperature of air, water and substrate at Erith.

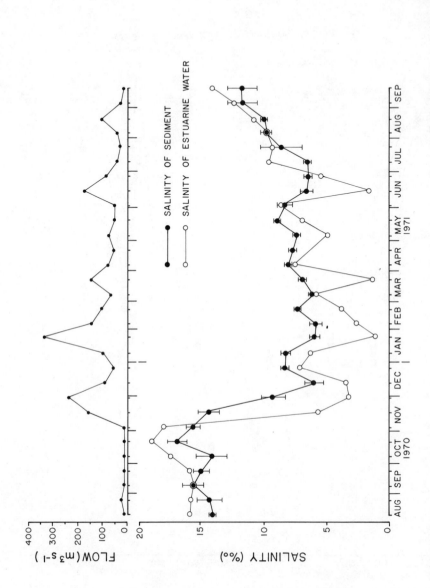

Figure 3. Seasonal changes in the freshwater flow entering the Thames estuary and variation in water and sediment salinities at Erith. (data courtesy of the Port of London Authority and Greater London Council).

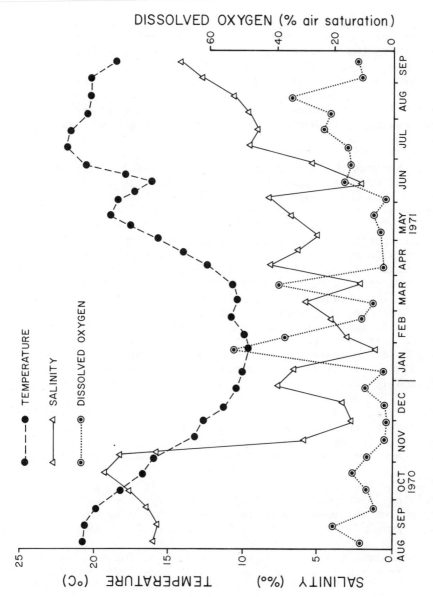

Figure 4. Seasonal changes in the dissolved oxygen, temperature and salinity of waters around Erith. (data courtesy of the Port of London Authority and the Greater London Council).

mean values (± standard deviation) recorded from 21 September 1970 to 20 September 1971 when the air temperature was $11.78\pm5.14^{\circ}$ C, the water temperature $15.55\pm4.28^{\circ}$ C and the temperature at 1 cm, 3 cm and 6 cm from the sediment surface $13.11\pm5.63^{\circ}$ C, $13.41\pm5.29^{\circ}$C and $13.36\pm5.11^{\circ}$ C respectively.

Since sampling was limited to periods of low water most of the records were taken in the morning during the late summer and autumn, and in the afternoon during the winter. Accordingly, higher temperatures would have occurred on emersion at low tide during early or mid-afternoon and for the same reason lower temperatures would have been encountered during emersion of the beach at night. Thus tubificids would be subject to higher and lower temperatures not only between sampling visits but also between tides.

The temporal change in water and sediment salinites and the rate of flow over Teddington Weir in 1970-1971 are given in Figure 3. From these data the following points emerge a) water salinity usually exceeds sediment salinity from August to mid-November (1970) and from July to September (1971) whilst in the intervening period the reverse situation occurs and b) the oscillations in sediment salinity are less pronounced than are those of water salinity which are manifested when freshwater flow over Teddington Weir increases or decreases. Increased freshwater flows reduce overall salinities in the estuary, while reduced flows operate conversely. Over the period of study the mean of the freshwater flow on sampling dates was $82.49\pm76.94m^3s^{-1}$ and the salinities of the water and the sediments were $8.08\pm5.20^{\circ}/oo$ and $9.28\pm3.38^{\circ}/oo$ respectively. The water content of sediment samples showed little annual fluctuation; the mean value being $50.35\pm4.16\%$ (by weight).

The variation in dissolved oxygen content of estuarine water around Erith beach over the 12 month period is given in Figure 4 together with the coincident temperature and salinity data. The data indicate the wide degree of variation which may occur over a short time. This sampling programme coincided in the autumn with a strike of local authority employees which brought about difficulties in disposing of sewage treatment waste into the estuary and whose overall effects are discussed more fully elsewhere (Sedgwick and Arthur, 1976, Sedgwick, 1978).

On 6 September 1970 the oxygen content was 20% air saturation and apart from minor oscillations was well under 10%, and even as low as 1.5% (30 November 1970), until 11 January 1971 (2.6%). On 26 January 1971 the level of oxygen in solution rose rapidly to 53.2% air saturation which coincided with the maximum freshwater flow of $336.7m^3s^{-1}$ recorded on any sampling occasion during the study. On 8 March 1971, 6.7% air saturation was recorded but

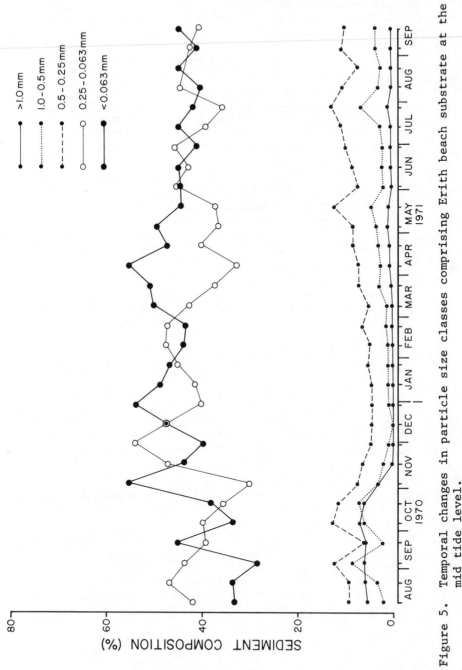

Figure 5. Temporal changes in particle size classes comprising Erith beach substrate at the mid tide level.

it rose to 33.6% air saturation on 22 March 1971. April and May
were characterized by concentrations of about 7% air saturation
but with the onset of increased freshwater flow to around 50 m^3s^{-1}
in early June levels rose to 15.5% air saturation by the middle of
the month. The dissolved oxygen content remained high and was at
34% air saturation on 22 August but thereafter reduced flows to
21 m^3s^{-1} (6 September) and 13.2 m^3s^{-1} (20 September) were reflected
in lowered oxygen saturation levels of 9.7% and 11.5% air saturation respectively. The mean dissolved oxygen level in the water
at Erith for each sampling period between 21 September 1970 to
20 September 1971 was 13.44±13.06% air saturation: the wide
variation serving to indicate the variability of this factor.

The temporal change in the percentage (by weight) of the
substrate fractions are given in Figure 5. The dominating particle
sizes at Erith were within the 0.25 to 0.063 mm and <0.063 mm
fractions and sizes in excess of 0.5 mm rarely attained 5% of
sediment composition. When sediments are subjected to increasing
freshwater flows the 0.25 to 0.063 mm particles increase and the
<0.063 mm particles decrease; a situation which is in a continual
state of flux.

From 10 August 1970 to 20 September 1971 the following mean
values were obtained:

Sediment fraction (mm)
>1.0   1.0-0.5   0.5-0.25   0.25-0.063   <0.063
Mean (%)
   1.74   3.35   8.59   41.77   44.71

The importance of determining not only the median size of
deposits but also the degree of their sorting (i.e. the proportion
of particles which are of similar diameter to that of the median
particle size) has biological implications in water retention of
the sediments, their stability and the capacity of intertidal
animals to burrow into them (Newell, 1970). The median particle
size was 3.8Phi or 0.070 mm, the sorting efficiency of the
"sedimentary agencies" or the Phi quartile deviation (QDØ), being
1.35. Perfect sorting would have a QDØ value of zero, and the
sediment collected at mean tide level was, accordingly, not
particularly well sorted. The degree of sorting above and below
the median, is represented by the Phi quartile skewness (SkqØ)
which was 0.15 suggesting that the efficiency of sorting of the
larger and smaller particles about the median was about the same
order.

The depths of the black ferrous sulphide layer are presented
in Figure 6. On 24 August 1970 the "reduced" layer was 31 mm
below the surface and, except for 21 September 1970 when it was
at 42 mm, it moved upwards to within 24 mm of the surface on

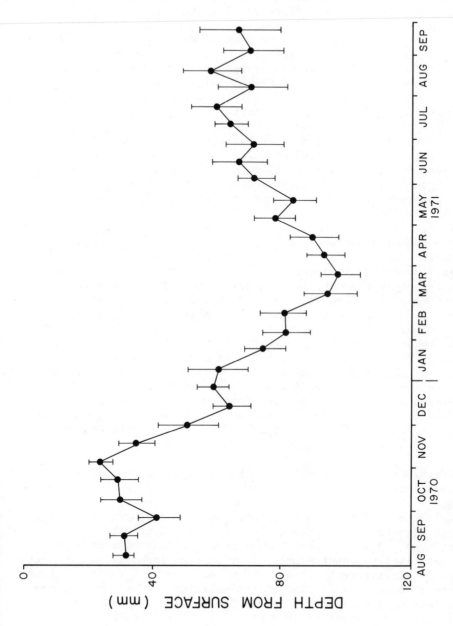

Figure 6. Temporal changes in the depth (±95% confidence limits of the mean) of the black layer in Erith beach sediments.

2 November 1970. After the onset of rains on 16 November and
higher freshwater flows the layer moved down to 35 mm and this
extended to 64.5 mm on 14 December where it remained for the
following month. In early February 1971 the depth from the
surface of the black layer increased with minor oscillations to
its maximum of 98 mm in the third week of March. From then to
the end of the survey the black layer became nearer to the
surface and on 20 September was at about 65 mm depth. The mean
depth over the sampling period was 65.53±20.19 mm.

b)   Correlations Between Environmental Factors

        To determine the association between factors, correlation
coefficients (and/or partial correlation coefficients) were
determined using the appropriate computer program of Dixon (1965).
Table 3 presents the correlation coefficients, and significance
level, for the relationship between the environmental factors.

        Initially environmental factors were relatively stable, but
most changed with the onset of heavy rain and the consequent
increase in freshwater flow. The correlation coefficients
between flow and environmental factors indicate the importance
of the former in the estuarine system. The high correlation
between freshwater flow and the 0.25 to 0.063 mm sediment
fraction might suggest that this fraction is transported seawards
in the estuary as an accompaniment of high flows with some settling
in the regions of Erith. Alternatively, it is conceivable that
the finer, upper layers of the sediment are removed resulting in
exposure of the coarser grades. The particle size fluctuations
throughout the sampling period do in fact suggest a changing
composition of the beach, the implications of which will be
considered later. The <0.063 mm fraction was correlated positively
with water content of the sediment and negatively with changes in
the 0.25 to 0.063 mm fraction.

        Freshwater flow was also significantly correlated with water
temperature for as the flow rate increased the water temperature
decreased. Nevertheless this is also a reflection of the time of
year when the high flows occur; late autumn and winter are the
times of high freshwater flows, and are coincident with low air
temperatures. Air and water temperature drops so does that of
the water, even though thermal pollution from industry may increase
the local temperature by about 4.5° C above ambient (Arthur, 1972).
The effect of freshwater flow on the salinity of the water in the
estuary would also be marked (H.M.S.O., 1964). A low freshwater
flow in the late summer of 1970 gave high salinities but heavy
inflow of freshwater in November reduced severely the salinity
in the estuary to show a strong negative correlation. Whereas
the water salinity changed rapidly with fluctuations in the flow
of freshwater, changes in the sediment were slower and this is

Table 3. The correlation between environmental factors recorded at Erith.

Parameters	Correlation co-efficient (r)	Significance level (P)
Freshwater flow with 250-63μ sediment fraction	+0.546	0.01
Freshwater flow with water temperature	-0.579	0.01
Freshwater flow with water salinity	-0.754	0.001
Freshwater flow with sediment salinity	-0.445	0.02
Freshwater flow with dissolved oxygen	+0.441	0.05
<0.063 mm sediment fraction with water content of the sediment	+0.471	0.02
<0.063 mm fraction with 0.250 to 0.063 mm sediment fraction	-0.427	0.05
Air temperature with water temperature	+0.800	0.001
Air temperature with sediment temperature: at 1 cm depth	+0.969	0.001
Air temperature with sediment temperature: at 3 cm depth	+0.952	0.001
Air temperature with sediment temperature: at 6 cm depth	+0.936	0.001
Water temperature with sediment temperature: at 1 cm depth	+0.922	0.001
Water temperature with sediment temperature: at 3 cm depth	+0.943	0.001
Water temperature with sediment temperature: at 6 cm depth	+0.962	0.001
Salinity of water with salinity of sediment	+0.833	0.001

reflected in the lower significance of the correlation (P=0.02).
Such a more stable saline environment may confer a greater survival
potential upon the tubificids.

A positive correlation exists between the dissolved oxygen
and freshwater flow but as may be seen from Figure 4, the initial
high flows which occurred in November and December 1970 did not
produce a marked effect on the dissolved oxygen level. The
dissolved oxygen level was low at this time, possibly reflecting
the adverse conditions which applied during the strike of local
authority workers, (Sedgwick and Arthur, 1976) and this could be
associated with the "flushing out" of the upper estuary of accumulated material during previous low flows. Scour of the sediment
also results in reduced dissolved oxygen levels due largely to
demands exerted by reduced sediments and fine, particulate organic
matter in suspension. That this situation probably occurred during
the late autumn and early winter (1970-1971) is shown by the high
oxygen levels which were recorded at the end of January and March
1971 when there was an increase in freshwater flow.

Air temperature was positively correlated with the temperature
at 1 cm, 3 cm and 6 cm in the sediment as was water temperature.
The correlations are very significant but it is noteworthy that
the correlation coefficients between air and temperature at 1 cm,
3 cm and 6 cm decreases in value with depth, whilst the opposite
occurs for the correlation with water temperature. This suggests
that at the time of collection (low tide) air temperature influenced
the temperature of the upper layers of sediment more than the lower
layers, and may be explained by the thermal capacity of the sediment. During the reported work the temperature of the water was
frequently higher than that in the sediment. On immersion the
temperature of the water and sediment would equilibrate slowly and
on emersion the air temperature and possibly evaporation would
cause the temperature of the surface layers to change quickly,
thus producing a temperature gradient. There is therefore a
"buffering" effect, the sediment temperature being intermediate
between water and air temperatures.

The depth of the black anoxic layer changed during the course
of the study of Erith beach, and substantiates the observations
of Perkins (1957) and of Brafield (1964) at Whitstable. Perkins
found that at Whitstable the black layer was about 80 mm from the
sediment surface for much of the winter period while during the
summer it was at about 10 mm from the surface. Temperature has
a significant effect here in its influence on chemical reactions
and on the abundance of bacteria. Brafield (1964) reports on the
association of the black layer with low oxygen concentrations in
the interstitial water, a high percentage of fine sand and poor
drainage. Unlike the Whitstable studies quoted, the depth of this
black layer in the Erith sediment did not return to its original

level during the summer of 1971 and remained much lower than in
the late summer of 1970. Fenchel and Riedle (1970) state that
the vertical movements of the black layer are correlated with
changes in four external parameters. These are: increase of
protection (e.g. by embayment) reducing permeability, and seasonal
changes related to temperature cycles, mixing of the sediment by
winter storms and the input of organic matter. Photosynthetic
activity of algae has also been implicated in lowering the black
layer during daylight, but it rises again at night (Fenchel and
Riedl, 1970). The involvement of organic matter may be considered
important, for in the autumn of 1970 coincident with low river
flows and the strike of local authority workers, the depth of the
black layer was close to the surface. It is likely then that there
was retention and the settling out of organic matter. With the
onset of high river flows, this accumulated organic material could
in part be removed, as is known for other periods in the Thames,
and this coincided with the reduction in depth of the black layer
from the sediment surface. Even so it is important to recall that
Ellis (1925) showed that pollution was not the cause of the blackening of sediment in the Clyde, nor is it likely that the sediment
at Whitstable is as "polluted" as that in the Thames estuary,
nevertheless, the black anoxic layer develops there and 'migrates'
to similar levels as at Erith.

c)   The Influence of Tubificids on the Depth of the Black
     Layer in Sediments

     The laboratory experiment to determine whether *T. costatus*
affected the depth of the black layer in sediments from Erith
revealed that after two days the tubificids which had been added
to a core of "reduced" sediment were within 20 mm of the surface,
whereas in the sediment without tubificids, aeration of the water
had produced a well defined 2 mm deep grey covering layer. The
tubificids continued to burrow in the black sediment bringing blue-
grey sediment to the surface layers, and after 9 days a well
defined black layer was present 18 mm below the surface. Moreover,
the tubificids were more aggregated. After 42 days the depth of
the black reduced layer was 23 mm from the sediment surface, but
in the vessel without tubificids the corresponding depth was
approximately 3 mm. The surface sediments in the container holding
T. costatus were prone to erosion by the water currents initiated
by aeration inferring the ability of the tubificids to loosen the
surface layers and promote erosion.

     In order to examine the possible effects of tubificids on the
depth of the black layer within Erith beach field data were
subjected to analysis. Correlation analysis was carried out
between the depth of the black layer and water and sediment
temperature, water and sediment salinities, dissolved oxygen,
number of tubificids, biomass of tubificids, life cycle stage

of tubificids and sediment fractions <0.063 mm and 0.25 mm to
0.063 mm. Initial analyses gave a positive correlation coefficient
of +0.565 (P=0.01) with tubificid biomass (which was also correlated with the number of mature and breeding *T. costatus*).
Furthermore, partial correlation analysis indicated that water
temperature was negatively correlated with the distance of the
black sediment from the surface (partial correlation coefficient
p-0.421; P=0.05) and that tubificid biomass was positively correlated (p+0.658; P=0.001). The stronger correlation with tubificid
biomass may well relate to the failure of the black layer to rise
close to the surface during the late summer 1971 and therefore the
overall assessment of the situation favoured tubificid biomass.
Experimental studies on the effect which tubificids have on the
depth of this black layer proved that they have the facility for
bringing material to the surface and for mixing the sediments.
In the field, Schumacher (1963) found a correlation between the
number of tubificids and the depth of the black layer in the River
Elbe. Edwards (1962) reported on the ability of tubificids to
mix sediments and concluded that tubificids were more effective
in transferring solid particles from lower layers than the larvae
of *Chironomus riparius*. By bringing deeper sediments of a "reduced"
nature to "oxygenated" surroundings, oxidation can proceed with
increased oxygen demand for biological and chemical activities.
The utilization of organic matter by bacteria at depth in the
sediment does not proceed as fast as that in the surface layers
(Zobell, 1939). In some oxygenated situations then, the utilization is likely to be greater and consequently the demand for
oxygen would be increased. The laboratory study described earlier
also illustrates the effect which the tubificids have on loosening
sediment whereby they render the surface layers susceptible to
scour. When this happens the fine sediment particles and organic
matter are carried into suspension and the resultant oxygen demand
is high (H.M.S.O., 1964; Edwards and Rolley, 1965).

The work of Alsterberg (1925) indicated that in 24 hours
T. tubifex and *L. hoffmeisteri* "displace" a quantity of mud four
times greater than their body weight; Appleby and Brinkhurst (1970)
found this to be greater at higher temperatures. Whilst their
results at 12° C for *T. tubifex* agree with those of Alsterberg
(1925) at higher temperatures, sediment displacement is about
8 times the body weight. Assuming comparable rates of sediment
passage in *T. costatus* and, using the mean biomass of 868gm^{-2}
for the sampling period at Erith, during one day about 7 kg m^{-2}
of sediment would be displaced or over one year this would amount
to about 2500 kg m^{-2}. Undoubtedly, this could influence the
oxygen resources of the environment.

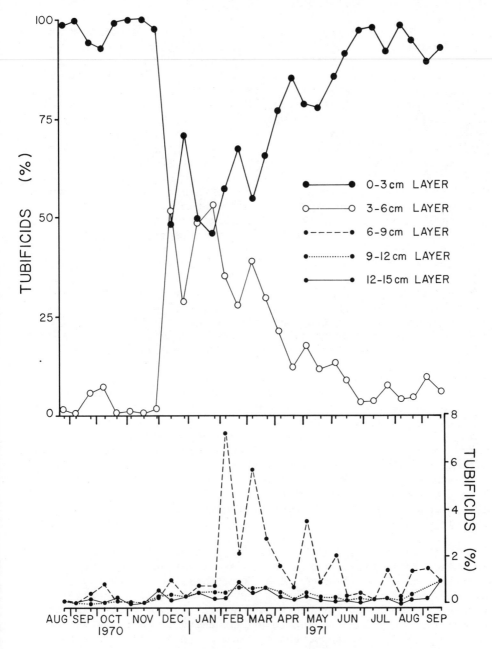

Figure 7.   Temporal changes in the vertical distribution
            of tubificids at Erith within 3 cm layers, to
            a depth of 15 cm.

d)   The Vertical Distribution of Tubificids

    The vertical distribution of tubificids in the sediment of
Erith beach at times of low water was studied from 24 August 1970
to 20 August 1971. Their distribution in 3 cm layers were expressed
as a percentage of the total tubificids in the cores as shown in
Figure 7. From these data it is evident that the proportion of
worms occupying sediments at depths exceeding 9 cm is at a level
of about 1% or less over the whole year, whilst maximal occupation
of the 6 cm to 9 cm level is around 7% in February. At the
beginning of this study (August 1970) the tubificid populations,
predominantly immature and maturing worms, were largely aggregated
at depths of 0-3 cm and this distribution persisted to the end of
November. A sudden, large reduction in this occupancy followed,
with a concomitant increase in the percentage of worms in the 3
to 6 cm layer. Thus, on the 14 December 1970, 47% were in the
0 to 3 cm layer and 51.4% in the 3 to 6 cm layer; corresponding
figures for 28 December were 70.7% and 28.3%, and on the 26 January
1971 the lowest of 45.6% in the 0-3 cm layer and the highest of
53% in the 3-6 cm layer were recorded. Subsequent to mid-August
the percentage of worms inhabiting the 0-3 cm layer showed an
overall increase, despite the oscillations. Conversely, the percentage of worms inhabiting both the 3-6 cm and the 6-9 cm layers
decreased. During August 1971 the proportion of worms in the 0-3
cm layer approximated to that found at the same time in the
previous year.

    This pattern of vertical distribution in the sediments based
only on one full year's survey may be subject to various interpre-
tations. It could, for example, be construed as an inherent
physiological behavioural mechanism whereby maturing and immature
worms occupy the surface 3 cm, whilst mature worms move to significantly greater depths preparatory to moving upwards as is shown
in breeding worms over this period. Alternatively, the pronounced
downward migration evident later in 1970 may have been induced
by increased freshwater flows entering the estuary, or by other
abiotic parameters.

    Correlation and partial correlation analyses were carried out
to investigate the effects of a number of variables on the vertical
distribution of tubificids. The factors considered were air, water
and sediment temperatures, water and sediment salinities, dissolved
oxygen, the depth of the black reduced layer, the percentage of
breeding *T. costatus*, of mature and breeding *T. costatus* and of
mature, breeding and postbreeding worms. The results of partial
correlation analysis revealed that the tubificids in the 0-3 cm
layer were correlated with water temperature (p+0.663; P=0.001)
and sediment salinity (p+0.448; P=0.05); those in the 3 cm to
6 cm layer with water temperature (p-0.683; P=0.001) and sediment
salinity (p-0.447; P=0.05) and those in the 6 cm to 9 cm layer

with mature and breeding *T. costatus* (p+0.445; P=0.05). Quite
obviously the influence of temperature and salinity is not as great
at depth in the sediment, and the data reflect the vertical migration
of the maturing and breeding tubificids into what was probably a
more stable environment in addition to the effects of temperature
and salinities, reductions of which were related to a downward
migration of the tubificids within the sediment. Hunter (1977)
found that *L. hoffmeisteri* deposited cocoons at depth in the sediment. It is likely that *T. costatus* may also deposit cocoons at
depth thus aiding population maintenance at a particular location.
Palmer (1968) thought that the black reduced layer in sediments
limited the vertical movements of tubificids and Vader (1964)
similarly thought this to be a factor in an intertidal area.
Observations at Erith beach indicated that *T. costatus* would
penetrate the black sediments and the depth of this layer was
not correlated with the vertical distribution of the tubificids
but with their biomass; thereby inferring the ability of the
oligochaetes to mix the upper sediment layers.

e)   Tubificid Species Recorded at Erith Beach

Population changes over the period of study are shown in
Figure 8 together with the 95% confidence limits based on logarithmic transformation of the data. Initially, the density of tubificids was 316.10^3m^{-2} of which 11.10^3m^{-2} were *Monopylephorus
rubroniveus*. A slight rise in the total density of worms occurred
thereafter to 2 November 1970. Within this period, on 5 October
1970 one specimen of *P. benedeni* was recorded giving a population
equivalent to 5.10^3m^{-2}, the rest being *T. costatus*. *M. rubroniveus* was recorded again at a density of $9.6.10^3m^{-2}$ on 19 October
1970 and small individuals were found on 16 November 1970 at
17.10^3m^{-2}, and on 28 December 1970 at 4.10^3m^{-2} before disappearing
from the fauna to reappear on 22 August 1971, at a density of
about 4.10^3m^{-2}. Later, young specimens were recorded on 6 September and 20 September 1971 representing densities of 14.10^3m^{-2}
and 18.10^3m^{-2} respectively. *L. hoffmeisteri* was first noted at
Erith on 1 June 1971 when the density was 4.10^3m^{-2} and small
individuals were further recorded on 28 June 1971 (8.10^3m^{-2}), and
again on 12 July 1971 and 26 July 1971 at densities of about
4.10^3m^{-2}. These species made only minor contributions to the
total tubificid populations on Erith beach which were almost
exclusively of *T. costatus*. The mean density of tubificids
recorded throughout the sampling year was approximately
$370\pm12.10^3m^{-2}$.

Population Studies and the Life History of *Tubifex costatus*

Tubifex costatus was the most abundant species of tubificid
recorded at Erith and for most of the study, the only one recorded.
Because of this, the following comments are considered generally
applicable to *T. costatus*.

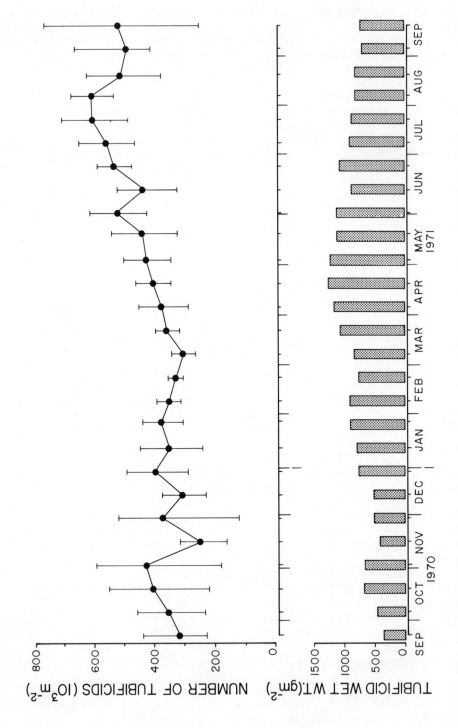

Figure 8.   Temporal changes in the population density (±95% confidence limits of the mean) and biomass of tubificids at Erith.

a)   Spatial Dispersion

A population is seldom randomly or regularly dispersed, but
frequently contagious when the variance is significantly greater
than the mean (Elliott, 1971). Alley and Anderson, (1969) found
that the curve relating the sample means and variance for oligochaetes, approximated to the negative binomial, a view supported
by Brinkhurst et al. (1969) for the distribution of oligochaetes
in Toronto Harbour. The variance and means based on the numbers
of tubificids recorded at Erith beach (Table 4) show that since
the variance is greater than the mean, there is contagion in the
population. This was substantiated by plotting the log of the
mean counts on each sampling occasion against the log of the
variance, in accordance with Taylor's Power Law (Taylor, 1961).
Its slope as stated by Taylor, gives a measure of dispersion,
varying continuously from 0, for a regular distribution, to infi-
nity, for a highly contagious distribution. The slope of the
line fitted to our data is 8.144 indicating a contagious distribution. The negative binomial distribution was considered to be
a suitable model and calculation of the exponent k of the negative
binomial distribution for the series of samples collected at Erith
gave a value of 8.046 (the method used for this calculation is
applicable, when there is no relationship between the mean ($\bar{x}$) or
log $\bar{x}$ and $\frac{1}{k}$ (Elliott, 1971) - as was so for these data). Brinkhurst et al. (1969) recorded a similar k value of 7.24 for tubificids in Toronto Harbour.

Changes in the degree of aggregation with respect to different
stages in the life history of animals have been reported and it is
conceivable that in tubificids, the degree of aggregation may alter
during the life cycle in response to biotic and abiotic environmental components. This was determined by use of the Lexis index

$$(\lambda = \frac{s}{\sqrt{\bar{x}}})$$

"which might be more appropriately named 'index of aggregation'"
(Debauche, 1962), whose significance can be tested by the Chi^2
test. The Lexis index (λ), calculated for each sampling occasion,
is shown in Table 4. The values for λ were significant at the
P=0.01 level, except for the samples of 22 February 1971 (2.30)
which was significant at the P=0.05 level. For a regular distribution λ approaches 0, and for a randomly distributed population
λ equals 1. λ indicating maximum aggregation would have a value
equal to $\sum x$, when x=value of the independent variate. From Table 4
it is also clear that the degree of aggregation changed throughout
the sampling period, the largest values of λ being in autumn 1970,
late spring and summer 1971, and the lowest values during winter
of 1971. Such a degree of contagion contrasts with the findings

Table 4.   The mean, variance and degree of aggregation (Lexis
            Index) for tubificids within 15.2 cm^2 samples
            collected from Erith in the Thames estuary.

Sampling Date		Mean	Variance	Lexis Index (λ)
21 September	1970	481	5668	3.43
5 October	1970	536	48591	9.52
19 October	1970	613	118467	13.90
2 November	1970	650	230096	18.81
16 November	1970	373	29316	8.87
30 November	1970	562	249100	21.03
14 December	1970	465	23187	7.06
28 December	1970	598	27011	6.92
11 January	1971	525	36917	8.39
26 January	1971	568	25863	6.75
8 February	1971	528	6707	3.56
22 February	1971	496	2572	2.30
8 March	1971	458	8737	4.37
22 March	1971	540	6215	3.39
5 April	1971	563	26027	6.80
19 April	1971	608	13081	4.64
3 May	1971	641	23391	6.04
17 May	1971	657	35566	7.36
1 June	1971	790	44045	7.47
14 June	1971	647	38430	7.71
28 June	1971	801	15354	4.38
12 July	1971	838	29075	5.89
26 July	1971	908	45046	7.04
9 August	1971	915	22124	4.92
22 August	1971	769	77403	10.00
6 September	1971	738	29129	6.28
20 September	1971	781	181724	4.82

of Brinkhurst (1964b) who found that *T. costatus* was under-
dispersed, perhaps reflecting the effects of different densities
and environmental factors between study locations.

In an attempt to determine which factors could affect the
degree of aggregation λ, consideration was given to sediment temperature and salinity, dissolved oxygen, the number of tubificids
and the percentage of maturing and breeding *T. costatus*. Between
the biological factors only the factor representing maturing and
breeding *T. costatus* was found to be significantly correlated
(P0.02; r =+0.449), and with the other factors sediment salinity
was found to be significantly correlated (P0.02; r=+0.486).
However, when subjected to partial correlation analysis the
significance levels decreased (p-0.2) but sediment temperature
(p-0.303) salinity (p+0.30) and the maturing and breeding *T. costatus* (p-0.298) were identified as possible factors affecting the
degree of aggregation. It is interesting to note that there was
no effect of the density of tubificids on λ. The vertical
distribution of the tubificids, associated with the response to
temperature and salinity changes and for breeding purposes could
be associated with the change in aggregation. During the winter
period the tubificids were recorded lower in the sediment (Figure 7)
and so occupied a larger volume of sediment than they did during
the summer and autumn periods. This could have resulted in a
more even distribution even though aggregation could be expected
at times of reproduction (Debauche, 1962).

b)   Population Studies

Changes in tubificid densities and biomass

In recognition of the spatial distribution of the tubificids
and the requirement to perform parametric tests, the population
data were normalized using a logarithmic function, log (x+1).
Figure 8 presents the temporal changes in the densities and
biomass of tubificids at Erith.

From 21 September 1970 to 2 November 1970 the tubificid
population at Erith beach increased progressively from $316 \cdot 10^3 m^{-2}$
to $427 \cdot 10^3 m^{-2}$. Subsequently, on 16 November 1970 the lowest
tubificid population in this annual study was recorded at
$245 \cdot 10^3 m^{-2}$, after which the population increased slightly and
oscillated at between $300 \cdot 10^3 m^{-2}$ and $400 \cdot 10^3 m^{-2}$ until 8 February
1971 before slowly declining to early March. A progressive
increase to $519 \cdot 10^3 m^{-2}$ on 1 June 1971 followed, before a temporary
reduction to $425 \cdot 10^3 m^{-2}$ on 14 June 1971, but by 28 June 1971 the
incidence was slightly in excess of that for 1 June 1971 ($526 \cdot 10^3$
m^{-2}). Thereafter the rise continued to $601 \cdot 10^3 m^{-2}$ in late July
and early August. Subsequently, on 22 August 1971 to 20 September
1971 the population level declined to $500 \cdot 10^3 m^{-2}$.

On 20 September 1970 the biomass (wet weight) of tubificids
was 350g m^{-2} and this increased to 675g m^{-2} on 2 November 1970.
A reduction in numbers of tubificids on 16 November 1970 was
reflected in the decline of biomass to 410g m^{-2}. Thereafter, and
except for minor decreases on 22 February 1971 and 8 March 1971,
the biomass increased to maximum of 1375g m^{-2} on 19 April 1971.
This was succeded by a progressive decrease, except for a more
rapid fall on 16 June 1971. On 20 September 1971 the biomass was
725g m^{-2}. Over the period of study, the mean biomass was 868.4
±333.1 g m^{-2} and the mean weight of tubificids was 2.04±0.65 mg.

The only other detailed study on the life history of *T. costatus* was undertaken by Brinkhurst (1964b) at Hale, Lancashire,
England, where this species represented almost 100% of the tubificid fauna. The maximum mean density recovered at Erith of
601·10^{3}m^{-2} contrasts with density recorded at Hale, of 40·10^{3}m^{-2},
the latter remaining almost constant over a 12 month period.
Brinkhurst explained this observation on the basis that as post-
breeding tubificids died the recruitment of young worms into the
population maintained relatively constant densities. At Erith,
recruitment was responsible for a major increase in the density
of tubificids to the maximum of 601·10^{3}m^{-2} on August 9, 1971, and
whereas population changes were marked, biomass changed little.
Such changes give support to the view that *T. costatus* has a two
year life cycle for in those tubificids which breed once or more
each year and then die (e.g. *T. tubifex, L. hoffmeisteri*) biomass
changes are extremely obvious (Hunter, 1977).

c)   Population Changes and Environmental Factors

The environmental parameters that were chosen for correlation
with changing tubificid populations were sediment fractions of
<0.063mm and 0.25 to 0.063mm, air, water and sediment temperatures,
water and sediment salinities and dissolved oxygen. The results
of the correlation and partial correlation analyses (Table 5) show
that certain environmental factors could influence the tubificid
populations.

It is probably coincidental that oxygen, temperature and
salinity were correlated with population changes, and the factor
which probably has most effect is that of beach movement (i.e.
particularly movement of the finer fractions <0.25mm). It is
unlikely that temperature and salinity markedly affected the
tubificid populations recognizing the tolerance of *T. costatus* to
these variables. In the partial correlation analysis dissolved
oxygen was positively correlated (albeit at a low level, P=0.2)
with post-recruitment densities, however, densities remained
relatively constant from 22 August 1971 to 20 September 1971 even
though the dissolved oxygen level decreased markedly.

Table 5.   Correlations between the density of tubificids and
environmental factors at Erith

	Correlation Coefficient (r)	Significance Level (P)
Density (over study period) with sediment temperature	+ 0.488	0.01
Density (over study period) with sediment salinity	− 0.324	0.1
Density before recruitment with sediment temperature	− 0.533	0.05
Post-recruitment density with sediment temperature	+ 0.753	0.02
Post-recruitment density with <0.063mm particle size	− 0.703	0.01
	Partial Correlation Coefficient (p)	
Density (over study period) with 0.25 to 0.063 mm particle size	+ 0.344	0.2
Density (over study period) with sediment temperature	+ 0.563	0.01
Density (over study period) with sediment salinity	− 0.595	0.01
Density before recruitment with <0.063mm particle size	− 0.671	0.02
Density before recruitment with 0.25 to 0.063 mm particle size	− 0.792	0.01
Density before recruitment with sediment temperature	− 0.706	0.02
Density before recruitment with sediment salinity	− 0.534	0.1
Post-recruitment density with <0.063mm particle size	− 0.568	0.2
sediment temperature	+ 0.742	0.05
sediment salinity	− 0.529	0.2
dissolved oxygen	+0.526	0.2

    Both from and before recuitment, particle size fluctuations
were correlated with population changes suggesting that this factor
could be important in the distribution of the tubificids. On 16
November 1970 there was a marked drop in the tubificid population
at a time when most of them were in the upper layer of the sediment
(based on low tide data). If scour occurred in response to the
high river flows, some tubificids would also be removed. Mud
scoured from the reaches close to Beckton and Crossness has, on
occasions, been deposited in Gravesend Reach (H.M.S.O., 1964).
Huddart noted a marked increase in the density of *T. costatus* at
Greenhithe from 12 to $40 \cdot 10^3 m^{-2}$ (unpublished data) and he attributed this population increase to tubificid transporation during
high freshwater flows. High river flows were also recorded after
this date, during January and March 1971, but no similar situation
occurred. This could be explained by the vertical distribution
of the tubificids. Assuming that the position of *T. costatus*
when exposed at low tide is similar to that when immersed, the
tubificids were deeper in the sediment and may not have been
scoured. High river flows were recorded in June 1971 when the
tubificids were closer to the sediment surface and again a drop
in their density was recorded indicating that scouring of the
beach had occurred. Young tubificids frequently occupy the upper
sediment layers, and therefore would be very susceptible to erosion.
Similarly, on 22 August 1971 high river flows occurred, relative
to the previous two weeks. The tubificids were in the upper sediment layers and again there was a marked reduction in the tubificid
population, which dropped from its maximum value of $601 \cdot 10^3 m^{-2}$ to
$505 \cdot 10^3 m^{-2}$. During the time when the population changes were
marked there were changes in the proportions of the 0.25 to 0.063mm
and <0.063 mm sediment fractions (these changes also occurred
during February 1971, but as stated above may not have had any
effect on the tubificid population). Thus particle size fluctuations could be associated with tubificid population changes.
George (1964) found that *Cirriformia tentaculata* (Montagu) juveniles were frequently in the upper sediment layers on the beach
at Hamble (Southampton). Sediment movement was correlated with a
reduction in the population of smaller worms. George concluded
that the distribution of *C. tentaculata* could not be correlated
with any environmental factor, but that the most significant
influence was periodic strong wave action. It is likely that
similar conclusions could be drawn from the studies at Erith where
the most significant factor associated with population changes
was water flow and its effect on beach scouring.

d)   Life History Studies on *Tubifex costatus* Growth Patterns

    It has been proposed that the life cycle of *T. costatus*
extends for over more than one year (Brinkhurst, 1964b, and
Birtwell, 1972) but the available evidence supporting this is
not strong. Length-frequency date (Figure 9) yields further

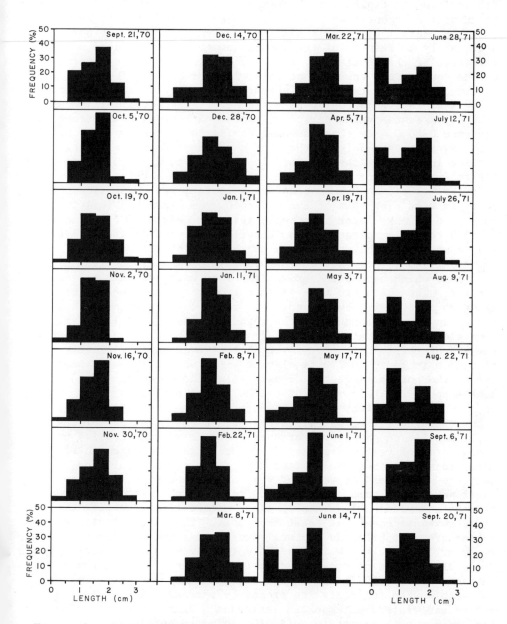

Figure 9. Temporal changes in the frequency of 0.5 cm. size classes in populations of *Tubifex costatus* at Erith.

interpretative information on the life cycle, but there are still
difficulties in explaining it entirely on the basis of 270 samples
taken over only one year. A basic assumption then has to be made,
that the overall pattern is similar from year to year but not
necessarily in respect of the details of timing nor indeed of the
magnitute of the population of each size class, for the vagaries
of the environment ensure that this will not be so. The productivity of an ecosystem will fluctuate from year to year above and
below what is considered to be a characteristic level. This can
only be assessed over a period of years. Accepting this assumption
then the 0 to 0.4 cm class assumes dominance in the population from
late May into June-July of the year n which coincides with the
decline in worms of the largest classes (2.0 to 2.4 and 2.5 to
2.9 cm) which having deposited cocoons a little earlier may now
be dying. The 0 to 0.4 cm class peak is followed about two months
later in year n by that of the 0.5 to 0.9 cm class and this in
turn 6 to 8 weeks later by a peak, again in year n, for producing
the 1.0 to 1.4 cm class. This latter class together with 1.5 to
1.9 cm worms possibly represents two generations of worms viz.
those derived by growth from the 0.5 to 0.9 cm class of year n
and those which have been derivatives from worms of year n-1 or
earlier. The postulate then is that growth from the 0-0.4 cm
class to the 1.0-1.4 cm of year n is rapid and covers about 4.5
months and that subsequently the worms of classes 1.0-1.9 cm grow
more slowly and remain in the sediments for about 12 months. At
the end of this time those of year n-1 have grown sufficiently
to become included in 2.0+cm classes which mature, breed, and pass
out of the population after deposition of the cocoons. There is
some support for this view for unlike the size classes <1 cm the
combined 1.0 to 1.4 cm and 1.5-1.9 cm classes persist in collections throughout the year at levels of not less than 32% of the
whole population sampled. Moreover at the time when 1.0-1.4 cm
class worms of year n are recruited into this population these
worms constitute between 64% and 84% of the total population, but
with the "transfer" of worms of year n-1 to the higher classes
their percentage composition thereafter, for the rest of the year,
fluctuates around 50%. Thus on these speculations the life cycle
of T. costatus covers at least two years. The vertical distribution of the tubificids may well support the speculation of a
two year life cycle. Whilst adult worms may move downwards in
the sediments the cocoons are deposited within 2-4 cm of the
surface (Hunter, 1977) and this would make them particularly
vulnerable when sediments are eroded by increased flow rates in
autumn and winter in temperate latitudes. Hence, there is a need
for rapid growth to attain sufficient size to enable them to
burrow more efficiently into deeper substrates (Arthur, 1965).
From the vertical distribution curve it is clear that initially
the tubificids are restricted to the 0 to 3 cm layer for the early
part of their lives but by January a substantial proportion are in

the 3 to 6 cm layer. Following this there is a gradual return to
the 0 to 3 cm layer and in general at the time of cocoon deposition
the vast proportion of the worms are relatively superficially
located in the sediments.

The changes in the mean dry weight and length of individuals
comprising the population of tubificids is shown in Figure 10.
While there was an increase in mean weight of *T. costatus* up to
19 October 1970, from this date until 14 December very little
growth occurred. This latter period coincided with low dissolved
oxygen levels in the estuary and it may not be coincidental that
growth was impaired. Hunter (pers. comm.) studying populations of
L. hoffmeisteri in the Thames estuary noted a reduction in growth
at the same time. The cessation of activity, as shown in laboratory experiments when *T. costatus* was exposed to low dissolved
oxygen levels and anaerobic conditions, may be associated with
the lack of growth. The mean dry weight of the tubificids varied
but did not drop below about 0.24 mg in contrast to those tubificids
with a one year life cycle such as *L. hoffmeisteri*. For this
species mean weight changes are very pronounced and much greater
than that shown for *T. costatus* (Hunter, 1977).

Biomass changes throughout the study period indicate a relationship with the number of tubificids. However, the correlation
between biomass and the number of tubificids was only significant
at the P=0.01 level (r= +0.496) while biomass with the percentage
of mature breeding and post-breeding *T. costatus* (r= +0.714) and
that between biomass and breeding *T. costatus* (r= +0.601) was at
the P=0.001 level of significance. The average wet weight of *T.
costatus* was about 2.0 mg (2.3 mg was given by Brinkhurst,(1964b)
for *T. costatus*) and that for breeding tubificids was about 3.0 mg.

Population Structure

An adequate assessment of population structure and the densities of the components on a seasonal basis requires information
on the ontogenetic changes through which the organisms pass. For
this purpose a number of stages in the life cycle of *T. costatus*
were identified. Their changes through time are shown in Figure 11.

On 21 September 1970 the population consisted largely of
immature worms (78% of the total population) which declined to 18%
in early November 1970. Subsequently, except for a slight increase
in November to about 24% and again in February 1971 to 16%, the
pattern was one of reduction in numbers in respect of these immature worms up to late March. Thereafter, numbers of immature worms
as a proportion of the whole population increased from 5.4%
(22 March 1971) to peak at 64.2% (26 July 1971), before falling
markedly to 21.3% of the total on 20 September 1971.

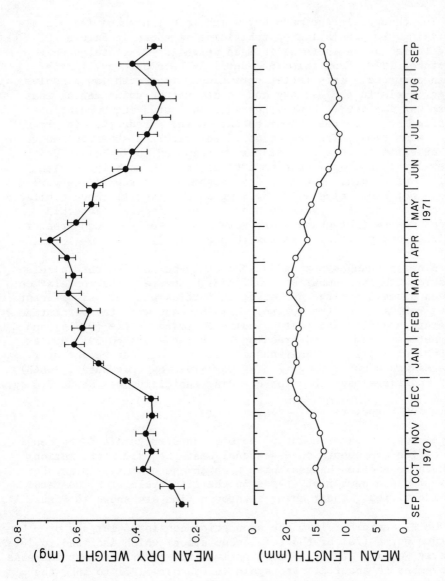

Figure 10. Temporal changes in the mean length and dry weight of *T. costatus* at Erith.

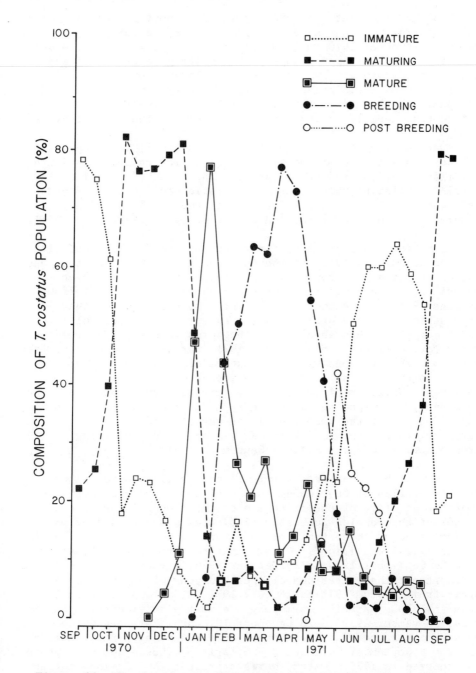

Figure 11. The incidence of 5 prescribed life cycle stages within populations of *T. costatus* at Erith from September, 1970 to September, 1971.

On the first date of sampling in 1970 about 22% of the worms
were maturing, which rose to around 80% over the next four or five
weeks (2 November 1970 to 28 December 1970). This was succeeded
by a rapid fall to approximately 48% in early January 1971 and to
13% by mid-January. Low levels of oscillating populations of
maturing worms (2-13%) occurred from February almost to the end
of June. At the end of this period these components of the population rose rapidly to about 79% at the end of the investigation.

Mature *T. costatus* to the extent of over 4%, first appeared
in the population in mid-December,1970, by the end of this month
was 11% and at the end of January 1971 had reached a maximum level
of 77%. Following which there was a rapid decline to 22 February
1971 when 26.6% were mature, this continued, with superimposed
fluctuations to zero at the end of the study (20 September 1971).

Breeding specimens were first encountered in the population
(7%) on 26 January 1971 and increased as a percentage of the population over the next ten weeks to attain a maximum of 77% on
5 April 1971. A rapid fall ensued to about mid-June, and from
14 June 1971 to 22 August only slightly more than 3% of the
population were in a breeding state except for samples collected
on 26 July 1971 when about 7% of the worms were so. No breeding
worms were collected in September 1971, nor were they evident from
September 1970 to January 1971. Expectedly, post-breeding worms
appear in the population as the incidence of breeding worms
declines. Whilst their role in the life history is uncertain they
represent, on morphological grounds, a distinctive phase. In mid-
May about 13% of the worm population was classified in this
category, increasing to 42% in early June and disappearing
progressively to zero, by late September 1971.

It will be clear from Figure 11 that whilst there is considerable overlap of developing phases of the tubificids over the
year the adoption of the criterion of their occurrence in excess
of 50% of the sampled population indicates a succession of phases
of dominance of those tubificids in varying morphological and
physiological conditions. Thus, immature worms were dominant
from 21 September 1970 to 19 October 1970, although this is likely
to represent the latter end of the 1970 peak of occurrence, and
again from 14 June 1971 to 26 July 1971, at a lower maximum level
than in 1970. Maturing worms are the most significant components
from 2 November 1970 to 28 December 1970 and there are strong
reasons for supporting the concept that the rise in their numbers
as from 9 September 1971 would similarly produce a dominant phase
as occurred in 1970. Mature worms represent >50% dominance from
28 December 1970 to 26 January 1971 (a relatively short period
when compared with maturing, breeding and immature worms). Whether
this is to be attributed to such coincidental factors as downward
movements into the sediments by tubificids and to significant

increased flow rates of freshwater (over 300m^{3}s^{-1} over Teddington Weir) on sampling occasions in 1970-71, or to real biological reasons cannot as yet be evaluated. Breeding worms predominate in the sample population from 22 February 1971 to the beginning of May 1971, since post-breeding worms are at no time pre-eminent elements of the population we do not propose to comment further on them at this time. The extension of low level populations of most of these phases into parts of the year when they are not dominant components is probably to be ascribed to late developing individuals of one generation being overlapped by early developing individuals of the next generation.

Comparison with other studies on the life history of *T. costatus* reveal some differences, but overall, similarity was found with the work of Brinkhurst (1964b) at Hale. Brinkhurst (1964b) chose the criterion for classifying "mature" *T. costatus* as the presence of genital ducts, accordingly there would be a temporal overlap between the time that mature worms were recorded at Hale and in the Thames estuary. At Hale about 10% of the tubificids were mature in January whereas in excess of 50% of the population was in this physiological state at Erith in the same month. Knöllner (1935) recorded mature worms throughout the year except in August and September, and Bülow (1957) at all times of the year. Breeding worms i.e. those with spermatophores in their spermathecae first appeared in the Thames in late January and dominated in the population from early February to May, whereas Knollner (1935) and Brinkhurst (1964b) recorded them in April-May and April-July respectively. This life cycle stage occurred at a maximum frequency of 77% in the Thames estuary but only at the 25% level at Hale (Brinkhurst, 1964b). An explanation of the differences in the duration of life cycle stages in different locations may well be related to temperature.

The influence of temperature on the periodicity and reproductive potential of tubificids has been studied. For example, in *Limnodrilus hoffmeisteri* by Poddubnaya (1959), Kennedy (1966) Aston (1973) and Wisniewski (1976); in *L. udekemianus* by Timm (1962) and in *T. tubifex* by Aston (1973) and Wisniewski (1976). In northern European lakes, for example, tubificids generally breed for only a restricted period in the summer when water temperatures exceed 15° C following the spring thaw. The temperature in the middle reaches of the Thames estuary is raised by 4.5° C above ambient due to thermal discharges (Arthur, 1972) and a positive correlation of sediment temperatures with water temperatures was determined. Furthermore, the percentage of mature and breeding *T. costatus* was found to be negatively correlated with water temperature (r= -0.601; P=0.01) and sediment salinity (r= -0.639; P=0.001) but from the recruitment of immature worms there was a positive correlation with temperature (P=0.02) which persisted when the data was subjected to partial correlation.

This suggests, that the temperature influences recruitment in this species and the effects of thermal pollution coupled with mild winters may advance the onset of breeding in the Thames estuary.

Brinkhurst (1964b) found that each cocoon of *T. costatus* contained 2 or 3 eggs and that in the survey he carried out at Hale the population remained constant. In the present study, however, the pre-recruitment population of $370 \cdot 10^3 m^{-2}$ increased to a maximum post-recruitment level of $601 \cdot 10^3 m^{-2}$. While this maximum population density infers a relatively large recruitment of *T. costatus* reference must be made to the duration of the breeding period (Table 6). The mean population of worms in the pre-recruitment period (30 November 1970 to 5 April 1971) was $350 \cdot 10^3 m^{-2}$ with relatively little variation around the mean. The changes taking place in this population were largely physiological; mature worms becoming breeding worms. In the period under discussion this change was initiated in late January 1971 (7% breeding worms), peaked in mid-April (77% breeding worms) before becoming almost negligible from June to September. At this time the population will constitute the number of non-breeding individuals plus those in the process of becoming breeding individuals (in addition to immigrants from other sites). Against this will be losses from natural mortality and/or removal from particular locations by physical factors.

In the post-recruitment period from early April to late September the mean number of worms was about $500 \cdot 10^3 m^{-2}$. The composition of this population (year n) will be the number of juveniles entering per unit time and is dependant on the number of worms breeding at some preceeding time. The time between breeding and emergence of the immature stages has not been established directly but circumstantial evidence suggests this may be between about 70-110 days. These figures are based on the time of the first appearance of breeding worms in the population to the first production of immature worms, and from the peak incidence of emerging young respectively. Additionally, there will be in this population worms which remain from year n-1 and will breed in April-June of year n. Coupled with these are post-breeding worms whose time of persistence and incidence of mortality are at present unknown. Post-recruitment losses from the population will arise from natural mortalities and from physical forces. The difference then in post-recruitment and pre-recruitment of worms is about $150 \cdot 10^3 m^{-2}$ equivalent to about 43% of the former, but presently the unquantifiable adverse conditions (generally of water flow encountered during the autumn and winter) may serve to reduce this further.

Table 6. Populations of breeding and non-breeding
          T. costatus at Erith, 1971.

	Total Population $(\cdot 10^3 \text{m}^{-2})$	Population of breeding *T. costatus* $(\cdot 10^3 \text{m}^{-2})$	(%)	Population of non-breeding *T. costatus* $(\cdot 10^3 \text{m}^{-2})$
26 January	373	26	7	347
8 February	347	149	43	198
22 February	326	163	50	163
8 March	301	196	65	105
22 March	355	224	63	131
5 April	370	285	77	85
19 April	399	291	73	108
3 May	421	227	54	194
17 May	432	177	41	255
1 June	519	88	17	431
14 June	425	14	3.3	411
28 June	518	17	3.3	501
12 July	547	12	2.2	535
26 July	593	44	7.3	549
9 August	601	20	3.3	581
22 August	501	17	3.3	484

ACKNOWLEDGEMENTS

We are grateful for the financial assistance provided by the Greater London Council in support of this research. We appreciate the assistance of S.J. Birtwell, M. Pedneault and S. Steele in the preparation of the manuscript and S. Prothero who prepared the figures.

REFERENCES

Alley, W.P. and R.F. Anderson, 1969, Small-scale patterns of spatial distribution of the Lake Michigan macrobenthos. Proc. 11th Conf. Great Lakes Rs. (1968) Intern. Ass. Great Lakes Res., 1-10.

Alsterberg, G., 1925, Arch. Hydrobiol. 15:291-338. Cited by M. Palmer in "Some aspects of the respiratory physiology of *Tubifex* in relation to its ecology" (1964) Ph.D. thesis Univ. of London.

Appleby, A.G. and R.O. Brinkhurst, 1970, Defecation rate of three tubificid Oligochaetes found in the sediment of Toronto Harbour, Ontario. J.Fish.Res. Bd. Can., 27:1971-1982.

Arthur, D.R., 1965, Form and function in the interpretation of feeding in lumbricid worms. Viewpoints in Biology, 4:204-251.

Arthur, D.R., 1972, Katabolic and Resource Pollution in Estuaries. Symposium on "Population and Pollution". Eugenics Society. London 65-83.

Aston, R.J., 1973, Field and Experimental Studies on the effects of a Power Station Effluent on Tubificidae (Oligochaeta, Annelida). Hydrobiologia, 42:2, 225-242.

Birtwell, I.K., 1972, Ecophysiological aspects of tubificids in the Thames estuary. Ph.D. thesis. Univ. of London.

Brafield, A.E., 1964, The oxygen content of interstitial water in sandy shores. J. Anim. Ecol., 33:97-116.

Brinkhurst, R.O., 1963a, A guide for the identification of British Aquatic Oligochaeta. Sci. Publ. Freshwat. Biol. Ass., No. 22

Brinkhurst, R.O., 1963b, Taxonomical studies on the Tubificidae (Annelida, Oligochaeta). Int. Rev. ges Hydrobiol/Syst. Beih., 2:1-89.

Brinkhurst, R.O., 1964a, Observations on the biology of lake-dwelling Tubificidae. Arch. Hydrobiol., 60:385-418.

Brinkhurst, R.O., 1964b, Observations on the biology of the marine oligochaete *Tubifex costatus*. J. Mar. Biol. Ass., U.K. 44:11-16.

Brinkhurst, R.O. and C.R. Kennedy, 1965, Studies on the biology of the Tubificidae (Annelida, Oligochaeta) in a polluted stream. J. Anim. Ecol., 34:429-443.

Brinkhurst, R.O., 1966, Taxonomic studies on the Tubificidae (Annelida, Oligochaeta) supplement. Int. Revue ges. Hydrobiol., 51:727-742.

Brinkhurst, R.O., K.E. Chua and E. Batoosingh, 1969, Modification
    in sampling procedure as applied to studies on the bacteria
    and tubificid oligochaetes inhabiting aquatic sediments.
    J. Fish. Res. Bd. Can., 26:2581-2593.
Bülow, T., 1957, Systematische studien un eulitoralen Oligochaeten
    der Kimbrishen Halbunsel. Kieler Meeresforsch., 13:69-116.
Dahl, I.O., 1960, The oligochaete fauna of three Danish brackish
    water areas. Medd. Dan. Fisk og Havunders, 2:1-20.
Debauche, H.R., 1962, The structural analysis of animal communi-
    ties in the soil. Progress in Soil Zoology (ed., P.W. Murphy)
    Butterworth: London 398p.
Dixon, W.J., 1965, Biomedical Computer Programs. Health Sciences
    Computing Faculty, Univ. Cal.,Los Angeles. 620p.
Edwards, R.W., 1962, Some effects of plants and animals on the
    conditions in freshwater streams with particular reference
    to their oxygen balance. Int. J. Air. Wat. Poll., 6:505-520.
Edwards, R.W. and H.L.J. Rolley, 1965, Oxygen consumption of
    river muds. J. Ecol., 53:1-19.
Elliott, J.M., 1971, Some methods for the Statistical Analysis
    of samples of Benthic Invertebrates. Sci. Publ. Freshwat.
    Biol. Ass., 25:144p.
Ellis, D., 1925, An investigation into the cause of blackening
    of the sand in parts of the Clyde estuary. J.R. Tech. Coll.
    Glasgow,144-156.
Fenchel, T.M. and R.H. Riedl, 1970, The sulphide system: a new
    biotic community underneath the oxidised layer of marine
    sand bottoms. Mar. Biol., 7, Nos. 3:255.
Finney, D.J., 1971, 3rd ed. Probit Analysis. Cambridge Univ.
    Press: Cambridge. 334p.
George, J.D. 1964, On some environmental factors affecting the
    distribution of Cirriformia tentaculata (Polychaeta) at
    Hamble. J. mar. biol. Ass. U.K., 44:383-388.
H.M.S.O., 1964, The effects of polluting discharges on the Thames
    estuary. W.P.R.L. Tech. Paper No. 11 London. 609p.
Huddart, R.H., 1971, Some aspects of the ecology of the Thames
    estuary in relation to pollution. Ph.D. Thesis, Univ. London.
Huddart, R.H. and D.R. Arthur, 1971a, Shrimps and whitebait in
    the polluted Thames estuary. Intern. J. Environmental
    Studies,2:21-34.
Huddart, R.H. and D.R. Arthur, 1971b, Shrimps in relation to
    oxygen depletion and its ecological significance in a
    polluted estuary. Environ. Pollut., 2:13-35.
Hunter, J.B., 1977, Some aspects of tubificids in the Thames
    estuary. Ph.D. Thesis Univ. of London.
Hunter, J.B. and D.R. Arthur, 1978, Some aspects of the ecology
    of Peloscolex benedeni Udekem (Oligochaeta: Tubificidae)
    in the Thames estuary. Estuarine Coastal Marine Science ,
    6:197-208.
Kennedy, C.R., 1966, The life history of Limnodrilus hoffmeisteri
    Clap. (Oligochaeta, Tubificidae) and its adaptive significance.
    Oikos ,17:158-168.

Knöllner, F.H., 1935, Ökologische und systematische untersuchungen über litorale und marine Oligochaeten de Kieler Bucht. Zool. Jb. (syst),66:425-512.

Ladle, M., 1971, The biology of Oligochaeta from Dorset chalk streams. Freshwat. Biol.,1:83-97.

Lumkin, P., 1971, Plankton of the Thames estuary. M. Phil. thesis Univ. of London.

Morgans, J.F.C., 1956, Notes on the analysis of shallow-water soft substrata. J. Anim. Ecol., 25:367-387.

Newell, R.C., 1970, Biology of intertidal animals. Logos: London 555p.

Palmer, M., 1964, Some aspects of the respiratory physiology of Tubifex in relation to its ecology. J. Zool. Lond., 154:463-473.

Perkins, E.J., 1957, The blackened sulphide containing layer of marine soils, with special reference to that found at Whitstable, Kent. Ann. Mag. nat. Hist.,10:25-35.

Poddubnaya, T.L., 1959, On the dynamics of the tubificid population (Oligochaeta, Tubificidae) in the Rybinsk reservoir (trans. title). Rep. Inst. Reservoir Biol. Borok,2:102-108.

Schumacher, A., 1963, Quantative Aspekte der Berziehung zwischen Starke der Tubificiden besiedlung und Schichtolichke der Oxydationszone in dem Susswasserwatten der Unterelbe. Arch. Fischwiss,14:48-50.

Sedgwick, R. and D.R. Arthur, 1976, A natural pollution experiment. The effects of a sewage strike on the fauna of the Thames estuary. Environ. Pollut.,11:137-160.

Sedgwick, R., 1978, The ecophysiology of fish and crustaceans in the Thames estuary. Ph.D. Thesis Univ. of London.

Smith, R.I., 1956, The ecology of the Tamar estuary. VII; Observations on the interstitial salinity of intertidal muds in the estuarine habitat of Nereis diversicolor. J. mar. biol. Ass. U.K., 35:81-104.

Stephenson, J., 1930, The Oligochaeta. Clarendon Press: Oxford.

Taylor, L.R., 1961, Aggregation, variance and the mean. Nature London,189:732-735.

Thames Migratory Fish Committee, 1977, Report of the Thames Migratory Fish Committee. H.M.S.O. 40p.

Timm, T., 1962, Uber die Fauna, Ökologie and Verbreitung der Subwasser-Oligochaeten der Estrischen S.S.R. Tartu Riiklika Ulikobli Toimetised Zoologica-Alaseid Toid,120:67-107.

Train, D. and P.S. White, 1971, Thames estuary: restoration and preservation of its quality. Soc. of Chem. Ind. 1251-1258.

Vader, W.J., 1964, A preliminary investigation into the reactions of the infauna of the tidal flats to tidal fluctuations in water level. Netherlands J. Sea Res., 2:189-222.

Wentworth, K., 1922, A state of grade and class terms for clastic sediments. J. Geol., 30:377-392.

Wisniewski, R.J., 1976, Effect of heated waters on biocenosis of
    the moderately polluted Narew River. Oligochaeta. Pol. Arch.
    Hydrobiol., 23:4:527-538.
Zobell, C.E., 1939, Occurrence and activity of bacteria in marine
    sediments. Am. Ass. Petroleum Geologists Oklahoma, 416-427.

SEASONAL MOVEMENTS OF SUBTIDAL BENTHIC COMMUNITIES

IN THE FRASER RIVER ESTUARY, BRITISH COLUMBIA,

CANADA

P.M. Chapman

E.V.S. Consultants Ltd.

ABSTRACT

The subtidal benthic fauna of the lower Fraser River, a salt-wedge estuary, was sampled monthly from June 1977 to August 1978 in mud substrates at six stations ranging from oligohaline to polyhaline. Subtidal interstitial salinities were also measured and were related to the seasonal distribution of the estuarine benthic fauna. Diurnal comparisons of sediment salinities with those of the immediately overlying water showed that interstitial salinites of sediments containing a high proportion of silt vary seasonally, not diurnally. Thus, there is a transition zone between salt and fresh interstitial water that is clinally shifted up and downstreams with seasonal variations in freshwater discharge. Because of this, seasonal shifts occur in the distribution and demography of benthic infaunal species. Seasonal shifts in relation to interstitial salinity variations are shown to occur in the distribution of the oligochaetes *Limnodrilus hoffmeisteri*, *Tubifex tubifex*, *Tubificoides gabriellae* and *Paranais litoralis* and of the polychaetes *Eteone longa*, *Amphicteis* sp. and *Polydora kempi japonica*. These seven species comprised over 25% of the total taxa collected and over 60% of the individuals collected. The data on other species distributions support the premise of cyclic changes related to seasonal interstitial salinities but this is not always shown clearly because sampling was

biased towards the most common oligochaete populations. The system
is stable and it is suggested that seasonal faunal shifts are a
feature of salt-wedge estuaries in general. In addition, vertical
salinity gradients exist in sediments and these change seasonally.
See Chapman, P.M. 1979, Seasonal movements of subtidal benthic
    communities in a salt-wedge estuary as related to interstitial
    salinites. Ph.D. Dissertation, University of Victoria,
    British Columbia, 222 pp.

TOLERANCE AND PREFERENCE REACTIONS OF MARINE

OLIGOCHAETA IN RELATION TO THEIR DISTRIBUTION

Olav W. Giere

Zoologisches Institut
Universität Hamburg
Martin-Luther-King Platz 3
2000 Hamburg 13
West Germany

ABSTRACT

Ecological studies on marine oligochaetes often stress the relevance of salinity, temperature, moisture and oxygen supply to their inhomogeneous distribution. However, laboratory experiments specifying the effect of single factors on the distributional pattern are scarce.

Hence, this paper reviews experimental tolerance and preference data and deals particularly with the reaction of some ecologically diverging and geographically disjunct populations of marine tubificids and enchytraeids. Differences between populations, attempts to modify their respective "home range", and comparison with data from other authors help to substantiate the often assumptive character of field studies rendering a more direct ecophysiological proof. These experiments can lead to a deeper understanding of the complicated microdistribution in marine oligochaetes on the basis of limitation, repellance and attraction within the net of abiotic factors. Considerations on the nature of the physiological disposition in these worms lead to speculations on their adaptive potential with respect to distributional consequences.

INTRODUCTION

Ecological investigations on marine oligochaetes have predominantly focussed on distribution studies, often supplemented by
monitoring an array of relevant abiotic parameters like salinity,
temperature and oxygen content. There is little experimentally
proved evidence, however, as to what factor or factorial combination
determines by attraction or repellance the occurrence of oligochaetes, both with respect to their microdistribution in a given
biotope and to more extended geographical aspects comparing syn-
specific populations in different areas. Hence, the need for
tolerance and preference studies has already been emphasized by
Jansson(1967a) and Kinne (1970, 1971).

RESULTS

Tolerance studies on marine oligochaetes have mainly dealt
with temperature and salinity, the most conspicuously fluctuating
ecofactors on the shores (Table 1).

From Table 1 it can be seen that many of the larger marine
oligochaetes, typically inhabiting the wrack beds of the seashore
(e.g. the enchytraeids *Enchytraeus albidus, Lumbricillus lineatus,
L. reynoldsoni* or the shallow bottoms of tidal estuaries and bights
(Tubifex costatus, Peloscolex benedeni, Paranais litoralis) possess
an extremely wide tolerance range which underlines their good
adaptation to the extreme and ever changing salinity and temperature
regime in these biotopes. Even species which preferably live in
freshwater habitats, like the enchytraeid *Lumbricillus rivalis,*
the tubificids *Limnodrilus hoffmeisteri* and *Ilyodrilus templetoni*
and the naidid *Paranais frici*, tolerate amazingly high salinities
rarely to be encountered in situ. The same expanded "tolerance
frame", considerably wider than the usual occurrence would suggest,
has been reported by Moroz (1974) for tubificids from a freshwater
lagoon with occasional influx of oligohaline water.

The tolerance data for typical brackish-water forms *(Monopylephorus irroratus, Marionina southerni)* exceeds by far the range
of their predominant occurrence. *M. southerni* populates mainly
the beaches of the Baltic Sea (Giere, 1976; Jansson, 1968b). In
this species, a similar resistance pattern is also found for the
extremes of temperature which extend up to 35°C. Even in deep-
water marine forms like *Tubificoides gabriellae* whose habitat
salinity fluctuations are certainly less than in intertidal species,
the tolerance range extends close to fresh water conditions.

In contrast to this extremely wide "zone of resistance adaptation" (Vernberg and Vernberg, 1972), the experimental data for
many of the typical meiobenthic species, often belonging to the
genus *Marionina*, indicates restriction to meso- or polyhaline

Table 1.   Experimental tolerance data of aquatic Oligochaeta.

Species	Salinity tolerance (°/oo S)	Temper.-tolerance (°C)	Area of origin and remarks	Author
Enchytraeus albidus adults:			western Baltic Sea (depending on substrate)	SCHONE, 1971
"		-11 to +34	western Baltic	SCHULZ, 1954
"		-13 to +36	western Baltic (depending on preacclimat.)	KAHLER, 1970
reproduction and development	0 - 40		western Baltic Sea	SCHONE, 1971
		1 - 25	England, sewage beds	REYNOLDSON, 1943
embryos in cocoons:	~3 - 21		western Baltic (depending on saison)	KAHLER, 1970
Lumbricillus lineatus	0 - 55		western Baltic Sea	GIERE, unpubl.
L. reynoldsoni	0 - 50		Irish Sea	TYNEN, 1969
L. rivalis adults:	0 - 25		Germany, sewage beds	GIERE, unpubl.
development:		5 - ~20	England, sewage beds	REYNOLDSON, 1943
Pontodrilus bermudensis	5 - 25		Indian shores	GANAPATI & SUBBA RAO, 1972
Tubifex costatus at 5°C:	0 - 34		Thames-Estuary	BIRTWELL & ARTHUR, 1979
at 20°C:	0.5 -34		(LC 50-values: sal.: 7 d	
		→ 32.4	temp.: 4 d)	continued:-

Species	Salinity tolerance (°/ooS)	Temper.- tolerance (°C)	Area of origin and remarks	Author
Peloscolex *benedeni*			Thames-Estuary (LC 50-values:	BIRTWELL & ARTHUR,
at 5°C:	2.8 – 34		sal.: 7 d	1979
at 20°C:	5.9 – 34	→28.5	temp.: 4 d)	
Tubifex *tubifex*				
at 5°C:	0 – 9.8			
at 20°C:	0 – 9.0	→33.9	"	"
Limnodrilus *hoffmeisteri*				
at 5°C:	0 – 10.7			
at 20°C:	0 – 14.7	→37.5	"	"
Limnodrilus *hoffmeisteri* mature,			Fraser River- estuary + environs of	CHAPMAN & BRINKHURST, 1979
accl.to 0°/ooS →10 immature,			Victoria, B.C.	
accl.to 0°/ooS →11 to 5°/ooS →14.5			(LC 50-values: 4 d; 10° C)	
Tubificoides *gabriellae* mature				
accl.to 10°/ooS 1- ?30 accl.to 20°/ooS 2- ?30			"	"
Paranais *frici* accl.to 15°/ooS 0.5 ?35			"	"
P. litoralis 1 – ?35			"	"
Monopylephorus *irroratus* 0 – ~25			"	"
Ilyodrilus *templetoni* 0 – ~7.5			"	"
Psammoryc-tides *barbatus* 0 – 8			Dnieper-Lagoon (Black Sea)	MOROZ, 1974

continued:-

Species	Salinity tolerance ($^o/oo$ S)	Temper.- tolerance (^{o}C)	Area of origin and remarks	Author
Potamothrix hammoniensis	0 - 5		Dnieper-Lagoon (Black Sea)	MOROZ, 1974
P. molda- viensis	0 - 5		"	"
Limnodrilus claparedea- nus	0 - 5		"	"
Marionina southerni	0 - 40	1 - 35	eastern Baltic Sea	ELMGREN, 1968
"	0 - 30		western Baltic	GIERE, unpubl.
M. subterra- nea	1.3 - 15		eastern Baltic Sea	JANSSON, 1968a
M. precli- tellochaeta	2.5 - 10		"	JANSSON, 1962
M. spicula	3 - 30	→30	French Atlantic coast	LASSERRE,1970
M. achaeta	3 - 25	→35	"	LASSERRE,1971
Aktedrilus monosperma- thecus	1.3 - 15		eastern Baltic Sea	JANSSON, 1962 1968a

brackish-water. This corresponds closely to the habitat conditions in the interstitial of the upper beach.

However, tentative low-temperature tests showed that many oligochaetes, at least those from the higher shore, can stand extremes far below 0° C which hardly occur in their microenvironment.

Compilation of tolerance results usually follows two interpretation lines: the first one referring to zonation patterns in a given area (e.g. Jansson, 1967a; Wieser et al, 1974), the second one dealing with distributional effects in a zoogeographical scale (Arndt, 1973; Theede, 1973).

In oligochaete studies, the first aspect was pointed out by Lasserre (1971) who found a close correspondence between habitat demands and oligochaete tolerance. He could relate high temperature-/low salinity-tolerance of *Marionina achaeta* to its occurrence in the upper beach whereas *M. spicula*, living around the high water (hw) line, displayed a good adaptation to higher salinities but attenuated extremes of temperature.

Temperature as a limiting distributional factor has been pointed out already relatively early by Reynoldson (1943) who related the semiterrestrial occurrence of *Enchytraeus albidus* to the high tolerance range of the species, whereas restriction of *Lumbricillus rivalis* to the more aquatic microhabitats corresponded with its less developed temperature resistance.

In contrast to this, the wide tolerance range of *Marionina southerni* (Elmgren, 1968; Jansson, 1968a) and *M. subterranea* (Elmgren, 1968) cannot be possibly related to their horizontal zonation on the beach where *M. southerni* was found near the water line, *M. subterranea* 1-3 m higher up. Here, other ecofactors seem to co-determine occurrence, indicated perhaps by results of Jansson (1966a) and Locy (in preparation) that *M. subterranea* possesses a marked preference for sediments of 200 to 250 µm grain size.

Besides these zonation aspects, Jansson was the first to discuss tolerance reactions of oligochaetes in the light of geographical distribution (1968a). He connected the occurrence of *M. preclitellochaeta* in the Baltic to its relatively restricted salinity capacity. He also monitored the salinity tolerance of the interesting tubificid *Aktedrilus monospermathecus* which, some 15 years later, became a main subject of our own studies in the field of oligochaete tolerance and preference reactions (Giere, 1977; Giere and Pfannkuche, 1978; Giere, Pfannkuche and Hauschildt, in preparation).

The relatively limited salinity range (1-20°/ooS) reported by
Jansson (1962) for populations in the eastern Baltic suggested
geographical restriction to this brackish-water basin. However,
occurrence of this species in North Sea shores and in highly saline
beaches of Bermuda posed the problem to what extent tolerance
ranges of wide-spread oligochaete species can vary in disjunct
populations from physiographically different biotopes.

Hence, we started comprehensive experimental studies including
tests for temperature, salinity, pH and oxygen supply and combinations of these parameters first with populations from Bermuda
(Giere, 1977), later from the North Sea and western Baltic (Giere
and Pfannkuche, 1978). To report the results in brief, it became
clear that in the subtropical intertidal of Bermuda the resistance
range of one single stress factor was amazingly wide (6-36°C;
5-40°/ooS; up to 9.5 ph) if oxygen supply was good and the concomitant parameters stayed "normal". Only the interaction of several
stressors and/or oligoxic conditions deteriorated viability considerably (Figure 3 in Giere, 1977).

These results match well with the field distribution of *A.
monospermathecus* in a tidal sand flat in Bermuda (Figure 1 in
Giere, 1977). Along the beach slope, where sufficient oxygen down
to more than 10 cm depth was available, the other recorded abiotic
parameters remained within the above frame of capacity resistance.
Here the density of *A. monospermathecus* ranged up to 200 ind./100 cm^{3}.
In the outer flat, at low tide, however, high salinity, often combined with high pH-values, prevailed at the surface. The worms could
not avoid this unfavourable environment by downward migration since
sediment below one cm was anoxic. Therefore, here, tubificid populations remained always suppressed.

The Bermuda worms showed a good survival in seawater of 35°/ooS
if temperature stayed below 20° C. These figures differ markedly
from Jansson's data (1962) (1-20°/ooS only) from brackish waters
in the eastern Baltic. This variance in physiological disposition
becomes even more apparent in combinations of high salinity with
another stress factor like high temperature (Figure 1). The
animals displayed best survival rates close to their averaged
"home salinity" (11°/ooS in the Baltic, 35°/ooS in Bermuda). The
North Sea-worms, not shown here, had best survival at 30°/ooS and
30° C, i.e. far higher a salinity than the Baltic population could
stand, but less than the Bermuda animals were used to. Similarly,
the populations from boreal areas were found to be less resistant
to temperatures above 35° C than the subtropical ones (Figure 1).

Adjustment of best resistance values to habitat conditions
was repeatedly reported also for polychaetes (Theede, 1973 - oxygen
decrease, Theede et al., 1973; Westheide and Basse, 1978 - low
temperatures.

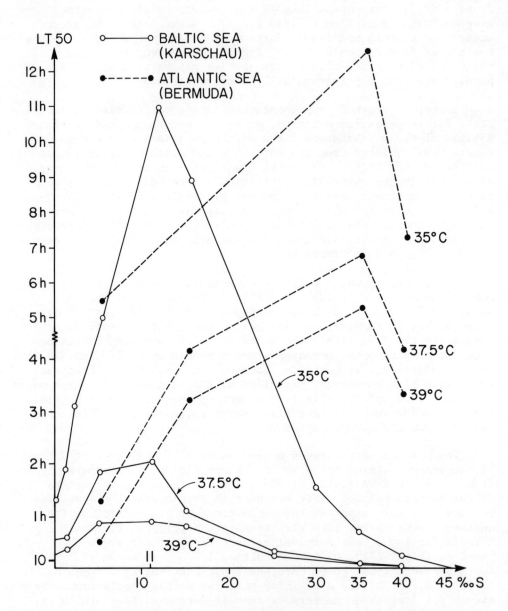

Figure 1. Salinity tolerance of *Aktedrilus monospermathecus*
Baltic and Bermudian populations); Survival (LT 50)
at various high temperatures.
(From Giere and Pfannkuche, 1978).

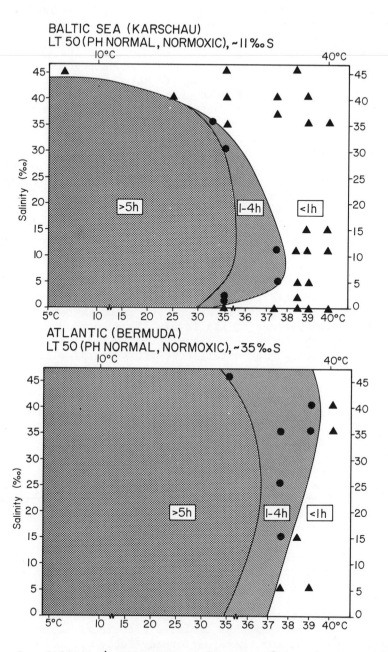

Figure 2.   Salinity/temperature tolerance of *A. monospermathecus*:
"Survival fields" in Baltic and Bermudian populations
(From Giere and Pfannkuche, 1978).

The shift in tolerance reactions of these populations towards divergent directions is an adaptational response to the different ecological situation which is especially apparent in Figure 2, where fields of corresponding LT-50 values are plotted along S/T-axes, again comparing the Baltic and Bermuda populations. The area of shortest survival for the Baltic *Aktedrilus* is particularly wide at higher salinities, whereas here, the subtropical worms from the Atlantic thrive well. In contrast to this, their short-survival field extends towards lower salinities.

To focus attention back to the question (how these differing ecophysiological dispositions reflect the microdistribution of the worms in their local habitats) one can state that vertical distribution in the Bermudian cove was basically defined by oxygen conditions (Figure 5 in Giere, 1977). Large populations could only establish in the slope where the animals were able to avoid the extremes of abiotic fluctuations at the surface by downward migration. This was rendered impossible in the flat by the anoxic milieu in deeper layers.

In the well-aerated temperate beaches of the Baltic (Figure 3), the worms preferably aggregated around the ground water layer following the fluctuations of its level. Thus, during a low water-level, they could be found near the water-line right at the surface (upper graph). During high water, however, *Aktedrilus* lived mainly below 10 cm depth.

In the North Sea shore, anoxic conditions (indicated by the redox potential-discontinuity layer) established in about the same depth as the groundwater layer and forced the population to move upwards with concentrations always above groundwater.

However, the factors which prevent this species from populating the upper horizons are still not fully understood especially at the backshore which offers good oxygen, salinity and temperature-conditions. Initial preference experiments showed that, here, moisture gradients might be involved.

In order to further differentiate the relation between distribution and ecophysiological response, preference experiments were carried out, in which the animals could adjust their macrodistribution in gradient fields, more comparable to the situation in the field. It is symptomatic for the poorly developed "state of the art" that only few and somewhat isolated preference tests on marine oligochaetes have, so far, been performed (Table 2).

From this table, it can be seen that the response of the tested worms usually matches their habitat conditions fairly well (see also Jansson, 1968a,b). Again, the considerable physiological accommodation of some typical wrack-bed forms to extremes of salinity and temperature is striking.

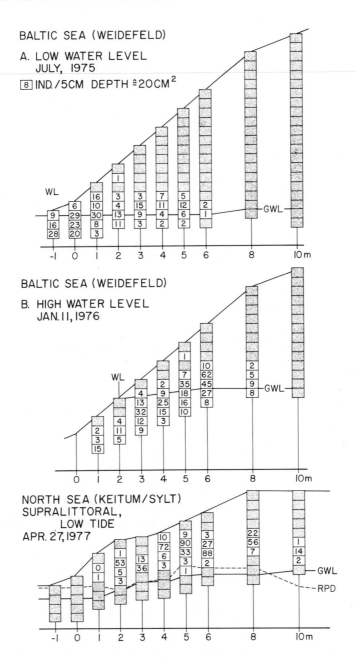

Figure 3.   Field distribution of A. *monospermathecus* in the Baltic
and North Sea (WL = water line; GWL = ground water line;
RPD = redox potential discontinuity).
(From Giere and Pfannkuche, 1978).

Table 2.   Experimental preference data of aquatic Oligochaeta.

Species	Preference	Area of origin	Author
Enchytraeus albidus	5-15°/ooS for development, reproduction	western Baltic Sea	SCHONE, 1971
"	3 - 10°/ooS	Baltic Sea	BACKLUND, 1945
"	no marked temp.-preference	western Baltic Sea	SCHULZ, 1954
"	5-15 % saturat. moisture prefer.	"	"
Lumbricillus lineatus	no salinity-preference	Irish Sea	TYNEN, 1969
"	no salinity-preference	Baltic Sea	BACKLUND, 1945
L. reynoldsoni	about 15°/ooS	Irish Sea	TYNEN, 1969
Marionina subterranea	125 - 500 µm	Eastern Baltic Sea	JANSSON, 1966a
"	200 - 250 µm sediment	Pacific coast of California	LOCY (in prep.)
"	16 - 20° C	"	"
"	5°/ooS	Eastern Baltic Sea	JANSSON, 1968a
M. southerni	1 - 5°/ooS	"	ELMGREN, 1968
M. preclitellochaeta	0.3^o/ooS	"	JANSSON, 1962
Aktedrilus monospermathecus	2.5 - 5°/ooS	"	"

We commenced our own preference studies with these larger
ubiquitous species, followed later by the interstitial *Marionina*
species *southerni* and *subterranea*. Most intensive, however, was
and still is the work on *Aktedrilus monospermathecus* which was
done in order to further clarify the different local distribution
patterns. By choosing species with disjunct populations, comparing
our results with those of other authors, and testing pre-acclimated
animals, we wanted information about the diversity and flexibility
of preference responses in some common marine oligochaetes.

The scarcity of preference studies might partly be due to
problems in experimental set-up. So, our experiments were preceeded
by the development of reliable and versatile devices for testing
preference in salinity, temperature and moisture gradients (Giere,
1979, in press) which are briefly described here.

Salinity gradients were established in sterile, sieved sand
(500 to 800 µm) in a perspex trough with a bottom of plankton gauze
which was tightly connected to a chambered "water trough" which
contained a series of differently saline sea water (o-50°/ooS).
Through cubes of foam rubber, inserted into the chambers, the water
moistened the overlying dry sand layer by capillary forces. This
device, the principle of which is based on Jansson's "alternative
chamber" (1962), provides smooth and well replicable salinity
gradients without any abrupt "bordering zone".

Regular and stable temperature gradients in moist sand of a
given average salinity were accomplished in a chromium-plated
copper trough connected to a copper bridge which dipped into a
heating bath at one end, a cooling bath at the opposite end.

Moisture gradients were set up in a long perspex trough
positioned at an oblique angle. By filling in sea water at the
lower end, a gradient of water content from complete saturation
to dryness in the layer of homogeneous sand was attained.

Combinations of temperature and salinity gradients can be
obtained replacing the perspex sea water chambers of the salinity
device by corresponding ones made of plated copper which, in turn,
are connected to the "temperature bridge."

All preference data are based on 2 to 5 replicates and
thorough statistical treatment (paired t-test, U-test according
to Wilcoxon, Mann and Whitney, see Massey and Dixon, 1969; 2 x 2
contingency table test). The results illustrated in the graphs
were selected only if they significantly (% probability-level)
corresponded to the parallel tests.

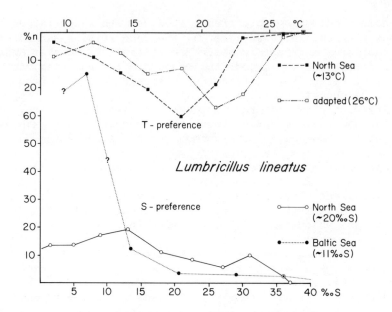

Figure 4. Salinity and temperature preference of *Lumbricillus lineatus*: North Sea and Baltic populations.

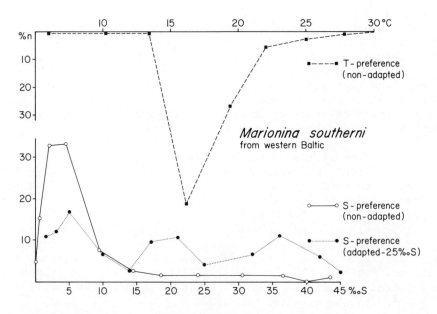

Figure 5. Salinity and temperature preference of *Marionina southerni* from the western Baltic Sea.

Our salinity experiments with *Lumbricillus lineatus* populations from the Baltic (Figure 4) showed a clear preference for the low domestic salinity, whereas the North Sea worms displayed a much more diffuse pattern, although a trend towards low salinities is recognizable. This is in contrast to the lack of any preference in the experiments of Tynen (1969) and Backlund (1945) (see Table 2). However, their use of artificial substrate in their devices might contribute to this contradiction (Giere, 1979) beside some possible variance in genetical background (see Discussion). Tynen stressed the correspondence between littoral zonation and response to salinity for *L. reynoldsoni,* a species from the upper shore with lower salinities, and for *L. lineatus* which lacked any marked salinity preference and duely was distributed all over the shore.

The temperature preference of *L. lineatus* (Figure 4, upper curve) reflects well the mean habitat conditions in their temperate climate. Long-term adaptation (> 30 d) shifted the preference significantly towards a higher range.

Conditions were more complicated in *Enchytraeus albidus,* the other dominant wrack form. Our salinity tests with Baltic populations resulted in a marked preference for salinities below $10^{\circ}/ooS$, particularly below $5^{\circ}/ooS$ which was confirmed by corresponding values in experiments with Baltic worms performed by Backlund (1945) and Schöne (1971). However, in our comparative North Sea-tests, an almost homogeneous distribution along the gradient from about 2 to $55^{\circ}/ooS$ was to be monitored (see Discussion).

In the interstitial enchytraeid *Marionina southerni,* coincidence between habitat conditions and experimental results was found in temperature (Figure 5, upper curve), but in salinity, the western Baltic populations with habitat water of about $10^{\circ}/ooS$ were always aggregated around $5^{\circ}/ooS$. Close correspondence to similar results by Elmgren (1968) with populations from the eastern Baltic (local salinity about $6^{\circ}/ooS$) might lead to the conclusion this species to have a stable, fixed salinity preference. However, by long-term adaptation, the preference peaks became somewhat smoothened with a clear trend toward higher salinities. Apparently, the pre-acclimation period Elmgren had chosen was too short to induce a clear shift of his recordings.

We tested temperature preference of another *Marionina* species, *M. subterranea* from the North Sea, for comparison with corresponding experiments carried out by Locy (in preparation) with populations from the California Pacific coast. Both populations had a well defined, but not significantly divergent, temperature preference. On the Pacific coast the preferendum was about 16 to 20° C, in the North Sea about 15 to 16° C, which could be slightly shifted after 5 weeks acclimation in 3° C to $12 - 14^{\circ}$ C.

For *Aktedrilus monospermathecus*, we had, as in the earlier
tolerance tests, the rare occasion to study disjunct populations
from physiographically differing habitats (Figure 6). Again, the
salinity preference corresponded fairly well to the "home conditions".
The North Sea animals reflected the wide tidal plus non-tidal salinity fluctuations in that shore line. Only the Baltic populations
exhibited a certain discrepancy between average (salinity) range
(about 11°/ooS) and preference data. Also temperature ranges could
be related to prevailing biotope conditions.

In the Bermuda populations, the preference range explains
the sub-optimal conditions in the surface layers of the sand flat
far clearer than tolerance data did. The preferred attenuated
temperature between 25 and 30° C and salinities of 35 %o S were
typical for the deeper layers of the hw-zone and the upper shore.

The restriction of *A. monospermathecus* to near-groundwater-
layers in the back shore was not yet explained. Since, in this
ubiquitous species, no substrate specificity seems to exist (judging
from sediment conditions at numerous sampling sites and experiments
reported by Jansson, 1968b), our studies focussed on moisture
gradients as a causative factor. First tests exhibited a marked
preference for very high water saturation (> 90 % of the inserted
worms preferred saturation between 80 and 100 %), thus strongly
suggesting a response to moisture differences to be well developed
(compare Schulz, 1954, for *Enchytraeus albidus*, Table 2). In
recent tests, however, this percentage has markedly decreased,
thus, necessitating further scrutinization of this problem.

DISCUSSION

Comparing the tolerance and preference results presented above,
it becomes evident that tolerance data respond more to ecological
extremes and represent "momentary situations" rather than a frame
relevant for the permanent distribution pattern. Moreover, they
are determined in highly unnatural mono-factorial stress situations
and certainly refer only to the "sterile expatriation area" (Ekman,
1953) in which adult animals might possibly survive for short
periods of time. Thus, the "zone of resistance" (Arndt, 1973),
defined by tolerance data, exceeds by far the area of regular
existence in which successful reproduction and embryogenesis are
accomplished (see Table 1). Kahler (1970) found mortality caused
by salinity stress in embryonic stages of *Enchytraeus albidus*
remarkably higher than in adults (Table 1). Comparison of data
for adults and embryos (Table 1; Reynoldson, 1943) prove the same
to be valid for temperature. Also, tolerance data disregard the
interaction in the array of natural ecofactors, e.g. between
temperature and salinity (see Table 1, Birtwell and Arthur, 1979).

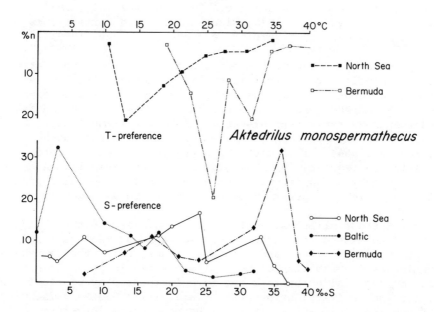

Figure 6. Salinity and temperature preference of *A. monosperma-thecus*: Baltic, North Sea, and Bermuda populations.

The tolerance thresholds closely match the extremes of salinity, temperature, pH and oxygen supply at the sand surface, especially in the subtropical *Aktedrilus* population. In the astatic intertidal in particular anoxia and H_2S often act as prevalent distributional controls (Caldwell, 1975) for many oligochaetes which usually display a well developed preference for rich oxygen supply (Hoven, 1975; Lasserre and Renaud-Mornant, 1973) (exception: *Peloscolex benedeni*).

The close alignment between field distribution and tolerance capacity, particularly in warmer climates, has been stressed already for other animal groups (Alderdice, 1972; Bodoy et al., 1977; Jansson, 1966b; Vernberg, 1975; Wieser, 1975) and certainly is of "economic advantage" by requiring less energy (Tynen, 1969).

The validity of the well known additive or even synergistic effect of abiotic parameters (Kinne, 1971; Remmert, 1967) became especially evident for oligochaetes in the low-temperature experiments which shifted tolerance thresholds for salinity and pH far beyond the usual range (see also data of Birtwell and Arthur, 1979, Table 1), probably due to a slow-down of metabolic processes (Percy, 1975; Schulz, 1954). This might also enable many oligochaetes to survive in ice (Kahler, 1970; Schulz, 1954; see Table 1). A marked cooling capacity is apparently even developed in some subtropical populations which are never exposed to temperatures below 10° C.

In temperate climates, most short-term tolerance thresholds were found to exceed the range of physiographical factors (again with the exception of H_2S and anoxia). Hence, here, the remarkably discontinuous distribution of most oligochaetes is probably not defined by physical parameters but more influenced by biotic factors (e.g. food selectivity; Brinkhurst et al., 1972; Giere, 1975; Giere and Pfannkuche, 1978; Gray, 1971) or by non-lethal long-term reactions (Southward, 1958), especially of the more sensitive juvenile stages (Percy, 1975).

Preference responses are the main basis for the horizontal and vertical orientation of endobenthos (Gray, 1965; Jansson, 1967b; Tynen, 1969). The choice or preference range, as a general ecological principle (Kinne, 1970, 1971) narrower than the tolerance limits, allows a more differentiated analysis of the animals distribution. However, it became evident that isolated, monofactorial preference experiments also represent only a limited approach to an adequate understanding of the field distribution.

It also has to be kept in mind that other kinds of preferences might influence the distribution pattern: Substrate selectivity probably plays a more important role than the few existing results (Table 2) would suggest (Conrad, 1975; Tynen, 1969). Moreover, attraction by food probably is a highly decisive distributional factor regarding the high ability of oligochaetes (so far tested) not only to discriminate among different food items (Chua and Brinkhurst, 1973; Giere, 1975; Giere and Hauschildt, 1979) but also to sense attractive food particles from certain distances (Hauschildt, 1978). Also seasonal preference variations might complicate the pattern (Bodoy et al., 1977; Percy, 1975; Theede, 1973).

Interaction of all these factors might explain the divergences between salinity/temperature preferendum and actual field situation or tolerance results found in some of our experiments (e.g. *Aktedrilus* - salinity preference in the Baltic).

The amazing range of capacity resistance in most of the investigated intertidal oligochaetes poses the question which osmophysiological regulatory mechanisms might be involved. This problem, although particularly interesting in marine invaders from primarily limnic group like oligochaetes, has only scarcely been studied. It should only be pointed out here that Kähler (1970) and Lasserre (1975) characterized their experimental species as fairly effective and rapid hyperosmotic regulators, thus being well adapted to environmental fluctuations. Brackish-water species often showed a transition from osmoregulation at moderate to osmoconformation at higher salinities (Lasserre, 1970; Schöne, 1971) which corresponds to the situation in some concomitant *Nereis* spp. (Oglesby, 1969, 1978; Smith, 1957). Contrasting to this, Tynen (1969) favoured the concept of poikilosmosis in *Lumbricillus reynoldsoni* with osmoregulation playing only a minor role.

It has repeatedly been found in polychaete (Jørgensen and Dales, 1957; Oglesby, 1978; Theede, 1973) and oligochaete annelids (Hohendorf, 1963; Kähler, 1970) that the physiological disposition of the species can be experimentally modified by the addition of specific ions (Na^+, K^+, Ca^{++}) or by pre-acclimation resulting in a shift of preference ranges for factors like salinity or temperature. Similar impacts also became evident in our acclimation experiments, both in macrofauna and meiofauna species (Figures 4, 5) and seem to be a general feature in oligochaetes of the sea shore. The contradictory results of Backlund (1945) on *Enchytraeus albidus*, of Elmgren (1968) and Jansson (1968a) on *Aktedrilus monospermathecus* which indicated a stable preference irrespective of the background conditions could not be repeated. They must probably be assigned to the relatively insignificant biotopic differences or an insufficient acclimation time and range (20 d; 20°/ooS) compared with the extremely euryecous nature of these species.

This flexibility of tolerance and preference ranges throws some light on the nature of adaptational responses in these oligochaetes. They seem to represent non-genetic adaptations (Kinne, 1964, 1971) accomplished by exposure to diverging ecological conditions though longer periods of time. These steady-state reactions

(Kinne, 1970) probably are superimposed to a rather stable, extremely wide genetically fixed physiological background which even might keep covering conditions never to be encountered by a local populations in its particular environment (e.g. low temperatures in Bermuda worms).

However, to generalize this modificatory interpretation might yet be premature regarding the widely occurring genetical inter-population variance in marine littoral species (see compilation in Battaglia and Beardmore, 1978), which was found also in polychaetes (Grassle and Grassle, 1974; Kerambrun and Guerin, 1978) and oligo-chaetes (Christensen et al., 1976; Christensen and Jelnes, 1976; Christensen et al., 1978). Our tentative electrophoretic studies on *Lumbricillus lineatus* and *Enchytraeus albidus* showed major differences in band position between North Sea and Baltic populations. Thus, different genetical disposition might account for their completely diverging preference behaviour reported earlier. Further studies in this field are required before conclusions on the establishment of physiological races or even species can be drawn. They are especially needed since cross-breeding experiments are hardly to be performed with these worms.

The concept of a non-genetic, fairly adaptable "opportunistic" adjustment might particularly refer to wide-spread coastal oligo-chaetes responding to the heavy fluctuations in these unstable biotopes which mostly favour r-strategists. Considering evolution-ary strategies, marine oligochaetes, being limnogenic invaders into the sea, had first to populate these astatic coastal habitats. Hence, especially for interstitial species with their restricted propagatory potential, maintenance of an unfixed pattern of physio-logical adaptation plus an euryecous potential on the basis of a wide, genetically fixed background, would be a significant pre-disposition for attaining a wide-spread distribution even in physiographically differing areas.

It is well possible that in more stable biotopes, e.g. the sublittoral or deep sea, genetic variability plus selection might result in a genetically fixed, more restricted, stenecous condition (Arndt, 1973) and, thus, lead a step further towards oligochaete speciation in the sea.

REFERENCES

Alderdice, D.F., 1972, Factor combination: responses of marine poikilotherms to environmental factors acting in concert. In: Marine Ecology, Vol. 1, Environmental factors, Part III O. Kinne, 1973, ed., pp. 1659-1722, Wiley, London.
Arndt, E.A., Ecophysiological and adaptational problems confronting animals living in brackish water. Oikos, Suppl.,15:239-245.

Backlund, H.O., 1945, Wrack fauna of Sweden and Finland. Ecology and chorology. Opusc. ent. Suppl., 5:1-237.

Battaglis, B., and J.A. Beardmore, (ed.), 1978, Marine organism. Genetics, ecology, and evolution. Proc. NATO Advance Research Institute. Venice, March 24 - April 4, 1977, NATO Conf. Ser., Ser. 4: Marine Science, Vol. 2, 768 pp. Plenum Publ. New York.

Bedoy, A., A., Dinet, H. Massé and C. Nodot, 1977, Incidence de l''acclimatation sur la tolerance thermique de Cerithium vulgatum (Mollusque Gasteropode) et d'Asellopsis duboscqui (Crustace Harpacticoide). Tethys, 8:105-110.

Birtwell, I.K. and D.R. Arthur, 1979, The ecology of tubificids in the Thames estuary with particular reference to Tubifex costatus. In "Aquatic Oligochaete Biology", R.O. Brinkhurst and D.G. Cook, ed., Plenum Publ. Co., New York.

Brinkhurst, R.O., K.E. Chua and N.K. Kaushik, 1972, Interspecific interactions and selective feeding by tubificid oligochaetes. Limnol. Oceanogr., 17:122-133.

Caldwell, R.S., 1975, Hydrogen sulfide effects on selected larval and adult marine invertebrates. Water Resour. Res. Inst., 31:1-27.

Chapman, P.M. and R.O. Brinkhurst, 1979, Salinity tolerance in selected aquatic oligochaetes. Intern. Revue ges. Hydrobiol. in press.

Christensen, B., U. Berg and J. Jelnes, 1976, A comparative study of enzyme polymorphism in sympatric diploid and triploid forms of Lumbricillus lineatus (Enchytraeidae, Oligochaeta). Hereditas, 84:41-47.

Christensen, B. and J. Jelnes, 1976, Sibling species in the oligochaete worm Lumbricillus rivalis (Enchytraeidae) revealed by enzyme polymorphisms and breeding experiments. Hereditas, 83:237-244.

Christensen, B., J. Jelnes and U. Berg, 1978, Long-term isozyme variation in parthenogenetic polyploid forms of Lumbricillus lineatus (nchytraeidae, Oligochaeta) in recently established environments. Hereditas, 88:65-74.

Chua, K.E. and R.O. Brinkhurst, 1973, Bacteria as potential nutritional resources for three sympatric species of tubificid oligochaetes. In: Belle W. Baruch Library in Marine Science, Vol. 1: Estuarine Microbial Ecology, Stephenson, L.H. and R.R. Colwell, ed., 513-517.

Conrad, J.E., 1976, Sand grain angularity as a factor affecting colonization by marine meiofauna. Vie Milieu, B., 26:181-198.

Dixon, W.J. and F.J. Massey Jr., 1969, Introduction to statistical analysis. 3 rd ed., 638 pp., McGraw-Hill, New York.

Ekman, S., 1953, Zoogeography of the sea. 417 pp., Sidgwick and Jackson, London.

Elmgren, R., 1968, Salthaltsresistens och salthaltspreferens hos Marionina southerni med nagra anmärkningar rörande dess volym och osmoregulation. 3-betygsuppsats i zoologi (unpubl.) 33 pp.

Ganapati, P.N. and R.V. Subba Rao, 1972, Salinity tolerance of a
    littoral oligochaete, *Pontodrilus bermudensis* Beddard. Proc.
    Ind. Natl. Sci. Acad., B, 38:350-354.
Giere, O., 1975, Population structure, food relations and ecologi-
    cal role of marine oligochaetes, with special reference to
    meiobenthic species. Mar. Biol., 31:139-156.
Giere, O., 1976, Zur Kenntnis der litoralen Oligochaetenfauna
    Sudfinnlands. Ann. Zool. Fennici, 13:156-160.
Giere, O., 1977, An ecophysiological approach to the microdistri-
    bution of meiobenthic Oligochaeta. I. *Phallodrilus monosperma-*
    thecus (Knollner) (Tubificidae) from a subtropical beach at
    Bermuda. In: Biology of benthic organisms, 11th European
    Marine Biology Symposium, Galway, Oct. 1976, Keegan, B.F.,
    P. O'Ceidigh, and P.J.S. Boaden, ed., pp.285-296, Pergamon
    Press, Oxford, New York.
Giere, O., 1979, Some apparatus for preference experiments with
    meiofauna. J. Exp. mar. Biol. Ecol.,(in press).
Giere, O., and D. Hauschildt, 1979, Experimental studies on the
    life cycle and production of the littoral oligochaete *Lumbri-*
    cillus lineatus and its response to oil pollution. In: Cyclic
    phenomena in marine plants and animals, Naylor,E and R.G.
    Hartnoll, ed., in press, Pergamon Press, Oxford, New York.
Giere, O. and O. Pfannkuche, 1978, An ecophysiological approach
    to the microdistribution of meiobenthic Oligochaeta. II.
    Phallodrilus monospermathecus (Tubificidae) from boreal
    brackish-water shores in comparison to populations from sub-
    trophical beaches. Kiel. Meeresforsch., Sdbd. 4, in press.
Giere, O., O. Pfannkuche and D. Hauschildt, in preparation, An
    ecophysiological approach to the microdistribution of meioben-
    thic Oligochaeta. III. Preference reactions in disjunct
    populations of *Aktedrilus monospermatecus*.
Grassle, J.F. and J.P. Grassle, 1974, Opportunistic life histories
    and genetic systems in marine benthic polychaetes, J. Mar.Res.,
    32:253-284.
Gray, J.S., 1965, The behaviour of *Protodrilus symbioticus* GIARD
    in temperature gradients. J. Anim. Ecol., 34:455-461.
Gray, J.S., 1971, Factors controlling population localization in
    polychaete worms. Vie Milieu, 22:707-722.
Hauschildt, D., 1978, Experimentelle Untersuchung zum Einfluß
    ausgewählter Okofaktoren auf Fortpflanzung, Entwicklung und
    Produktion mariner Oligochaeten. Dipl.-Arb. Univ. Hamburg
    (unpubl.).
Hohendorf, K., 1963, Der Einfluß der Temperatur auf die Salz-
    gehaltstoleranz und Osmoregulation von *Nereis diversicolor*
    O.F. Müller. Kieler Meeresforsch., 19:196-218.
Hoven, W.V., 1975, Aspects of the respiratory physiology and
    oxygen preferences of four aquatic oligochaetes (Annelida).
    Zool. Afr., 10:29-46.

Jansson, B.-O., 1962, Salinity resistance and salinity preference
of two oligochaetes *Aktedrilus monospermathecus* Knöllner and
Marionina preclitellochaeta n. sp. from the interstitial fauna
of marine sandy beaches. Oikos, 13:293-305.

Jansson, B.-O., 1966b, On the ecology of *Derocheilocaris remanei*
Delamare and Chappuis (Crustacea, Mystacocarida). Vie Milieu,
17:143-186.

Jansson, B.-O., 1967a, The importance of tolerance and preference
experiments for the interpretation of mesopsammon field
distributions. Helgoländ. wiss. Meeresunters., 15:41-58.

Jansson, B.-O., 1967b, The availability of oxygen for the interstitial fauna of sancy beaches. J. exp. mar. Biol. Ecol.,
1:122-143.

Jansson, B.-O., 1968a, Quantitative and experimental studies of
the interstitial fauna in four Swedish sandy beaches. Ophelia,
5:1-71.

Jansson, B.-O., 1968b, Studies on the ecology of the interstitial
fauna of marine sandy beaches. Thesis Univ. Stockholm: 16 pp.

Jørgensen, C.B. and R.P. Dales, 1957, The regulation of volume and
osmotic regulation in some nereid polychaetes. Physiol. comp.
Oecol., 4:357-374.

Kähler, H., 1970, Uber den Einfluß der Adaptationstemperatur und
des Salzgehaltes auf die Hitze- und Gefrierrestenz von *Enchytraeus albidus* (Oligochaeta). Mar. Biol., 5:315-324.

Kerambrun, P., and J.-P. Guérin, 1978, Mise en évidence, par
électrofocalisation, de l'heterogénéité de la population de
Scolelepis (Malacoceros) fuliginosa de Cortiou (Bouches du
Rhones). C.R. Hebd. Séances Acad. Sci. Paris, Ser. D, 286
(16):1207-1210.

Kinne, O., 1964, Non-genetic adaptation to temperature and salinity.
Helgoländ. wiss. Meeresunters., 9:433-458.

Kinne, O., 1970, Temperature: Animals - Invertebrates. In:
Marine Ecology, Kinne, O., ed., Vol. 1: Environmental Factors,
Pt. 1(3):407-514, Wiley, London.

Kinne, O., 1971, Salinity: Animals - Invertebrates. In: Marine
Ecology, Kinne, O., ed., Vol. 1: Environmental Factors, Pt.2
(4):821-995, Wiley, London.

Lasserre, P., 1970, Action des variations de salinite sur le
metabolisme respiratoire d'oligochetes euryhalins du genre
Marionina Michaelsen. J. exp. mar. Biol. Ecol., 4:150-155.

Lasserre, P., 1971, Donees ecophysiologiques sur la repartition
des Oligochetes marins meiobenthiques. Incidence des para-
metres salinite, temperature, sur le metabolisme respiratoire
de deux especes euryhalines du genre *Marionina* Michaelsen
1889 (Enchytraeidae, Oligochaeta). Vie Milieu, Suppl. 22:
523-540.

Lasserre, P., 1975, Metabolisme et osmoregulation chez une
Annelide Oligochaete de la meiofaune: *Marionina achaeta*
Lasserre. Cah. Biol. Mar., 16:765-798.

Lasserre, P., and J. Renaud-Mornant, 1973, Resistance and respira-
    tory physiology of intertidal meiofauna to oxygen deficiency.
    Neth. J. Sea Res., 7:290-302.
Locy, S., in preparation, Some preference and tolerance experiments
    using *Marionina subterranea*, a common interstitial oligochaete
    from the Pacific coast.
Moroz, T.G., 1974, Effects of water salinity on survavl and repro-
    ductive capacity of oligochaetes of the family Tubificidae in
    the Dnieper-Bug lagoon (USSR). Gidrobiol. Zh., 10:102-104.
Oglesby, L.C., 1969, Salinity stress and desiccation in intertidal
    worms. Amer. Zool., 9:319-331.
Oglesby, L.C., 1978, Salt and water balance. In: Physiology of
    Annelids, Mill, P.J., ed., 555-658, Academic Press, London,
    New York, San Francisco.
Percy, J.A., 1975, Ecological physiology of arctic marine inver-
    tebrates. Temperature and salinity relationships of the
    amphipod *Onisimus affinis* H.J. Hansen, J. exp. mar. Biol.
    Ecol., 20:99-117.
Remmert, H., 1967, Physiologisch-ökologische Experiments an *Ligia
    oceanica* (Isopoda). Z. Morph. Okol. Tiere, 59:33-41.
Reynoldson, T.B., 1943, A comparative account of the life cycles
    of *Lumbricillus lineatus* Müller and *Enchytraeus albidus*
    Henle in relation to temperature. Ann. appl. Biol., 30:60-66.
Schöne, C., 1971, Über den Einfluß von Nahrung und Substratsali-
    nität auf Verhalten, Fortpflanzung und Wasserhaushalt von
    Enchytraeus albidus Henle (Oligochaeta). Oecologia (Berlin),
    6:254-266.
Schulz, W., 1954, Zur Biologie von *Enchytraeus albidus* (Henle).
    I. Okologie, Z. wiss. Zool., 158:31-78.
Smith, R.I., 1957, A note on the tolerance of low salinities by
    nereid polychaetes and its relation to temperature and repro-
    ductivity. Coll. Int. Biol. Mar. Stat. Biol. Roscoff 1956,
    Annee Biol., Ser. 3, 33:93-107.
Southward, A.J., 1958, Note on the temperature tolerance of some
    intertidal animals in relation to environmental temperatures
    and geographical distribution. J. mar. biol. Ass. U.K., 37:
    49-66.
Theede, H., 1973, Resistance adaptations of marine invertebrates
    and fish to cold. In: Effects of temperature on ectothermic
    organisms, Wieser, W., ed., 249-269, Springer, Berlin,
    Heidelberg, New York.
Theede, H., J. Schaudinn, and F. Saffe, 1973, Ecophysiological
    studies on four *Nereis* species of the Kiel Bay. Oikos,
    Suppl., 15:246-252.
Tynen, M. J., 1969, Littoral distribution of *Lumbricillus reynold-
    soni* Backlund and other Enchytraeidae (Oligochaeta) in relation
    to salinity and other factors. Oikos, 20:41-53.
Vernberg, W.B., 1975, Multiple factors effects on animals. In:
    Physiological adaptation to the environment. Vernberg, F.J.
    ed., 521-537, Intext, New York.

Vernberg, W.B. and E.J. Vernberg, 1972, Environmental physiology
    of marine animals. Springer, Berlin, Heidelberg, New York,
    346 pp.
Westheide, W. and M.V. Basse, 1978, Chilling and freezing resis-
    tance of two interstitial polychaetes from a sandy tidal
    beach. Oecologia (Berl.), 33:45-54.
Wieser, W., 1975, The meiofauna as a tool in the study of habitat
    heterogeneity: Ecophysiological aspects. A review. Cah.
    Biol. Mar., 16:647-670.
Wiesser, W., J. Ott, F. Schiemer and E. Gnaiger, 1974, An ecophy-
    siological study of some meiofauna species inhabiting a sandy
    beach at Bermuda. Mar. Biol., 26:235-248.

OLIGOCHAETA COMMUNITY STRUCTURE AND FUNCTION IN

AGRICULTURAL LANDSCAPES

Krzysztof Kasprzak

Polish Academy of Sciences
Department of Agrobiology
Swierczewskiego 19
60-809 Poznan, Poland

ABSTRACT

Considering the great importance of Oligochaeta in energetic transformations and circulation of matter in aquatic and land-ecosystems an essential problem is the estimation of the role of these animals not only in selected environments but also on the area of whole landscapes. The annual average biomass of Oligo-chaeta as well as other groups of benthos (Mollusca, Chironomidae) in a canal (8, 8 g fresh wt m^{-2}) is a dozen or so times greater than in a eutrophic lake (0,7 g fresh wt m^{-2}). The fauna of Oligochaeta in the eutrophic lake shows considerably greater diversity than in the drainage canal. Great changes of species composition and density of Oligochaeta in aquatic-reservoirs of the agricultural landscape, especially in mountain and piedmont rivers, are caused by the flow of mineral salts as well as municipal sewage and wastes of the food industry.

Agroecosystems are inhabited chiefly by species Lumbricidae and Enchytraeidae of a wide range of occurence and great adaptive ability. Due to transformation of natural environments in agroeco-systems, distinct changes in the typical composition and the relations of the domination of communities occurred. In comparison with Enchytraeidae the fauna of Lumbricidae shows greater quanti-tative impoverishment on fields under cultivation than in arable lands of grasslands. Organic and mineral fertilization spray irrigation as well as mechanical soil cultivation, exert great influence on changes of the density of Oligochaeta. The cultiva-tion of some plants, especially Papilionaceae increases the density

of Lumbricidae. The cultivation of root crops, however, exerts a
positive influence on the development of the population of Enchytraeidae.

INTRODUCTION

The Oligochaeta are a group found both in terrestrial biotopes
and in water-reservoirs. In many natural and anthropogenic biotopes
they are one of the most numerous and most often found animals.
They are characterized by a number of specific biological features
and anatomical-physiological properties, determining the important
role of Oligochaeta in biotope, especially in processes of the
self-purification of water-reservoirs during the waste water
degradation of organic matter, formation of the soil structure and
the increase of its fertility as well as during the course of
processes concerning the displacement of sediment in water-reservoirs.
The greater part of the investigations on Oligochaeta were until
recently carried out in natural ecosystems of different types, and
only recently have these animals been the object of detailed and
more complex ecological investigations carried out in rural agricultural areas entirely given over to human economy. Much attention
has been paid to investigations of agricultural landscapes, or areas
of polyfunctional forms of land use with the supremacy of farming
(Kasprzak, 1977), the accurate knowledge of which is very important
from the economic point of view.

This paper presents information concerning the occurrence of
Oligochaeta in the water-reservoirs and cultivated soils of Poznań,
their role in agroecosystems as well as the influence of agricultural
methods of different kinds on changes of their fauna. The estimation
of the role of Oligochaeta in aquatic and land ecosystems within
the agricultural region of an intensively conducted farm operation
is a part of complex investigations of the Department of Agrobiology
of the Polish Academy of Sciences in Poznań aiming at the repre-
sentation of the model of energetic changes and the circulation
of matter in the whole agricultural landscape as well as the determination of the influence of agriculture on the whole of the fauna
of ecosystems.

AQUATIC ECOSYSTEMS

Information concerning the occurrence of Oligochaeta in field
water-reservoirs are the result of complex hydrobiological investigations (Banaszak and Kasprzak, 1980) carried out in the eutrophic
(polytrophic) Lake Zbechy and in a drainage canal Rów Wyskoc, which
is connected with a fine net of melioration ditches. Both water-
reservoirs are situated within the agricultural landscape of the
central part of the Wielkopolska-Kujawy Lowland situated about
45 km south of Poznan, a very intensive farming district (Figure 1).

Table 1. Domination (in %) of the Oligochaeta in eutrophic
Lake Zbechy and drainage canal (Banaszak and
Kasprzak, 1980).

Species	Lake	Drainage canal
Potamothrix hammoniensis (Mich.)	54,6	41,9
Stylaria lacustris (L.)	16,0	-
Psammoryctides barbatus (Grube)	10,4	-
Psammoryctides albicola (Mich.)	6,8	-
Limnodrilus hoffmeisteri (Clap.)	4,7	51,5
Others: Nais pardalis Pig., N. christinae N. barbata O.F. Müll., Uncinais uncinata (Oerst.), Vejdovskyella comata (Vejd.), Chaetogaster diaphanus (Gruith.), Tubifex tubifex (O.F. Müll.), Limnodrilus udekemianus Clap., L. claparedeanus Ratzel, L. profundicola (Verrill), Peloscolex ferox (Eis.), Lumbriculus variegatus (O.F. Müll.)	7,7	6,6

The lake Zbechy is a post-glacial lake, actually devoid of a hypo-limnion and with completely treeless banks, which are chiefly occupied by meadows and plough-lands.

Species Composition and Density

In Lake Zbechy the Tubificidae make up the greatest number and biomass of oligochaetes, populating the whole lake basin. Potamothrix hammoniensis (55% of the whole of population)(Table 1), characteristic chiefly of eutrophic lakes (Brinkhurst, 1964; Grigyalis, 1974), is the most abundant species. The average annual (1977) biomass of Oligochaeta in the lake amounted to 0,7 g fresh wt m^{-2}. The low density of Oligochaeta in comparison with other lakes as well as of the remaining groups of benthos is a proof of intensively progressing eutrophic processes in Lake Zbechy. This is the result of fertilization of the lake due to an excessive flow of mineral salts. The greatest density of Oligochaeta and greatest number of species was situated in the littoral, in the zone of a depth of 0,5 - 1,5 m (Figure 2). However, in the zone of the lake of the greatest depth (5 m) amounting to 56% surface of the whole lake, only immature individuals of Tubificidae and Potamothrix hammoniensis occurred (0,04 g fresh wt m^{-2}) (Table 2). Relatively high numbers of Chironomidae (Diptera) (0,2 g fresh wt m^{-2}), exert an influence on the low density of Potamothrix hammoniensis in the profundal (Jonasson and Thorhauge, 1972; Thorhauge, 1976).

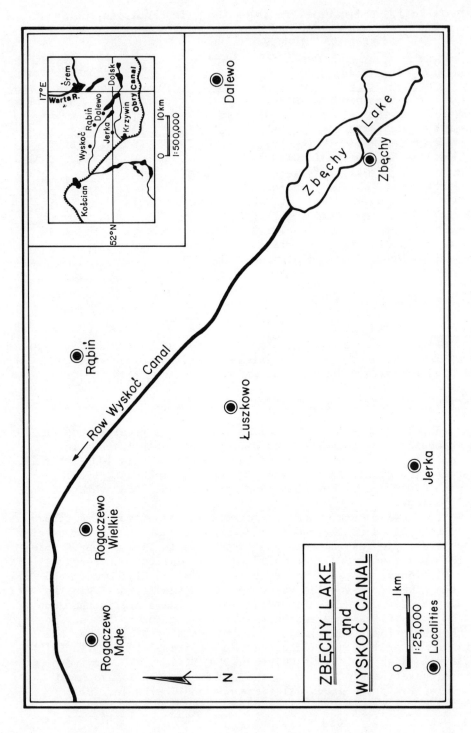

Figure 1. Investigation territory.

Table 2. Comparison of the density of Oligochaeta in
eutrophic Lake Zbechy and drainage canal
(Banaszak and Kasprzak, 1980)

| Depth m | Lake | | | Drainage canal | |
	Surface area of zone ha	Numbers individuals m^{-2}	Biomass mg fresh wt m^{-2}	Numbers individuals m^{-2}	Biomass mg fresh wt m^{-2}
0 - 1,5	19,3	191,1	569,6		
1,5 - 3,5	27,8	40,2	116,8	2466,5	8830,1
3,5 - 5,5	61,8	10,3	36,8		

The differentiations concerning the size of biomass between
the profundal and littoral of the lake is connected chiefly with
the role of the trophic factor. For that reason in the littoral
to which flow additional quantities of feed are noted considerably
higher values of biomass (Kajak and Dusoge, 1976). Many investigations show that, in different water-reservoirs, the chief factor
exerting the decisive influence on the total numbers of benthos is
the trophic situation (Kajak, 1966). In many cases dependences
between the number of benthos and different environmental factors
(oxygen, content of organic matter) cannot be discerned because
of the great complexity of conditions exerting an influence on
bottom organisms. It has been stated many times that there is an
immediate relation between the density of benthos and the quantity
of food flowing to the water-reservoir. The proof of this lies,
among other things, in the positive correlation between the density
of zoobenthos and the density of bacterial microbenthos and phytoplankton, as well as regional experiments with the organic fertilization of lakes causing a repeated increase of the numbers of
benthos.

Oligochaeta in the drainage canal are represented chiefly by
Tubificidae and Lumbriculidae (*Lumbriculus variegatus*), whereas
Naididae are almost completely absent. In contrast to the lake,
the drainage canal shows a considerable increase of the numbers
of *Limnodrilus hoffmeisteri* - a species which occurs in water-
reservoirs of different genesis and trophy. It is characterized
by very great abilities to adjust to changing environmental conditions, chiefly the content of oxygen and the quantity of organic
matter. This species dominates in the drainage canal, chiefly
during the winter-spring (February-March) and autumn, whereas
during the summer period (July) *Potamothrix hammoniensis* dominates.
The annual average biomass of Oligochaeta in the canal (8,8 g fresh
wt m^{-2}) was more than eleven times greater than the average biomass
in the Lake Zbechy (Table 2). Similar differences between the

Table 3. Values of Shannon–Weaver indices (H') of diversity
          for the Oligochaeta in eutrophic Lake Zbechy and
          drainage canal (Banaszak and Kasprzak, 1980).
          P = probability.

Lake: depth	Drainage canal	P
0-5, 5 m		
H' = 1,90	H' = 1,12	$(t_{185}) > t_{0,05}$
2,5-5,5 m		
H' = 0,43	H' = 1,12	$(t_{79}) > t_{0,05}$

density of benthos in the drainage canal and Lake Zbechy can also
be observed in the case of the Chironomidae (Diptera) and the
Spheriidae (Mollusca, Bivalvia) (Banaszak and Kasprzak, 1980).
This is proof that in the drainage canal very favourable conditions
occur for the development of saprophages and filtrators.

Diversity

        In the drainage canal, a smaller number of species than in the
eutrophic Lake Zbechy occur, as shown by the values of the Shannon-
Weaver indices (H') (Table 3). Differences occurring between the
numerical values of H' for the whole of the lake and canal are not
caused by errors of chance and are statistically significant. This
is stated by the verification of the results obtained by means of
the parametric test of significance based on the statistics which
has Student's distribution t of a number of degrees of freedoms
determined by the formula taking into consideration the variation
of calculated values H' (Poole, 1974). The shallow littoral zone
exerts an influence on the greater variety of the lake fauna of
Oligochaeta. This is shown by the comparison of the index H' calculated for Oligochaeta occurring in the profundal of lake and in the
canal. The differences between the H' values are statistically
significant. Thus the oligochaete fauna in the canal shows greater
variety than the fauna of these animals occurring in the profundal
zone of the lake. A decrease of the value of the diversity index
H' was observed both in the lake and in the canal by the distinct
increase of the numbers. This is caused by the domination of
individual species, namely by very unequal distribution of individuals on all species.

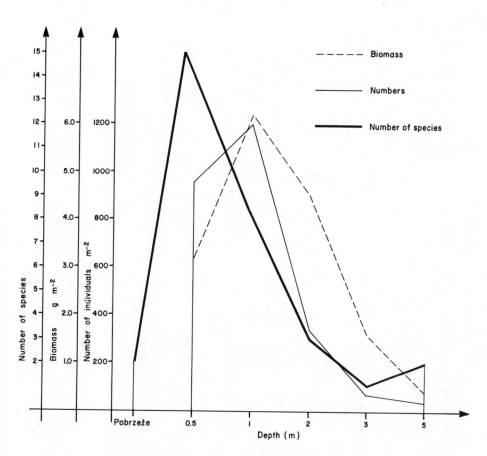

Figure 2.   Dependence of the number of species and density of
             Oligochaeta on depth in eutrophic Lake Zbechy (after
             Banaszak and Kasprzak, 1980).

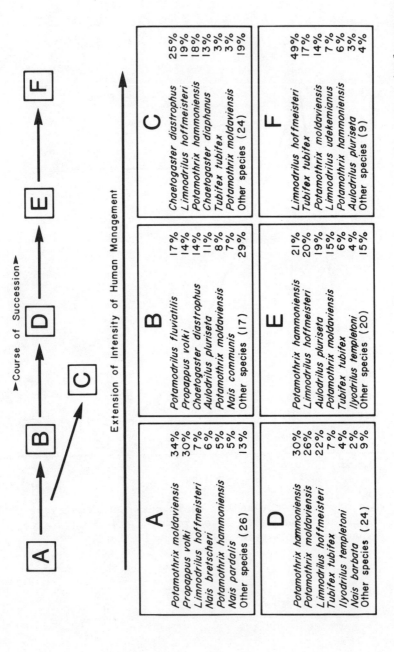

Figure 3.   Succession of the aquatic Oligochaeta community of bottom environments in lower part of the River Welna; A. sandy-gravel bottom in quick current, B. sandy bottom, C. sandy-bottom with putrid plant fragments, D. sandy-muddy bottom, E. muddy bottom in plant beds, F. accumulations on bank (after Kasprzak, 1976b).

Influence of Pollution

The flow of mineral salts from the reception basin as well as municipal wastes and organic wastes of the food industry exerts a great influence on changes of the typical composition and density of Oligochaeta in water-reservoirs. This pollution causes severe changes in the oligochaete fauna especially in mountain and piedmont rivers (Kasprzak, 1976a; Kasprzak and Szczesny, 1976). Strong pollution with allochthonous sewage is the cause of an increase of the numbers of pelophile species of Tubificidae as well as mass appearances of some species (especially *Nais elinguis*). The flow of the allochthonous matter to rivers causes an increase in the silting of the river bed, which directly affects changes of the typical species composition and relations of the domination of the communities of Oligochaeta (Kasprzak, 1976b). In the case of the communities of species of the River Welna, flowing through agricultural regions, the general outline of the course of succession of these communities according to the increase of the silting of the bottom of river bed can be observed. In the presented successive series (Figure 3) a succession is first of all the distinct decrease of the domination and participation of psammorheophile species (*Propappus volki* Mich., *Potamodrilus fluviatilis* (Last.), *Chaetogaster krasnopolskiae* Last.) and simultaneous increase of the numbers and the frequency of the occurrence of species being elements of pelorheophile communities (*Potamothrix hammoniensis*, *Limnodrilus hoffmeisteri*, *Tubifex tubifex* and *Ilyodrilus templetoni*).

Land-Ecosystems

Among all invertebrates living in soil the earliest investigations were carried out on earthworms of the family Lumbricidae. The occurrence of these animals in soils of different type and their influence on the fertility of soils was discussed in the nineteenth century by Ch. Darwin, V. Hensen and P.E. Müller. In consideration of the great importance of Lumbricidae in the formation of the soil structure as well as their essential participation in processes of the decomposition of organic matter, considerable increase in the interest in this group lead to many investigations and a comprehensive world literature concerning the systematics, taxonomy, biology and ecology of Lumbricidae. Considerable less attention is however given the family Enchytraeidae, the biology and ecology of which has been known somewhat better in the last 20 years. The intensification of investigations concerning this group of animals was connected with a systematic revision of this family (Nielsen and Christensen, 1959, 1961, 1963) as well as the elaboration of effective methods of the extration of Enchytraeidae from soil samples (Nielsen, 1953; O'Connor, 1955).

Most past investigations on Lumbricidae were carried out first of all in different natural ecosystems, chiefly forests, meadows, swamps, moorlands and mountains soils. Though agroecosystems in many countries occupy a considerable percentage of the area, they have been least investigated. This must be specially emphasized, because information concerning this environment included in many former works, is rather random and treated by some authors incidentally in comparison with considerably more abundant material originating from natural ecosystems. Most investigations were simply faunistic works, but detailed estimations of the role of Oligochaeta, especially Enchytraeidae in agroecosystems are absent. The influence of agrotechnical measures of different types on the density and species composition of their fauna is not known.

Soil Oligochaeta are represented in Poland by the Lumbricidae and the Enchytraeidae and in soils in agricultural cultivation they are one of the most common groups of the edaphon. Widely known is the fact that these animals are of great importance in the formation of the proper structure and increase of the fertility of soil. This is connected with the occurrence in general of great densities of these animals in soils of different types as well as with some of their specific physiological properties. As Lumbricidae and Enchytraeidae are saprophagous, fulvic acids and humic acids are produced in the gut by the action of micro-organisms. Because of this, these animals exert an influence on the process of humification of organic matter (Kurceva, 1971; Kozlovskaya, 1976)

Species Composition

In general, in the soils of meadows, pastures and fields in Poland, 20 species and forms of Lumbricidae have been listed (Table 4), which is 54% of all species of these animals being found till now in Poland (Kasprzak and Ryl, 1978). In meadows and pastures the number of species of Lumbricidae are the greatest. This must be explained by the considerably smaller number of agro-technical measures on meadow soils as well as a great quantity of attainable food. In meadow soil many species occur which are the same as those of adjoining soils of fields and forests, together with species characteristic for only some types of meadows (Plisko, 1965). The comparison of the Lumbricidae occurring in soils of fields and forests shows the distinct influence of agriculture on the change of the community of species. Plough-lands are inhabited by species of a wide range of occurrence and great adaptive abilities (Table 5). Other species lived on present plough-lands before the transformation of the natural environment into an agroecosystem, or have partly perished due to the changed ecological conditions (for example, species connected with forest soils) or have been superceded by species of greater adaptive abilities (Plisko, 1965). The simplification of animal communities in

Table 4.   List of species of Lumbricidae fauna of the other soils
in cultivated areas of Poland (Kasprzak and Ryl, 1978).

Taxons	Meadows and pastures	Fields
Eisenia foetida (Sav.)	+	
Dendrobaena octaedra (Sav.)	+	
Dendrobaena rubida (Sav.) f. *typica*	+	
Dendrobaena rubida (Sav.) f. *subrubicunda* (Eis.)	+	+
Dendrobaena rubida (Sav.) f. *tenuis* (Eis.)	+	
Dendrobaena platyura (Fitz.) f. *montana* (Cern.)	+	+
Dendrobaena platyura (Fitz.) f. *depressa*(Rosa)	+	
Octolasion lacteum (Oerley)	+	+
Octolasion cyaneum (Sav.)		+
Allolobophora caliginosa (Sav.)	+	+
Allolobophora rosea (Sav.)	+	+
Allolobophora chlorotica (Sav.)	+	+
Allolobophora georgii (Mich.)	+	+
Allolobophora antipai (Mich.) f. *tuberculata* (Cern.)	+	
Eiseniella tetraedra (Sav.)	+	+
Lumbricus rubellus Hoffm.	+	+
Lumbricus castaneus (Sav.)	+	
Lumbricus baicalensis Mich.	+	
Lumbricus polyphemus (Fitz.)	+	
Lumbricus terrestris L.	+	+

regions being within the domain of the human economy is known in
the case of many different systematic groups. This is a proof of
the decrease of the general number of species, the value of diversity indices, the increase of domination of one or several species
in the community as well as decrease of accessory species.

In cultivated soils of Poland there are about 30 species of
Enchytraeidae (Table 6). The number of species on meadows and
pastures as well as plough-lands of different type is similar.
According to the geographic situation and local soil conditions
certain differences in the type composition can however occur.
On meadows, the type composition of the fauna of Enchytraeidae as
well as the structure of domination depends to a great degree on
the character of the economy. On meadows mown regularly and
manured with fertilizers as well as on fields with annual cultivations species of the genus *Fridericia* Mich. of great adaptive
abilities dominate (Table 7).

Table 5. Domination (in %) of the Lumbricidae species of the other
          soils in cultivated areas of Poland.

Species	Rye and potato light sandy-clay soil (Jopkiewicz, 1972)	Cultivation pasture; peat-gley, sandy loam soil (Atlavinyte, 1976)
Allolobophora caliginosa (Sav.)	19,0	43,6
Dendrobaena octaedra (Sav.)	3,0	–
Lumbricus terrestris L.	78,0	–
Allolobophora rosea (Sav.)	–	11,5
Octalasion lacteum (Oerley)	–	1,5
Lumbricus rubellus Hoffm.	–	37,8
Allolobophora chlorotica (Sav.)	–	5,6

Density

     The mean density of Lumbricidae in agroecosystems oscillates
from several to several dozens of individuals per m^2 (Table 8).
The numbers of these animals are distinctly lower in annual cultivations, especially on cultivations of rye, where the fewest
animals are found. Higher numbers of Lumbricidae occur in soils
cultivated for perennial plants, cultivated meadows, and pastures.
In comparison with the Enchytraeidae the Lumbricidae undergo a
greater quantitative impoverishment on plough-lands than in utilized
grasslands. Their density decreases in direct proportion to the
intensity of the loosening of the soil (Dzangaliev and Belousova,
1969). In agroecosystems Lumbricidae as well as Protozoa are one
of the most important groups of animals in respect to biomass, and
their participation in the biomass of adephon in different ecosystems
amounts to 30 - 80% (Ryszkowski, 1978). Estimates of the density
of edaphon performed in the region of the Agroecological Station
in Turew (situated about 40 km south of Poznan) show that, where
there is a decrease in the density of Lumbricidae an important
compensation of the general biomass of edaphon is composed chiefly
of Protozoa as well as to a lesser degree Enchytraeidae and Nema-
toda (Ryszkowski, 1978). On the grounds of these investigations it
can be stated that the average biomass of all animals in the region
Turew amounts to about 6 g dry wt m^{-2} (Ryszkowski, 1974; Golebiowska
and Ryszkowski, 1977) and the biomass of edaphon is almost 95% of
the whole biomass of all animals living in investigated agroecosystems. This is caused chiefly by the occurrence of Lumbricidae
and Protozoa, the biomass of which amounts almost to 84% of the
whole.

     The numbers of the Enchytraeidae in agroecosystems is
considerably higher and oscillates in limits from several to

Table 6. List of species of Enchytraeidae fauna of the other soils
          in cultivated areas of Poland (Kasprzak and Ryl, 1978;
          Kasprzak, unpublished.

Species	Meadows and pastures	Fields
Buchholzia appendiculata (Buchh.)	+	+
Buchholzia fallax Mich.	+	
Henlea similis Niel. et Christ.	+	+
H. ventriculosa (d'Udek.)	+	+
H. perpusilla Friend	+	+
H. nasuta (Eis.)	+	+
Enchytraeus buchholzi Vejd.	+	+
E. albidus Henle		+
E. parva Niel. et Christ.	+	+
Achaeta bohemica (Vejd.)		+
A. camerani (Cog.)	+	+
A. seminalis Kasp.	+	
Hemifridericia parva Niel. et Christ.	+	+
Fridericia bulboides Niel. et Christ.	+	+
F. bulbosa (Rosa)	+	+
F. bisetosa (Lev.)	+	+
F. leydigi (Vejd.)	+	+
F. galba (Hoffm.)	+	+
F. alata Niel. et Christ.	+	+
F. perrieri (Vejd.)	+	+
F. callosa (Eis.)	+	+
F. ratzeli (Eis.)	+	+
F. striata (Lev.)		+
F. hegemon (Vejd.)		+
F. connata Bret.	+	+
F. paroniana Issel	+	+
F. maculata Issel	+	
F. tubulosa Dozsa-Farkas		+
F. semisetosa Dozza-Farkas	+	
F. gracilis Bülow	+	
Marionina argentea (Mich.)	+	+

several dozen thousands of individuals m^2 (Table 9). In general
these numbers are lower than those occurring in natural ecosystems.
They are dependent to a considerable degree on the type of cultivation (Ryl, 1977). In cultivations of root crops numbers of
Enchytraeidae above two times higher than in soils of cultivations
of cereals are observed. On pastures, cultivated meadows and other
grasslands densities of the population of Enchytraeidae reach a
distinctly higher level.

Table 7.   Domination (in %) of the Enchytraeidae species of the
           light sand-clay soil in cultivated areas of Poland
           (Ryl, 1977; unpublished).

Species	Rye	Potato	Alfalfa
Enchytraeus buchholzi Vejd.	43,3	24,4	29,3
Fridericia bulboides Niel. et Christ.	13,6	20,9	-
F. bisetosa (Lef.)	17,6	14,0	10,2
Henlea ventriculosa (d'Udek.)	-	-	18,7
Fridericia gracilis Bülow	-	-	26,3
Others	25,5	40,7	15,5

Table 8.   Average annual density of the Lumbricidae in other types
           of cultivations in Poland.

Type of cultivation	Density		Authors
	numbers individuals m^{-2}	biomass g fresh wt m^{-2}	
Oat	30,0	16,0	} Atlavinyte, 1975
Potato	2,3	1,3	} Jopkiewicz, 1972
	15,0	6,3	
Alfalfa	15,2	7,7	} Jopkiewicz, 1972
	~18,7	9,6	
Pastures	445	60,0	} Nowak, 1975
	91	20,6	

Table 9.   Average annual density of the Enchytraeidae in other
           types of cultivations.

Type of cultivation	Density		Authors
	numbers individuals m^{-2}	biomass g fresh wt m^{-2}	
Rye (Poland)	9600	~0,9	} Ryl, 1977
Potato (Poland)	23000	~2,4	} Ryl, 1977
Grasslands (Sweden)	~24000	4,7	} Persson and Lohm, 1977
Pastures (Denmark, Germany)	~28000	~2,0	} Nielsen, 1955
	~80000	~5,0	
	~25000	~3,7	} Möller, 1969

Respiration

The comparison of the biomass and respiration of Lumbricidae and Enchytraeidae obtained on fields under cultivation (rye, potato) in the region of Turew shows that the general participation of Enchytraeidae in the mineralization process of organic matter is considerably greater than the participation of Lumbricidae. The relation of the mean dry biomass of Enchytraeidae to the mean dry biomass of Lumbricidae amounts to 1 : 10, however, the relation of the respiration (kcal m^{-2}) is inverse and amounts to 3.4 : 1 (Golebiowska and Ryszkowski, 1977). Information obtained by Persson and Lohm (1977) show that the mean annual dry biomass of Lumbricidae in unused meadows in Sweden is about seven times greater than the dry biomass of Enchytraeidae, however, the respiration increases only 1,8 times (Table 10).

The participation of Lumbricidae in the degradation of organic matter is dependent also on its quantity. On meadows with a covering layer of organic matter the most important process stimulated by Lumbricidae is mineralization, however, on meadows without accumulated organic leavings the delay of the decomposition of organic matter by its storage in Lumbricidae (Nowak, 1975) is important.

Influence of Agricultural Methods

Organic fertilizers and mineral fertilizers, spray irrigation as well as mechanical cultivation of the soil, exert a strong influence on the changes of densities of Lumbricidae. Organic fertilization and small or mean quantities of mineral fertilizers (about 100 kg ha^{-1}) have a stimulating influence on the increase of the population of Lumbricidae and Enchytraeidae. Great quantities of mineral fertilizers (more than 600 kg ha$^-$), especially nitrogen fertilizers, cause a decrease of the density of these animals of about 30%. The influence of artificial spray irrigation on the changes of the density of the population Lumbricidae, and probably Enchytraeidae, are seen mainly in summer months at high temperature and low soil humidity. Artificial spray irrigation of cultivated lands, in the case of Lumbricidae, exerts an even greater influence than mineral fertilizers (Budziszak, 1975). It also exerts an influence on the species composition. Mechanical cultivation of the soil reduces the density of Lumbricidae. The increased density of Lumbricidae sometimes observed under the influence of rotary ploughs is connected with a better and more exact mixing of organic matter in the soil by rotary cultivator, the level of which on cultivated fields is a factor reducing the numbers of these animals. A similar reducing influence on the fauna of Lumbricidae are pesticides. Moderate doses (1,6 kg ha^{-1}) of some pesticides (Geratop, Tenoran) cause no distinct decrease of density, and may even create a stimulating influence on the development of the Lumbricid fauna as shown by Lipa,(1958a, 1958b).

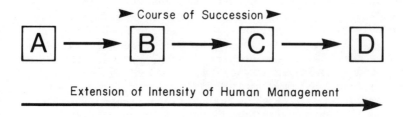

A		B	
Buchholzia appendiculata	27%	*Fridericia* sp. juv.	47%
Fridericia sp. juv.	24%	*Buchholzia appendiculata*	25%
Bryodrilus ehlersi	14%	*Fridericia galba*	14%
Achaeta eiseni	5%	*Fridericia bulbosa*	4%
Mesenchytraeus glandulosus	5%	*Fridericia paroniana*	3%
Fridericia paroniana	4%	*Fridericia maculata*	3%
Other species (20)	21%	Other species (6)	4%

C		D	
Fridericia sp. juv.	58%	*Fridericia* sp. juv.	56%
Fridericia paroniana	9%	*Enchytraeus buchholzi*	9%
Fridericia bisetosa	8%	*Fridericia galba*	7%
Fridericia bulbosa	7%	*Fridericia alata*	5%
Henlea similis	4%	*Henlea similis*	4%
Other species (9)	14%	Other species (9)	19%

Figure 4.   Succession of the Enchytraeidae community of terrestrial
environments in the Pieniny Mts.   (A) beech wood (*Fagetum
carpathicum typicum*); (B) natural meadows (association
Veratrum lobelianum and *Laserpitium latifolium*); (C)
artificial meadows (association *Anthylli-Trifolietum*);
(D) fields. (After Kasprzak, 1979a.)

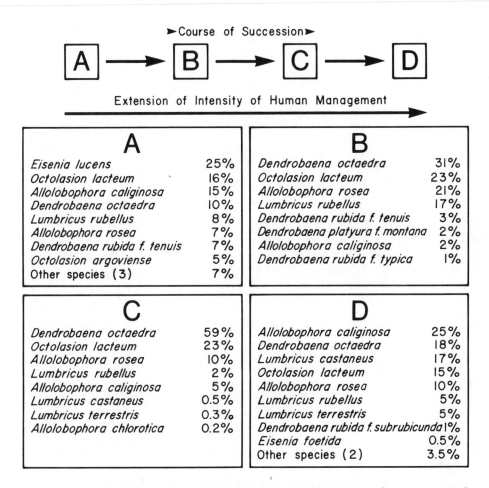

Figure 5. Succession of the Lumbricidae community of terrestrial
environments in the Pieniny Mts. (A) beech wood (*Fagetum
carpathicum typicum*); (B) natural meadows (association
Veratum lobelianum and *Laserpitium latifolium*); (C)
artificial meadows (association *Anthylli-Trifolietum*);
(D) fields. (After Kasprzak, 1979b.)

Table 10. Comparison of biomass and respiration of Enchytraeidae and Lumbricidae in other types of
cultivation.

Taxons	Biomass mg dry wt m^{-2}	Respiration kcal m^{-2}	Authors
Enchytraeidae	199,4	18,7	Golebiowska and
Lumbricidae	2000,0	5,5	Ryszkowski, 1977
Enchytraeidae	852	42,3	Persson and
Lumbricidae	5930	75,2	Lohm, 1977

Influence of the Type of Cultivation

     Some differences concerning the species composition, relations
of domination and density of Oligochaeta, occurring in different
agroecosystems, are related to the type of cultivation because of
changes of the soil property under the influence of human economy.
The numbers and biomass of Lumbricidae is low in the soil of annual
crops but distinctly higher in the soil of perennial crops and
pastures. Cultivation of some plants of the family Papilionaceae,
especially clover, exerts a distinct influence on the density of
Lumbricidae. The numbers of Lumbricidae are also conditioned by
the situation of the field in relation to the field afforestation.
The greater number of species of these animals occur in the woodlands, and average density distinctly decreases as distance from
the woods increases. A more positive influence on development of
the population of Enchytraeidae is exerted by root crops rather
than cereals. Changes of the whole species composition of the
lumbricid fauna under the influence of human economy was indicated
by the results of faunistic-ecological investigations carried out
on Oligochaeta in Pieniny Mountains. The agricultural economy of
man in these areas has caused a change of the whole species composition of the Lumbricidae and Enchytraeidae in soils of semi-
natural and anthropogenic environments developed on primary forest,
considerable reduction of the occurrence of specialized species
connected with forest soils, as well as an increase of the number
and density of eurytopic species of great adaptive abilities
(Kasprzak, 1978, 1979a, 1979b) (Figures 4 and 5).

REFERENCES

Atlavinyte, O.P., 1975, Ekologija dozdevych cervej i ich vlijanie na plodorodie pocvy v Litovskoj SSR Vilnius, Mokslas Publishers, 200 pp.

Atlavinyte, O.P., 1976, The effect of land reclamation and field management on the change in specific composition and densities of Lumbricidae. Pol. Ecol. Stud., 2:147-152.

Banaszak, J., and K. Kasprzak, 1980, Ocena wystepowania i zageszczenia Oligochaeta, Mollusca i Chironomidae w osadach dennych zbiornikow wodnych krajobrazu rolniczego. Pol.Ecol. Stud., in press.

Brinkhurst, R.O., 1964, Observations on the biology of lake-dwelling Tubificidae. Arch. Hydrobiol., 60:385-418.

Budziszak, I., 1975, Proba ustalenia zalesnosci pomiedzy wplywem deszczowania i nawozenia gleby a wystepowaniem w niej fauny Lumbricidae. Zesz. Nauk. AR Wroclaw - Rolnictwo, 31:241-249.

Dzangaliev, A.D., and N.K. Belousova, 1969, Dozdevye cervi y orosajennych sadach rasnych sistem soderzanija. Pedobiologia, 9:103-105.

Golebiowska, J., and L. Ryszkowski, 1977, Energy and Carbon Fluxes in Soil Components of Agroecosystems (Ed. U. Lohm and T. Persson, Soil Organisms as Components of Ecosystems). Ecol.Bull., 25:274-283.

Grigyalis, A.I., 1974, Struktura populacij dominirujuscich bentosnych organismov oz. Dusja. Tr.Akad.Nauk Lit. SSR 4:77-83.

Jonasson, P.M. and F. Thorhague, 1972, Life cycle of *Potamothrix hammoniensis* (Tubificidae in the profundal of a eutrophic lake. Oikos, 23:151-158.

Jopkiewicz, K., 1972, Zageszczenie i przeplyw energii przez populacje dzdzownic (Ed. L. Ryszkowski, Ekologiczne efekty intensywnej uprawy roli). Zesz. Nauk.IE PAN, 5:226-236.

Kajak, Z., 1966, Obfitosc i produkcja bentosu oraz czynniki na nie wplywajace (In: Produktywnosc ekosystemow wodnych). Zeszyty Problem. "Kosmosu", 13:69-92.

Kajak, Z., and K. Dusoge, 1976, Benthos of Lake Sniardwy as compared to benthos of Mikolajskie Lake and Lake Taltowisko. Ekol. Pol., 24:77-101.

Kasprzak, K., 1976a, Materials to the fauna of Oligochaeta of the Ojcow National Park and its vicinity - the Pradnik-Bialucha stream. Acta Hydrobiol., 18:277-289.

Kasprzak, K., 1976b, Badania nad skaposzczetami (Oligochaeta) dolnego biegu rzeki Welny. Fragm. Faun. 20, 24:426-467.

Kasprzak, K., 1977, Pojecie krajobrazu w naukach przyrodniczych. Wiad. Ekol., 23:119-131.

Kasprzak, K., 1978, Wplyw gospodarki czlowieka na zmiany glebowej fauny skaposzczetow (Oligochaeta) Pieninskiego Parku Narodowego. Chronmy Przyr. Ojczysta, 34, 5:65-69.

Kasprzak, K., 1979a, Skaposzczety (Oligochaeta) Pienin, I. Wazonkowce (Enchytraeidae). Fragm. Faun., 24, 1:7-56.

Kasprzak, K., 1979b, Skaposzczety (Oligochaeta) Pienin, III.
    Dzdzownice (Lumbricidae). Fragm. Faun., 24, 3:81-95.
Kasprzak, K., and B. Ryl, 1978, Wplyw rolnictwa na wystepowanie
    skaposzczetow (Oligochaeta) w glebach pol uprawnych. Wiad.
    Ekol., 24, 4:333-366.
Kasprzak, K., and B. Szczesny, 1976, Oligochaetes (Oligochaeta)
    of the River Raba. Acta Hydrobiol., 18, 1:75-87.
Kozlovskaya, L.S., 1976, Rol bespozvonocnych v transformacii
    organiceskogo vescestva bolotnych pocv. Leningrad, Nauka
    Publishers, 210 pp.
Kurceva, G.F., 1971, Rol pocvennych zivotnych v razlozenii i
    gumifikacii rastitelnych ostatkov. Moskva, Nauka Publishers,
    155 pp.
Lipa, J.J., 1958a, O roli dzdzownic w glebie i wplyw heksachloro-
    cykloheksanu (HCH) na ich populacje oraz plony roslin
    uprawnych. Post. Nauk. Rol., 1:39-53.
Lipa, J.J., 1958b, Effect on earthworms and Diptera populations
    of BHC dust applied to soil. Nature, 181:863.
Moller, F. 1969, Okologische Untersuchungen an terricolen
    Enchytraeidenpopulationen. Pedobiologia, 9:114-119.
Nielsen, C.O., 1953, Studies on Enchytraeidae. 1. A technique
    for extracting Enchytraeidae from soil samples. Oikos, 1952-
    1953, 4:187-196.
Nielsen, C.O., 1955, Studies on Enchytraeidae. 2. Field Studien.
    Nat. Jutland., 4:5-58.
Nielsen, C.O., and B. Christensen, 1959, The Enchytraeidae,
    critical revision and taxonomy of European species. Nat.
    Jutland., 8-9:1-160.
Nielsen, C.O., and B. Christensen, 1961, The Enchytraeidae,
    critical revision and taxonomy of European species. Supple-
    ment 1. Nat. Jutland., 10:1-23.
Nielsen, C.O., and B. Christensen, 1963. The Enchytraeidae,
    critical revision and taxonomy of European species. Supple-
    ment 2. Nat. Jutland, 10:1-19.
Nowak, E., 1975, Population densities of earthworms and some
    elements of their production in several grassland environment.
    Ekol. Pol., 23:459-491.
O'Connor, F.B., 1955, Extraction of Enchytraeidae worms from a
    coniferous forest soil. Nature, 175:815-816.
Persson, T., and U. Lohm, 1977, Energetical Significance of the
    Annelids and Arthropods in a Swedish Grassland Soil. Ecol.
    Bull., 23:1-211.
Plisko, J.D., 1965, Materialy do rozmieszczenia geograficznego
    i ekologii dzdzownic w Polsce (Oligochaeta, Lumbricidae).
    Fragm. Faun., 12:57-108.
Poole, R.W., 1974, An Introduction to Quantitative Ecology.
    McGraw-Hill Book Company, New York, 532 pp.
Ryl, B., 1977, Enchytraeids (Oligochaeta, Enchytraeidae) of the
    rye and potato fields in Turew. Ekol. Pol., 25:519-529.

Ryszkowski, L. (ed.), 1974, Ecological Effects of Intensive
    Agriculture, Warszawa, Polish Scientific Publishers, 84 pp.
Ryszkowski, L., 1978, Wplyw intensyfikacji rolnictwa na faune
    Zesz. Probl. Post. Nauk Rol., 233: in press.
Thorhauge, F., 1976, Growth and life cycle of *Potamothrix hammo-
    niensis* (Tubificidae, Oligochaeta) in profundal of eutrophic
    Lake Esrom. A field and laboratory study. Arch. Hydrobiol.,
    78 (1):71-85.

OLIGOCHAETE COMMUNITIES IN POLLUTION BIOLOGY: THE EUROPEAN:

SITUATION WITH SPECIAL REFERENCE TO LAKES IN SCANDINAVIA

Göran Milbrink

Institute of Zoology
Uppsala University and Institute of
    Freshwater Research
Drottningholm, Sweden

ABSTRACT

Indicator species of communities are often the biological
units of choice for describing the trophic situation in a body of
water or a situation of intermittent pollution. Communities of
oligochaete species within the families Tubificidae and Lumbriculidae usually describe the current trophic situation with sufficient accuracy. For several oligochaete species which are known
to be cosmopolitan or near to it, the ecological requirements do
not seem to differ much from one part of the world to another.
The ecological demands of immigrants also seem to remain unchanged
in their new localities.

There is, however, some hesitation as to the ecological
attributes of some species possibly as an effect of interspecific
competition. The eventuality of sibling species is considered.
The present paper summarizes the known ecological demands of about
25 tubificid and lumbriculid species rather frequently occurring
in the profundal of lakes in Scandinavia. References are also
made to recent experience gained in this field in lakes and
running water habitats in other parts of Europe outside Scandinavia.

Specific morphological reactions of individual specimens to pollutants in the environment are noted. The obvious ease with which some tubificid species are capable of colonizing even very distant localities is discussed as is the possibility of predicting the arrival of characteristic species in situations of drastically changed local conditions.

The frequent use of diversity indices in efforts to describe environmental changes or trophic conditions is commented upon. A suggested improvement of the Howmiller - Scott environmental index based on oligochaete community structure is briefly presented.

INTRODUCTION

Indicator species or indicator communities are often the biological units of choice for describing the trophic situation in a body of water or a situation of intermittent pollution. Benthic macro-invertebrates have proved to be particularly indicative in this context (cf. Milbrink, 1973). The community concept is here generally preferred, rather than relying upon individual indicator species. The underlying assumption, often verified empirically, is that interactions between biotic and abiotic components result in characteristic assemblages of organisms.

Whereas chemical surveys indicate conditions of one kind or the other only at the time of sampling, profundal benthic communities are considered to make an integral measure of autotrophic and heterotrophic lake processes and also of disturbances into these processes, which makes such communities suitable for various kinds of environmental monitoring. Sessile forms within the bottom fauna are more or less confined to macro-habitats where autochtonous and allochtonous material continuously accumulate and where heterotrophic processes are particularly effective. Accordingly, it is not surprising to find that sessile chironomid larvae and oligochates - despite considerable taxonomical difficulties have been the most frequently used indicator organisms in any pollution detection schemes or lake-typologies (cf. Milbrink, 1978).

LAKE-TYPOLOGIES AND SAPROBIEN SYSTEMS, GENERAL BACKGROUND

In Europe knowledge about tolerance and preferences of so-called indicator species or communities of different taxocenes within the bottom fauna early developed into the lake-type system proposed by Thienemann. The trophic concept introduced by Naumann soon after was accepted by Thienemann and by 1925 the typology was a rather comprehensive classification scheme, though with obvious limitations. Several experienced hydrobiologists have later been involved in the further development of the lake-typology to fit any lake - or even running water situations. Among these numerous authors may be specially mentioned Lundbeck and Brundin, but also

Valle, Lenz and others have made valuable contributions. The
lake-type system was early criticized for various reasons, for
instance by Lastockin and by Wesenberg-Lund (all of the above
in Brinkhurst, 1974).

     Characteristic chironomid species, in particular, have been
utilized for describing the properties of each lake-type. The
known ecological requirements of other insects, oligochaetes,
molluscs, etc. gradually complemented knowledge about the former
group.

     Another approach to the problem of classification - and
closely related to the former system - is the Saprobien System
(Kolkwitz and Marson, 1902, 1908, 1909; in Hellawell, 1978) which
has been widely used in its modified forms, especially in parts
of central and eastern Europe. The system was originally based
upon experience from self-purification gradients in running-waters,
but was later extended into lakes and whole catchment areas. The
interest has to a large extent been focused on the observed
changes in the populations of dominant bottom organisms in relation to organic pollution, i.e. organically enriched sediments and
the depletion of oxygen near the bottom as a consequence of organic breakdown (Brinkhurst, 1974). Although this kind of approach
in a more or less modified form has led to many important studies
of polluted rivers (for instance Hynes, 1960) and whole river
catchment areas (for instance Liebmann, 1951), and the original
system has grown in sensitivity but also in complexity (cf. Sladecek, 1973), it now seems to be of rather limited use, at least
in western Europe. Among those who have contributed with severe
criticism of the Saprobien System idea may be mentioned Hynes
(1960) and Elster (1966).

     Despite the fact that more and more information could be
included into the original lake-typologies and Saprobien Systems
thanks to, among other things, progress in the taxonomical understanding of different invertebrate groups, the basic idea seems
never to have received the popularity among limnologists and
hydrobiologists which was originally hoped for. The obvious
misuse by several authors in describing phenomena for which the
typologies were never intended has resulted in these systems
falling more and more into disreputation (Wiederholm, in manuscript).

     Chironomid species have traditionally played a dominant part
in most classification schemes of the above kind - in Europe and
elsewhere. The reason for this dominance may today seem a bit
difficult to understand since it is recognized that the taxonomy
of chironomid larvae has not been worked out to any large extent,
especially not for the first larval stages. The obvious periodicity

with which chironomids occur in the bottom fauna is also a
complicating factor – a phenomenon which may also characterize
oligochaetes to some extent.

   Another way of working with lake-classification and yet avoiding the slight confusion and often semantic discussion about what
is general pollution biology, what is "trophism" or "saprobity"
etc. (cf. Sladecek, 1978), is to concentrate on one entire benthic
group, such as chironomids or oligochaetes, considering the ecological attributes of the constituent species as proposed by Brundin
(1949, 1956), Wiederholm (1976), Saether (1975) and others for
chironomids, and by Brinkhurst with co-workers in a series of works
over the last 15 years (Brinkhurst, 1974), Milbrink (1973, 1978),
Lang (1978a), Howmiller and Scott (1977) and others for oligochates.
Under normal conditions freshwater benthic communities are usually
completely dominated by these two groups. There is, however, one
obvious disadvantage with the last kind of approach, since it is
recognized that there are many difficulties in identifying chironomid larvae to species (see above) and very few biologists,
indeed, are capable of determining oligochaetes, especially
immature tubificid worms, which are normally in the vast majority
and which really have to be determined correctly in order to avoid
misinterpretations (Milbrink, 1978). Most information about the
occurrence or the ecological attributes of indicator species or
communities within these two groups is thus largely confined to
a limited number of specialists. There is an urgent need for the
information to be transformed into a general, easily assimilated
form comparable to other environmental data (see below). The
need for reliable numerical indices is apparent.

   Communities of oligochaetes of the families Tubificidae and
Lumbriculidae could tell the initiated observer a good deal about
the current trophic situation, about a situation of environmental
stress etc. provided a minimum of background data are available.
Due to taxonomical difficulties nearly impossible to overcome by
early bottom faunists working with lake-typologies or sapro-
biology, oligochaetes in most instances have merely been referred
to as just oligochaetes or Tubificidae, *Tubifex* sp., *Limnodrilus*
sp. etc. Sometimes names of species such as *Tubifex tubifex* have
also been used, even if the species identity often seems to have
been nothing but a qualified guess. The *"Tubifex tubifex* concept"
is also an important link in several of the most popular pollution
and biotic indices, such as in the Biotic Score (Chandler), the
Trent Biotic Index (Woodwiss), the Empirical Biotic Index (Chutter),
the Index of Pantle and Buck (Pantle), the Saprobity Index of
Zelinka and Marvan, the Relative Purity Index of Knöpp, the Species
Deficit Index of Kothé (in Hellawell, 1978).

   It is not the intention of the present review to mention all
authors who have used the above broad and rather non-specific
nomenclature. Several early studies have successfully used the

relative abundance of worms in the macro-benthos as a measure of pollution, thereby utilizing the empirical fact that small shifts in the balance between the different entities of the fauna often reflect slightly changed environmental conditions (cf. Brinkhurst and Jamieson, 1971; Brinkhurst, 1974; Milbrink, 1973).

In the last mentioned paper, the present author made an attempt to compile data on the known ecological demands of those aquatic oligochaete species which could be obtained in the profundal of Scandinavian lakes - in fact the majority of worms within the above families commonly found in waters all over Europe. Brinkhurst (1974) has contributed with further information on the subject, as have for instance, Pfannkuche (1977), Milbrink (1978) and Lang (1978a). For more detailed information on the historical background to what we now know about the taxonomy, faunistics, autecology, relationships between different species in terms of interactive segregation for available resources, etc., the reader is referred to the above works and to other contributions in the present volume.

The following will rather closely follow the contents in Milbrink (1973, 1978) and Brinkhurst (1969, 1974) and refers mainly to the profundal of lakes where conditions are normally fairly stable over extended periods (cf. Brundin, 1956). Less stress is here laid upon littoral and riverine situations, where the oxygen factor, for instance, is less likely to be limiting. Ecological notes from detailed river surveys, however, have here been included as well (e.g. Brinkhurst and Kennedy, 1965; Dzwillo, 1967; Wachs, 1968; Ladle, 1971; Timm, 1970). The ranking of species below rather closely follows the scheme presented by Zelinka and Marvan (1961) with the addition of some species not yet found in Scandinavia (Table I). Strictly littoral species which do not reflect overall changes in the environment to the same extent have been omitted from the scheme. Species considered more or less endemic in their distribution, for instance from the Balkans and partly from eastern Europe have likewise been omitted. Due to considerable difficulties in obtaining relevant ecological information on various species occurring in eastern Europe, mostly because of language barriers, this part of Europe is definitely under-represented in all reviews of this kind. This is probably inevitable despite considerable efforts carried out to remedy this by the All-Soviet Oligochaete Association and by individual specialists in eastern Europe.

PRESENTATION

A correct definition of the term pollution is in reality very difficult to formulate. In most instances this concept means something totally negative such as the emission of poisonous substances into the environment, adverse effects of eutrophication etc. It may also, however, cover a more positive side, implying mild eutrophication causing, for instance, a rise in fish production. A more

correct term for the latter would be nutrient enrichment. Margalef
(1975) has pointed out that pollution is hardly a scientific concept,
rather a comparative procedure. Most of what we know about the ecological demands of aquatic oligochaetes refer to certain trophic
situation, less so to situations of poisoning.

We now seem to know quite a good deal about the preferences
and tolerances of a fair number of species, especially those which
are cosmopolitans or near to it, such as *Limnodrilus hoffmeisteri*
and *Tubifex tubifex* (Brinkhurst, 1969; Milbrink, 1973, 1978). The
ecological requirements of species of that kind do not seem to differ
much from one part of the world to another. Others such as a number
of species of the genus *Potamothrix, Limnodrilus cervix, Branchiura
sowerbyi* etc., which have obviously rather recently conquered areas
across the Atlantic or mountain barriers or other migration obstacles
seem to keep their ecological demands unchanged in their new habitats.
There are, however, species which appear to react slightly differently
in different parts of the world. One of these is presumably *Peloscolex ferox* (see below).

Table 1 lists most of the species discussed in pollution
biology in Europe. Each species has received ten points totally
allotted among the three main lake-categories in which the species
are usually encountered, i.e. the oligotrophic, the mesotrophic
and the eutrophic lake-types, here defined as holding less than
20 µg/1, up to 40 µg/1 and above 40 µg/1, respectively of total-
phosphorus (cf. Milbrink, 1978; see above).

There is little doubt that the tubificids *L. hoffmeisteri,
T. tubifex, Limnodrilus claparedeanus* and *Potamothrix hammoniensis*
belong to the eutrophic end of the scheme. They may appear in
extremely dense abundance – sometimes in monocultures – given the
ideal conditions (Brinkhurst, 1970, 1974; Milbrink, 1973; Pfannkuche, 1977). From the hyper-trophic Toronto Harbour area in Canada
there are even observations of "balls" exclusively composed of worms
rolling over the bottom surface (Brinkhurst, personal communication).

Even if *Limnodrilus hoffmeisteri* may be present in freshwater
of any quality its progressive dominance of the benthic community
is clearly correlated with the degree of organic pollution (Brinkhurst, 1969). The species is often associated with *T. tubifex*
(see discussion below) in severely polluted waters (Brinkhurst and
Kennedy, 1965; Hiltunen, 1969; Timm, 1970; Aston, 1971; Milbrink
1973; Brinkhurst, 1974). By mass occurrence this association indicates low oxygen concentration and organic pollution (Wachs, 1964;
Saether, 1970; Aston, 1971). Järnefelt (1953) describes the tremen-
dous increase in population densities of this species below winter
roads on frozen lakes in Finland due to animal droppings. Hiltunen
(1969), Saether (1970) and others have called the species saprophilous. Further references to the ecological demands of *L. hoffmeisteri* may be found in Milbrink (1973).

The following species *Tubifex tubifex* after some hesitation
now having been inserted into the scheme, has also been extensively
treated in Milbrink (1973).   In a recent paper (Milbrink, 1978) the
present author had arrived at the conclusion that *T. tubifex* should
be kept outside this list, since its distribution pattern very
clearly indicates that it has a habit of turning up wherever other
oligochaetes (or other bottom animals, for instance chironomid
larvae) are few or where the predation pressure from fish and other
predators is very low.   Many authors have observed and commented
upon this peculiarity of the species, especially Brinkhurst in a
series of works (in Milbrink, 1973).   Among recent reflections of
this kind may be mentioned Särkkä (1978).   From the references
given in Milbrink (1973) it is apparent that, in Europe, the species
may dominate the oligochaete fauna in lakes or running waters of
any trophic quality, size, sediment texture, etc.   It may be a main
component in the oligochaete fauna of small mountain lakes very poor
in nourishment as well as in deep and large oligotrophic lakes in
Switzerland, Italy and Scandinavia and in eutrophic or even heavily
polluted lakes and rivers all over Europe.

T. tubifex is mostly found in the outer margins of the trophic
scale (Brinkhurst, 1969).   Studies by, for instance, Berg et al.
(1962), Palmer (1968) and Aston (1971 have shown that it can tolerate
very low oxygen concentrations for extended periods.   Many reports
have witnessed that of all aquatic oligochaetes *T. tubifex* has the
greatest resistance against oxygen deficiency, H_2S, methane gases
etc. (cf. Wachs, 1964, 1968; see Milbrink, 1973).   The reason why
the species is so characteristic of deep oligotrophic lakes in
Europe is not entirely clear.   It may be a consequence of interactive
competition with other species such as *P. hammoniensis* (see discussion below), of decreasing light intensity with depth (Särkkä, 1978)
or of a maintained, sufficient reproductive capacity in relation to
other species at the low temperatures prevailing in the deep waters
(Aston, 1971).   Poddubnaya (this volume) has suggested the existence
of two distinct forms of the species, one limnophilous form in the
deep parts of lakes and one riverine form tolerating variable conditions.   Already Thienemann (1920, in Timm, 1970) distinguished one
small and thin form, which he called *Tubifex filum (?)* - which is
possibly the same variety of *T. tubifex* as Timm found in the profundal of lakes in Estonia.   This form - externally very suggestive of
Tubifex ignotus - is regionally common in the profundal of eutrophic
lakes in Sweden (Milbrink, unpublished).   In accordance with the
above, *T. tubifex* has been given five points in each of the oligotrophic and eutrophic lake-types.

Potamothrix hammoniensis is probably the commonest tubificid
species in eutrophic lowland-lakes in Europe.   The species is generally associated with pollution (see references in Milbrink, 1973).
Särkkä (1978) defines the optimum habitats of the species in Finland
as shallow lakes which are either eutrophic or polluted.   The species

Table 1. Occurrence of freshwater oligochaetes in different lake-types with special reference to Scandinavian conditions. For explanations, see text.

	Oligotrophy	Mesotrophy	Eutrophy
1. *Stylodrilus heringianus* Claparède	9	1	
2. *Rhynchelmis limosella* Hoffmeister	8	2	
3. *Bythonomus lemani* Grube	8	2	
4. *Peloscolex velutinus* (Grube)	8	2	
5. *Peloscolex ferox* (Eisen)	8	2	
6. *Lamprodrilus isoporus f. variabilis* Svetlov	(8)	(2)	
7. *Limnodrilus profundicola* (Verrill)	7	3	
8. *Psammoryctides barbatus* (Grube)	5	5	
9. *Rhyacodrilus falciformis* Bretscher	(5)	(5)	
10. *Rhyacodrilus coccineus* (Vejdovský)	4	6	
11. *Psammoryctides albicola* (Michaelsen)	4	5	1
12. *Aulodrilus pigueti* Kowalewski	3	5	2
13. *Bothrioneurum vejdovskyanum* Stolc	2	5	3
14. *Aulodrilus limnobius* Bretscher	2	5	3
15. *Tubifex ignotus* (Stolc)	2	4	4
16. *Potamothrix moldaviensis* (Vejdovský and Mrázek)	2	4	4
17. *Limnodrilus udekemianus* Claparède	1	4	5
18. *Potamothrix bavaricus* (Oschmann)		5	5
19. *Potamothrix bedoti* (Piguet)		5	5
20. *Potamothrix vejdovskyi* (Hrabê)		5	5
21. *Ilyodrilus templetoni* (Southern)		5	5
22. *Potamothrix heuscheri* (Bretscher)		4	6
23. *Aulodrilus pluriseta* (Piguet)		4	6
24. *Limnodrilus cervix* Brinkhurst		4	6
25. *Limnodrilus claparedeanus* Ratzel	1	3	6
26. *Potamothrix hammoniensis* (Michaelsen)	1	3	6
27. *Tubifex tubifex* (Müller)	5		5
28. *Limnodrilus hoffmeisteri* Claparède	1	2	7

occurs in dense mono-cultures in the markedly eutrophic Lake Esrom in Denmark (Jonason and Thorhauge, 1972). Lang (1978a) on the other hand suggests an intermediate position of the species between tolerant and sensitive species (see discussion below). According to Cekanovskaya (1965, in Särkkä, 1978) *P. hammoniensis* has a preference for a slow currents. Several authors have commented upon the apparent inverse correlation of the species with *T. tubifex* (for instance Lang, 1978a; Särkkä, 1978) especially with depth. Among genuine freshwater tubificids *P. hammoniensis* is the most euryhaline, tolerating salinites up to around 6°/oo in the Baltic archipelagos (Timm, 1970; Särkkä, 1978) - especially in case of nutrient enrichment (Milbrink, unpublished material from the eastern coast of Sweden).

Limnodrilus claparedeanus is more local and patchy in its distribution pattern than *L. hoffmeisteri* in England (Kennedy, 1965) as well as in Sweden (Milbrink, 1973). The former author found that *L. claparedeanus* demands high concentrations of oxygen in organically polluted water, whereas Dzwillo (1967) and Pfannkuche (1977) found the species to be one of the most abundant species in stagnant parts of the Elbe estuary. It is often found in the same communities as *L. hoffmeisteri* and *L. cervix,* especially at shallow depths (cf. Howmiller and Scott, 1977). The species may locally occur in very rich monocultures down to depths of about 30 meters in western Lake Mälaren (Milbrink, unpublished).

Limnodrilus cervix certainly shares many of the characteristics of the preceeding species with which it possibly also hybridizes (Howmiller and Beeton, 1970; Brinkhurst and Jamieson, 1972). It is likely that this species is a recent immigrant into European waters from North America (see below). It is mostly found together with the other *Limnodrilus* species, preferably at shallow depths (Milbrink, unpublished). It is supposed to spread from harbours and river-mouth areas (Hiltunen, 1967, 1969). In Europe little is known about the ecological requirements of *L. cervix.* In Mälaren it very locally co-exists with *L. claparedeanus,* (the "hybrid *cervix/claparedeanus",* see below), *L. hoffmeisteri* and *Branchiura sowerbyi* (Milbrink, unpublished). *L. cervix* together with *B. sowerbyi* and *P. multisetosus* are particularly good indices of organic pollution in the Great Lakes of North America (Brinkhurst, 1969).

Aulodrilus pluriseta is considered to be the most tolerant species of the genus, but more sensitive to organic pollution than, for instance, *P. hammoniensis* (Brinkhurst, 1964) or about as tolerant, even if it is more sporadically distributed (Juget, 1958; Milbrink, 1973). For further references, for instance concerning its distribution in North America, see the latter work.

Potamothrix heuscheri. Wherever in Europe this species has reached some sort of equilibrium with the more indigenous worm fauna it seems to be a reliable indicator of eutrophy (Milbrink, 1973;

Lang, 1978a). It is otherwise known to be extremely tolerant to
anoxic conditions in brackish or moderately alkaline bottom water
(Gitay, 1968; Milbrink, 1973). Since this species has comparatively
recently reached central and northern Europe ecological notes in the
early literature are few. Apart from the above works there is still
little known about the species (see below). *P. heuscheri* may be
found in the littoral of the Baltic (Särkkä, 1978; Milbrink, unpublished material from the eastern coast of Sweden).

Ilyodrilus templetoni is, no doubt, an easily over-looked tubificid species, and probably often mistaken for *T. tubifex* or *P.
hammoniensis*. Accordingly, ecological observations remain scarce
(Milbrink, 1973). Like *L. hoffmeisteri* the species is saprophilous
(Hiltunen, 1969), but often characteristic of improved conditions
in a polluted area (Howmiller and Beeton, 1970). In Swedish waters
I. templetoni is generally associated with *P. hammoniensis* and sometimes *L. hoffmeisteri* in the profundal of moderately eutrophied
waters (Milbrink, 1973).

Potamothrix vejdovskyi is indicative of mildly polluted or even
markedly eutrophic conditions in Sweden - as well as in North America
(Milbrink, 1973). Like most of the other so-called Ponto-Caspian
species it is generally found in fairly polluted habitats or in the
middle of the trophic scale (Brinkhurst, 1964, 1969). According to
Lang (1978a) *P. vejdovskyi* may be looked at as an indicator of oligotrophy in the Lake of Geneva - but also inversely proportional to
P. hammoniensis in the same lake. Specimens found in lakes may
differ from those found in rivers - Le Rhone (Juget, personal information).

Potamothrix bedoti was once considered a rather doubtful species
with morphological characteristics indistinct from those of a *P.
bavaricus* (see Milbrink, 1973). However, this matter has now been
settled (Spencer, 1978). In Europe both the morphology and the ecology of the two species have appeared very different. In Estonia
Timm (1970) found *P. bedoti* to be particularly characteristic of
springs and *P. bavaricus* or river mouths in brackish water. *P.
bedoti* - although always scarce - and *P. vejdovskyi* seem to have
similar patterns of distribution in Lake Mälaren and most likely
very similar ecological requirements.

Potamothrix bavaricus (see the former species) is a species
which has not been much referred to in pollution biology. In Johnson
and Brinkhurst (1971) its ecological requirements are suggestive of
those of *P. vejdovskyi*.

Limnodrilus udekemianus tolerates oxygen deficiency, but does
not usually inhabit grossly polluted waters (Brinkhurst, 1964c;
Kennedy, 1965; Dzwillo, 1967, in Milbrink, 1973). In the latter
paper reference is also made to several works with records of the

species from severely polluted rivers - often in great abundance -
together with *L. hoffmeisteri, L. claparedeanus* and *T. tubifex*
(Wachs, 1964; Timm, 1970; Howmiller and Scott, 1977) - river-mouths
in lakes and from littoral areas of eutrophic basins of Lake Mälaren
and other eutrophic lakes in Scandinavia (Milbrink, 1973 and unpublished material). The species is generally sporadic in its distribution, but may be locally common (cf. Kennedy, 1965).

Potamothrix moldaviensis belongs to the tubificid species about
which we know very little of the ecology. There is ample evidence
in the literature, however, that it is regionally common in running
water (for references, see Milbrink, 1973). In Lake Mälaren it may
also be locally common in littoral areas exposed to moving water
masses and on steep bottom sides, where spring water flows out. The
species is nearly always found on sandy substrates (Ibid.).

Tubifex ignotus is generally associated with pollution and most
often found together with *P. hammoniensis* in Swedish waters (Milbrink,
1973). Lastockin (1927), however, found *T. ignotus* to be characteristic of dystrophic lakes in Russia (in Milbrink, 1973). On the
whole, records of the species in Europe remain few, which may partly
be attributable to the fact that immature worms are difficult to
distinguish from *T. tubifex* (see above).

The following species, *Aulodrilus limnobius, Bothrioneurum
vejdovskyanum, Aulodrilus pigueti* and *Psammoryctides albicola* are
all fairly irregular in their distribution pattern and have been
little used in pollution biology. For references, see Milbrink
(1973).

Rhyacodrilus coccineus has proved to be a very useful indicator
organism. In Swedish waters - and preferably on sandy substrates -
the species is particularly characteristic of zones of transition
from eutrophic conditions or even gross organic pollution on to
oligotrophy (Milbrink, 1973, 1978). It seems to be able to withstand slight pollution (cf. Dzwillo, 1967), but is more sensitive
than, for instance, *P. hammoniensis, Psammoryctides barbatus* and
L. udekemianus (Brinkhurst, 1962, in Milbrink, 1973). In the classification of oligochaetes by Howmiller and Scott (1977) *R. coccineus* is referred to the group of species which are largely restricted to oligotrophic conditions. For further information on this
species, see Milbrink (1973).

Rhyacodrilus falciformis is a little known species from the
extreme profundal of deep oligotrophic lakes (Bretscher, 1901;
Piguet, 1913; Ekman, 1915, in Milbrink, 1973), but it is ubiquitous
in Lake Annecy and the Lake of Geneva (Juget, 1958).

Psammoryctides barbatus, although being one of the species most sensitive to oxygen deficiency, may still occur in great abundance in grossly polluted streams and in the littoral of eutrophic lakes provided turbulence is sufficient to produce local re-aeration (see reference list in Milbrink, 1973). *P. barbatus* was a characteristic species already in Thienemann's "Sauerstoffreichen *(Tanytarsus)* Seen" and physiological works (Berg et al., 1962) would seem to confirm field observations as to the sensitivity of the species in comparison with, for instance, *T. tubifex* and *P. hammoniensis* (Milbrink, 1973). *P. barbatus* may be a dominant oligochaete species in unstratified organically enriched lakes (Ibid.). Särkkä (1978) considers the species to be an indicator of oligotrophy.

Limnodrilus profundicola is likewise, when occurring in the profundal of a lake, a good indicator of oligtrophic conditions (Milbrink, 1973, 1978; Howmiller and Scott, 1977; Lang, 1978a) and particularly characteristic of the deepest lakes in Europe (Brinkhurst, 1964). Several works, however, have verified that this species may also be found in eutrophic situations – especially in the littoral – provided the water is well oxygenated (Milbrink, 1973).

The following species (at the top of Table 1) are all considered to be very reliable indicators of oligtrophy.

Lamprodrilus isoporus f. variabilis is a little known and possibly easily over-looked lumbriculid species from the east (Timm, 1970; Brinkhurst and Jamieson, 1971; Särkkä, 1978). Särkkä has equalized the ecological demands of the species with those of *Stylodrilus heringianus, Peloscolex ferox* and *P. barbatus* in Finnish waters.

Peloscolex ferox belongs to the most reliable indicators of oligotrophy, often mentioned in the early literature describing the bottom fauna of mountain lakes, large and deep oligotrophic lakes, humic lakes etc. (Ekman, 1915; Piguet, 1919, Brundin, 1949, in Milbrink, 1973). Although sometimes present also in the littoral of mesotrophic and eutrophic lakes and in organically enriched rivers with sufficient oxygen concentrations, it is a main component in the profundal bottom fauna of most oligotrophic lakes in Scandinavia (Milbrink, 1973, 1978; Särkkä, 1978). Juget (1958), however, found *P. ferox* to be ubiquitous and Lang (1978a) even regarded this species as indicative of eutrophic conditions (see discussion below; for further references, see Brinkhurst, 1964 and Milbrink, 1973). Bonomi (1967) has described the gradual disappearance of *P. ferox* and *Bythonomus lemani* in large and deep lakes in Italy under eutrophication, e.g. the now mesotrophic Lago di Maggiore. According to Della Croze (1955) these species are antogonists, even at great depths.

Peloscolex velutinus is a species particularly characteristic
of very deep, oligotrophic lakes (Juget, 1958). Lang (1978a) describes the ecological attributes of the species as close to those
of *P. vejdovskyi* and *S. heringianus*.

Bythonomus lemani is, like the former species, a dominant
species in very deep, oligotrophic lakes, such as in the Lake of
Geneva and Lake Annecy (Juget, 1958) and in Lago de Maggiore (Della
Croze, 1955). See also *P. ferox* above.

Rhynchelmis limosella is a stenothermous cold-water species.
In Sweden it may be found in the large and oligotrophic Lakes Vättern
(Ekman) and Vänern (<u>in</u> Milbrink, 1973).

Stylodrilus heringianus is considered to be the most reliable
indicator of oligotrophic conditions in the palaearctic region (e.g.
Juget, 1958; Brinkhurst in a series of works; Hiltunen, 1967; Howmiller and Beeton, 1970; Timm, 1970, <u>in</u> Milbrink, 1973; and more
recently Howmiller and Scott, 1977; Lang, 1978a; Särkkä, 1978).
S. heringianus, however, may also be a common species in well-oxygen-
ated small streams (Timm, 1970; Ladle, 1971).

Some species have deliberately been omitted from this list for
different reasons; the lumbricid *Eiseniella tetraedra* because we do
not fully understand what is really limiting the distribution of the
species. It has a habit of turning up in the extremes of the trophic
scale (cf. Milbrink, 1978); the tubificid *B. sowerbyi* because its
presence in Europe seems to be fundamentally linked to warm water
effluents; other species are not (yet) used in pollution biology at
all either because of their very vague ecological attributes or that
they are mainly littoral or scarce in any locality – some of them
even endemic in their distribution (see above).

DISCUSSION

In limnology and hydrobiology as well as in other branches of
biology there was an early need for reducing masses of empirical
descriptive data into simple units of classification. It is rather
obvious that if all available information about causal relationships
in inland waters could be condensed into a logic scheme of classification this would certainly be of great predicative value. In pollution biology it is also of the greatest importance that the biologist
communicate with technicians and decision makers. A great number of
indices – be it pollution, biotic or diversity indices – have been
suggested and tested in a variety of situations (Hellawell, 1978).
Despite considerable efforts to accomplish this, to date there seems
to be few – if any – such schemes covering enough for the objective
characterization of inland waters.

Promising results, however, seem to have been obtained with a
variety of ranking methods and graphic methods - including cluster
analyses (Hellawell, 1978; see also Brinkhurst, 1974; Lang, 1978b).
Ahl and Wiederholm (1976) and Wiederholm (1976) suggested the use
of a "Benthic Quality Index" based on the ecological requirements
of species of either chironomids or oligochaetes.   The index is
plotted versus the concentration of total-phosphorus over the middle
depth as suggested by Ryder et al. (1974, in Milbrink, 1978) in
their presentation of the "morpho-edaphic index."

In a recent paper Howmiller and Scott (1977) proposed an alternative system of classification of lakes in North America.   In
short, their "Environmental Index" is based upon knowledge about
the ecological demands of the constituent oligochaete species.
Even if the system is to some extent dominated by species more or
less confined to North America, it could easily be transfered to
European conditions (Milbrink, in manuscript).   The present author
has tested the sensitivity of this index - although in a slightly
modified form - in a variety of situations in Scandinavian lakes
and has also, like Howmiller and Scott, made comparisons with the
kind of information some of the most popular diversity indices may
provide under the same conditions.   As these authors have demonstrated, indices of this kind - such as Shannon's and Simpson's indices
(cf. Hellawell, 1978) - only utilize information about community
structure, i.e. about how many species there are in a community with
no reference to actual levels of abundance and neglecting the autecology of the species involved.   The same fairly low diversity
values are usually obtained at either ends of the trophic scale,
which means that in a gradient of organic enrichment there is little
difference numerically between values obtained from a situation of
heavy dominance of the tolerant species *L. hoffmeisteri* and *P.*
hammoniensis and a situation characterized by sensitive species such
as *S. heringianus* and *P. ferox* in low abundance.   Relatively high
values of diversity, on the other hand, are commonly obtained in
situations of transition, such as when organically enriched water
masses flow into an oligotrophic lake.   In contrast to diversity
estimates the "environmental index" changes rather smoothly along
a gradient of enrichment (cf. Table II; partly from Milbrink, in
manuscript).

The following modification of the Howmiller-Scott index has
been suggested in order to increase its sensitivity in Europe and
possibly also in North America:

$$\text{Modified Environmental Index} = c \cdot \frac{\frac{1}{2} \sum n_0 + \sum n_1 + 2 \sum n_2 + 3 \sum n_3}{\sum n_0 + \sum n_1 + \sum n_2 + \sum n_3}$$

where c is a coefficient of abundance; $c = 1$ when the abundance of
macro-zoobenthos exceeds 3000 specimens/ m^2 (cf. Milbrink, 1978),
$c = 3/4$, $1/2$, $1/4$ and 0 when the lower limits of abundance are 1200,

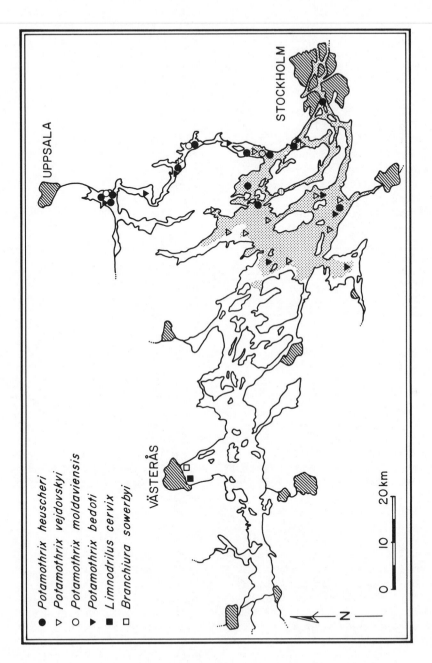

Figure 1. Distribution of immigrant oligochaete species in Lake Malaren. The full extension of brackish water in the profundal in the thirties (Dr. S. Vallin, personal information) is indicated in grey.

Table 2.   Percentage composition of oligochaetes along a
hypothetical transect extending from a center of gross organic
pollution to oligotrophic conditions in one of the large lakes of
southern Sweden. *T. tubifex* (alt. I) means dominance of species
at the bottom of Table I and (alt. II) means a marked influence
of species at the top of the table.

Stations	1	2	3	4	5	6
Limnodrilus hoffmeisteri	90	30	10			
Tubifex tubifex (alt. I)	10	30				
Potamothrix hammoniensis		40	60	10		
Rhyacodrilus coccineus			20			
Psammoryctides barbatus			10	20		
Tubifex tubifex (alt. II)				20	20	10
Peloscolex ferox				40	30	10
Stylodrilus heringianus				10	50	80
Abundance of macro-benthos (specimens/ m^2 (x 100)	52	24	11.6	8	2	1.2
Shannon's diversity index (base e)	.33	1.09	1.09	1.47	1.03	.64
Modified Environmental Index	3.00	1.95	.90	.38	.13	0

400, 160 and 0, respectively. Group n_0 comprises species 1 - 5
(see Table I) plus *T. tubifex* (alt. II). Group n_1 comprises species
7 - 14, group n_2 species 15 - 26 and group n_3 finally species 27 and
28 plus *T. tubifex* (alt. I; cf. Table II). The maximum index value
is 3.

It has proved to be difficult to define the requirements of
several of those oligochaete species more or less constantly invad-
ing new water systems which inevitably means the confrontation with
already established bottom fauna associations. One example of this
may be the gradual expansion of the Ponto-Caspian species to the
west and the north-west in Europe (Timm, this volume; see also
Cekanovskaya, 1962).

Several of the Ponto-Caspian species *P. heuscheri, P. vejdovskyi, P. moldaviensis, P. bedoti* and *P. bavaricus* have obviously
reached the eastern half of Lake Mälaren - directly or indirectly
via littoral bottoms of the Baltic (Figure 1). Their main centers
of distribution coincide fairly well with profundal bottom areas
reached now and then by weakly brackish water from the Baltic as
late as in the thirties and the forties. Some of these species
were found by the present author in the bottom material collected
all over the lake by Dr. G. Alm in 1915 and 1916 and preserved to
this day (Milbrink, unpublished). There are indications that they

are not yet in balance with the established oligochaete fauna in Lake Mälaren and will possibly spread further to the west in the lake (see below).

In the western part of Lake Mälaren are two other very recent immigrant species. One of them, *B. sowerbyi*, was recorded by Wiederholm (1970, in Milbrink, 1973) from the warm-water effluent of the stem power-station in Västeras, and the other species, *L. cervix* (as well as the possible "hybrid *cervix/claparedeanus*") is also locally common in the same region close to the harbour of the city. It is likely that the above species have all reached their new localities as a result of shipping activities and that they will extend their range considerably in the near future, possibly also changing the species composition characteristic of the western part of the lake.

Another method of detecting enrivonmental changes, possibly new for oligochaetes, is the identification of specific morphological reactions of individual specimens to pollutants in the environment (Milbrink, in manuscript). In the latter paper, specimens of *P. hammoniensis* with characteristic deformities are described from four bays of Lake Vänern, southern Sweden. These worms are easily recognized by their aberrant chaetae, which are more or less split-sometimes to the "root". Bifid chaetae are often enlarged distally in a seemingly grotesque way, others are strongly serrated etc. (Figure 2 is an example of this). It is apparent that these deformities are caused by pollutants, since the limited areas of the lake where the phenomena occur have long been under severe stress from industries and hence mercury compounds is here specially emphasized – but also other heavy metal ions and chlorinated organic compounds are present in high concentrations; synergistic effects cannot be excluded. Exceptionally high concentrations of mercury in the sediments – up to 10.000µg/g dry weight in some places – have been recorded.

One factor which may significantly influence the composition of oligochaetes in a community and hence our opinion about the ecological demands of various species is the reproductive strategy of each species. It is a well established fact that some species are more or less opportunistic (with a reproductive "r-strategy"). In the absence of competitors they respond very effectively to situations of environmental stress – having wiped out the majority of the other species – either by producing very large broods or by repeated reproduction with very short time intervals. In more mature communities these opportunists are often more or less outcompeted by species with lower reproductive potential, but with superior qualities of other kinds (for further details, see Aston, 1973 and reviews in Pfannkuche, 1977 and Lang (1978b).

Figure 2. Dorsal crotchet of *Potamothrix hammoniensis*
(segment 5) gravely deformed as a consequence of pollutants
in the environment. The bay of Kattfjorden (station no.
39), Lake Vänern, 1971.

Interspecific Competition between Oligochaete species

Several of the early workers on the bottom fauna have stressed
the importance of abiotic factors such as the quality of the sediment, e.g. its texture, its organic contents etc. Here may be
mentioned Ekman, Valle, Lundbeck, Brundin, Della Croze, Järnefelt,
Wachs and Timm (in Milbrink, 1973). Others such as Alsterberg
(1922), Berg et al. (1962) and Palmer (1968) have found the oxygen
factor to be of decisive importance for the composition of species.

Currently, many authors stress the over all importance of
interspecific competition between the species involved and their
environment. The distribution of species can rarely be exactly
correlated with the variation of chemical or physical factors one
by one, which is hardly comprising. A full analysis of the situation, however, including the combined effects of abiotic and biotic
factors may be extremely laborious to make and therefore seldom
seen in the literature. Most works have concentrated on key-factors,
such as the oxygen or temperature regime, and the level of nutrient

enrichment. In the light of recent experience that aquatic oligochaetes may interact very strongly with each other for different
resources, there is naturally a great risk that the occurrence of
one species or the other in a particular situation is mostly attributable to the simultaneous occurrence of other species and thus
not necessarily to abiotic factors alone.

The most obvious example of interspecific competition and
resource partitioning between sympatric oligochaete species has
been presented by Brinkhurst and co-workers (Brinkhurst et al. 1969;
Brinkhurst et al., 1972; Wavre and Brinkhurst, 1971, in Brinkhurst,
1974) having analysed in detail the *T. tubifex, L. hoffmeisteri,
P. multisetosus* community in the Toronto Harbour. It seems quite
clear that *L. hoffmeisteri* very actively searches the sediments for
the faeces of *T. tubifex* containing the favourite bacteria of the
former species in concentration. The faeces, no doubt, make up the
most nutritious components of the mud. In a mixed culture *P. multisetosus* appears to stimulate growth of the other two species quite
dramatically (Brinkhurst, 1974). The mechanisms behind this are
not fully comprehended. It would seem to be more correct to label
these inter-relationships "positive interactions" rather than
"competitive interactions". Another expression which has been
suggested is trophic symbiosis.

Resource partitioning between sympatrically occurring bottom
animals, however, is not a new discovery even if the most challeng-
ing results have been achieved in the last decade. A fairly large
number of benthic studies have shown the spatial co-existence of
species either closely related taxonomically or nutritionally -
their feeding habits make them seem to occupy the same niche. In
some cases the explanation may be a temporal succession of species
with very similar ecological requirements (cf. Brinkhurst, 1974).

Many rather vaguely formulated interactions of this kind may
be found in the chironomid literature. An interesting case of
niche separation between different *Chaoborus* species has been presented by Stahl (1966). Reynoldson and Bellamy (1970) studied the
establishment of interspecific competition between triclad flat-
worms. Fenchel (1976) described the close partitioning of common
resources between three congeneric snails in the Ise Fjord in
Denmark. Brinkhurst and Kennedy (1965) were aware of the existence
of interspecific interactions between species in mixed laboratory
cultures of tubificid oligochaetes and these authors also recommended that the food requirements of the species involved should
be investigated. Preliminary observations along this line had
been made already by Poddubnaya (1961).

Coler et al. (1967) and Giere (1975) have discussed the selective feeding of oligochaetes on bacteria and Wavre and Brinkhurst
(1971) and Whiteley and Seng (1976) have been able to separate

several species of heterotrophic bacteria from the intestines of
tubificids. Interspecific competition between species may be the
mechanism behind the occurrence of *P. ferox* in mesotrophic or even
eutrophic profundal areas of the Lake of Geneva. It has been
suggested that *P. vejdovskyi* is a superior competitor in unpolluted
parts of that lake (Lang, 1978a) excluding the former species from
such trophic situations in which we are used to find it. In the
large, deep lakes of northern Italy, however, *P. ferox* is again an
indicator of oligotrophy (see above). Of course, the existence of
sibling species cannot be excluded in the Lake of Geneva, but as
long as there are no electrophoretic tests proving that is the case,
this is not necessary as an explanation.

In order to explain the distribution pattern and the homoge-
neity in abundance between sympatric tubificid species with similar
ecological demands (*T. tubifex*, *L. hoffmeisteri* and *L. claparedeanus*) species-selective feeding on microorganisms must be considered
a most important mechanism (Pfannkuche, 1977). It is likely that
several of the *Potamothrix* species occurring in the eastern parts
of Lake Mälaren interact with each other very closely. *P. hammoniensis* and *P. heuscheri* simultaneously reach very dense populations
in the polluted basin of Ekoln in the north-eastern part of this
lake. The sympatric occurrence of *P. hammoniensis*, *P. vejdovskyi*,
P. moldaviensis and *P. heuscheri* in the basin of Görväln in eastern
Lake Mälaren is another example from Scandinavia. It seems highly
unlikely that these congeneric species would simultaneously reach
extraordinarily high abundance values or even seek the company of
each other without certain mutual nutritional advantages.

The mere balance between different groups of bottom animals can
reveal whether there is excessive organic enrichment in the water,
or if there are high concentrations of poisonous substances such as
heavy metal ions or DDT and derivatives thereof. Situations of
heavy silting, mobility of the bottom material and a heavy predation
pressure from fish could likewise be detected by analysing abundance
and biomass of oligochaetes and other components of the bottom fauna.
The full potential of benthic studies for environmental assessment,
however, can only be realized by considering existing information
on the autecology of the constituent species and also by learning
more about interactions between species.

REFERENCES

Ahl, T. and T. Widerholm, 1976, "Svenska Vattenkvalitetskriterier.
    Eutrofierande ämnen," Statens Naturvardsverk PM 918, 124pp.
Alsterberg, G., 1922, Die respiratorischen Mechanismen der Tubifi-
    ciden, <u>Lunds Univ. Arsskr. N.F. Avd.</u>, 18:1-175.
Aston, R.J., 1971, The effects of temperature and dissolved oxygen
    concentrations on reproduction in *Limnodrilus hoffmeisteri*
    (Claparéde) and *T. tubifex* (Müller), C.E.R.L.Lab. Note RD/L/M
    312.

Aston, R.J., 1973,   Field and experimental studies on the effects
    of a power station effluent on Tubificidae (Oligochaeta, Anne-
    lida), Hydrobiologia,42:225-242.
Berg, K., P.M. Jonason and K.W. Ockelmann, 1962, The respiration
    of some animals from the profundal zone of a lake, Hydrobio-
    logia,19:1-39.
Bonomi, G., 1967, L'evoluzione recente del Lago Maggiore rivelata
    dalle conspicue modificazioni del macrobenton profondo, Mem.
    Ist.Ital.Idrobiol., 21:197-212.
Brinkhurst, R.O., 1964, Observations on the biology of lake
    dwelling Tubificidae (Oligochaeta), Arch. Hydrobiol., 60:385-
    418.
Brinkhurst, R.O., 1969, The fauna of pollution, in: "The Great
    Lakes as an environment," Gt.Lakes Inst.Univ. Toronto,P.R. 39.
Brinkhurst, R.O., 1970, Distribution and abundance of tubificid
    (Oligochaeta) species in Toronto Harbour, Lake Ontario, J.
    Fish.Res.Bd.Canada, 27:1961-1969.
Brinkhurst, R.O., 1974, "The Benthos of Lakes", The MacMillan
    Press Ltd., London, 190 pp.
Brinkhurst, R.O. and C.R. Kennedy, 1965, Studies on the biology of
    the Tubificidae in a polluted stream, J.Anim.Ecol., 34:429-443.
Brinkhurst, R.O. and B.G.M. Jamieson, 1971, The Aquatic Oligochaeta
    of the World", Oliver and Boyd, Edinburgh, 860pp.
Brundin, L., 1949, Chironomiden und andere Bodentiere der Südschwe-
    dischen Urgebirgsseen, Rep.Inst.Res.Drottningholm,30:914pp.
Brundin, L., 1951, The relation of O_2-microstratification at the
    mud surface to the ecology of the profundal bottom fauna,
    Rep.Inst.Freshw.Res.Drottningholm, 32:32-42.
Brundin, L., 1956, Die Bodenfaunistischen Seetypen und ihre Anwend-
    barkeit auf die Sudhalbkugel. Zugleich eine Theorie der
    produktionsbiologischen Bedeutung der Glazialen Erosion.
    Rep.Inst.Freshw.Res.Drottningholm,37:186-235.
Cekanovskaya, O.V., 1962, The aquatic Oligochaete fauna of the
    USSR, Opred.Faune SSSR, 78:1 (in Russian).
Coler, R.A., H.B. Gunner and B.M. Zuckermann, 1967, Selective of
    Tubificids on Bacteria, Nature,216:1143-1144.
Della Croze,N., 1955, The conditions of the sedimentation and
    their relationship with Oligochaeta populations of Lake
    Maggiore, Mem.Ist.Ital.Idrobiol.Suppl., 8:39-62.
Dzwillo, M., 1967, Untersuchungen über die Zusammensetzung der
    Tubificidenfauna im Bereich des Hamburger Hafens, Verh.
    Naturw.Ver.Hamburg, N.F. Bd XI:101-116.
Elster, H.J., 1966, Uber die limnologischen Grundlagen der biolo-
    gischen Gewässer-Beurteilung in Mitteleuropa, Verh.Internat.
    Verein.Limnol.,16:775-785.
Fenchel, T., 1976, Evidence for exploitative interspecific compe-
    tition in mud snails (Hydrobiidae), Oikos,27:367-376.
Giere, O., 1975, Population Structure, Food Relations and Ecolo-
    gical Role of Marine Oligochaetes, with Special Reference to
    Meiobenthic Species, Marine Biology,31:139-156.

Gitay, A., 1968, Preliminary data on the ecology of the level-
    bottom fauna of Lake Tiberias, Israel Zool., 17:81-96
Hellawel, J.M., 1978, "Biological Surveillance of Rivers," Water
    Research Center, Stevenage, 331 pp.
Hiltunen, J.K., 1967, Some oligochaetes from Lake Michigan, Trans.
    Amer.Microsc.Soc., 86(4):433-454.
Hiltunen, J.K., 1969, Distribution of Oligochaetes in Western Lake
    Erie, 1961, Limnol.Oceanogr., 14(2):260-264.
Howmiller, R.P. and A.M. Beeton, 1970, The Oligochaete Fauna of
    Green Bay, Lake Michigan, Proc.13th Conf.Great Lakes Res.,
    15-46.
Howmiller, R.P. and M.A. Scott, 1977, An environmental index based
    on relative abundance of oligochaete species, Water Poll.
    Control Fed., 49:809-815.
Hynes, H.B.N., 1960, "The Biology of Polluted Waters," Liverpool.
    U.P. pp. XIV+202.
Järnefelt, H., 1953, Die Seetypen in Bodenfaunistischer Hinsicht,
    Ann.Soc.zool-bot.fenn Vanamo, 15:1-38.
Johnson, M.G. and R.O. Brinkhurst, 1971, Associations and Species
    Diversity in Benthic Macroinvertebrates of Bay of Quinte and
    Lake Ontario, J.Fish.Res.Bd.Canada,28:1683-1697.
Jonason, P.M. and T. Thorhauge, 1972, Life cycle of Potamothrix
    hammoniensis (Tubificidae) in the profundal of a eutrophic
    lake, Oikos,23:151-158.
Juget, J., 1958, Recherche sur la faune de fond du Léman et du
    lac d'Annecy, Annls.Stn.Cent.Hydrobiol.appl.,7:9-96
Kennedy, C.R., 1965, The distribution and habitat of Limnodrilus
    Claparéde (Oligochaeta:Tubificidae), Oikos,16:26-38.
Ladle, M., 1971, The Biology of Oligochaeta from Dorset chalk
    streams, Freshwat.Biol., 1:83-97.
Lang, C., 1978a, Factorial correspondence analysis of oligochaeta
    communities according to eutrophication level, Hydrobiologia,
    57(3):241-247.
Lang, C., 1978b, Approche multivariable de la détection biologique
    et chimique des pollutions dans le Lac Léman (Suisse), Arch.
    Hydrobiol., 83(2):158-178.
Liebmann, H., 1951, "Handbuch der Frischwasser und Abwasserbio-
    logie," Bd 1, 1.Aufl., Verlag Oldenbourg, Munich.
Margalef, R., 1975, External factors and ecosystem stability,
    Schweiz.Z.Hydrobiol., 37:102-117.
Milbrink, G., 1973, On the Use of Indicator Communities of Tubifi-
    cidae and some Lumbriculidae in the Assessment of Water
    Pollution in Swedish Lakes, Oikos,1:125-139.
Milbrink, G., 1978, Indicator communities of oligochaetes in
    Scandinavian lakes, Verh.Internat.Verein.Limnol., 20:2406-2411.
Palmer, M.F., 1968, Aspects of the respiratory physiology of
    Tubifex tubifex in relation to its ecology, J.Zool.Lond.,154:
    463-473.
Pfannkuche, O., 1977, Ökologische und systematische Untersuchungen
    an naidomorphen Oligochaeten brackiger und limnischer Biotope,

Dissertation der Universität Hamburg, 138 pp.

Poddubnaya, T.L., 1961, Observations on the nutrition of the
    common species of tubificids in the Rybinsk basin. Trudy.
    Inst.Biol.Vodokhr.Akad.Nauk SSSR, 4:219-231. (In Russian).

Reynoldson, T.B. and L.S. Bellamy, 1970, The establishment of
    interspecific competition in field populations with an
    example of competition in action between *Polycelis nigra*
    (Mull.) and *P. tenuis* (Ijima)(Turbellaria, Tricladida),
    Proc.Adv.Study Inst.Dynamics Numbers Popul., (Oosterbeek 1970):
    282-297.

Saether, O.A., 1970, A survey of the bottom fauna in lakes of the
    Okanagan Valley, British Columbia, Fish.Res.Bd Canada, Tech-
    nical Rep. no. 196.

Sládecek, V., 1973, System of water quality from the biological
    point of view, Arch.Hydrobiol., (Ergebn.Limnol.) 7:1-218.

Sládecek, V., 1978, Relation of saprobic to trophic levels, Verh.
    Internat.Verein.Limnol., 20:1885-1889.

Spencer, D.R., 1978, Oligochaeta of Cayuga Lake, New York with a
    description of *Potamothrix bavaricus* and *P. bedoti*. Trans.
    Amer.Microsc.Soc., 97:139-147.

Stahl, J.B., 1966, Characteristics of North America *Sergentia*
    lake, Gewässer und Abwässer,41/42:95-122.

Särkkä, J., 1978, New records of profundal Oligochaeta from
    Finnish lakes, with ecological observations, Ann.Zool.Fennica,
    15:235-240.

Timm, T., 1970, On the fauna of the Estonian Oligochaeta,
    Pedobiologia,10:52-78.

Wachs, B., 1964, Beitrag zur Oligochaeten-Fauna eines schiffbaren
    Flusses, Z.angew.Zool., 51:179-191.

Wachs, B., 1968, Die Bodenfauna der Fliessgewässer in Beziehung zu
    den bedeutendsten Substrattypen, Wässer und Abwässer-Forschung,
    63:1-11.

Wiederholm, T., 1976, Chironomids as indicators of water quality
    in Swedish lakes, Naturvardsverkets Limnologiska Undersökning,
    Information 10, 17 pp.

Wavre, M. and R.O. Brinkhurst, 1971, Interactions between some
    tubificid oligochaetes and bacteria found in the sediments
    of Toronto Harbour, Ontario, J.Fish.Res.Bd.Canada,28:335-341.

Whitley, S.L. and T.N. Seng, 1976, Studies on the bacterial flora
    of tubificid worms, Hydrobiologia,48:79-83.

Zelinka, M. and P. Marvan, 1961, Zur Prazisierung der biologischen
    Klassifikation der Reinheit fliessender Gewässer, Arch.Hydro-
    biol., 57:389-407.

STRUCTURE OF TUBIFICID AND LUMBRICULID WORM COMMUNITIES,

AND THREE INDICES OF TROPHY BASED UPON THESE COMMUNITIES,

AS DESCRIPTORS OF EUTROPHICATION LEVEL OF LAKE GENEVA

(SWITZERLAND)

Claude Lang
Barbara Lang-Dobler

Conservation de la Faune
César Roux 16
1005 Lausanne
Switzerland

ABSTRACT

In 1977, the nearshore zone of Lake Geneva (Switzerland) between isobaths 10 and 40 m was divided into 17 areas. In each of these, 19 to 26 sediment cores were collected, the tubificid and lumbriculid worms quantitatively sampled and then pooled together. The subsequent biological analyses based on 17 pooled samples were undertaken to describe the eutrophication level of this zone. Four basic worm communities, each characterized by a different species, were singled out: (1) oligotrophic *Stylodrilus heringianus*; (2) mesotrophic *Potamothrix heuscheri*; (3) eutrophic *Potamothrix hammoniensis*; and (4) eutrophic *Peloscolex ferox*. Communities (3) and (4) were related to heavy inputs of allochthonous organic matter which also increases the worm density in these areas. Species were divided into three groups according to their value as trophic level indicators established in a previous study. Three indices of trophy based upon these results were calculated for each community: (1) an index of oligotrophy

characterized by the relative abundance in the community of
Stylodrilus lemani, S. heringianus, Peloscolex velutinus, and
Potamothrix vejdovskyi; (2) an index of mesotrophy based mainly
on *Potamothrix heuscheri, Aulodrilus pluriseta,* and *Limnodrilus
hoffmeisteri*; (3) an index of eutrophy based mainly on *Potamothrix
hammoniensis, Tubifex tubifex,* and *Peloscolex ferox.* According to
these indices, most communities were mesotrophic (65%), 29% were
eutrophic, and 6% oligotrophic. Total nitrogen and organic carbon
in the sediment were negatively correlated with the index of oligotrophy, positively with the index of mesotrophy. The index of
eutrophy was positively correlated with total phosphorus and heavy
metal content in the sediment.

INTRODUCTION

     Man-made eutrophication of lakes has become a worldwide process
which may be evaluated from different points of view. Until now,
water chemistry and phytoplankton ecology have been applied more
successfully than benthos to the study of this process. However,
as shown by Brinkhurst (1974), the trophic typology of lakes can
also be based on benthic communities. Tubificid and lumbriculid
worms seem the most useful tool to fulfill this goal. Zahner (1964)
and Milbrink (1973) in Europa, and several authors in North America
(reviewed by Cook and Johnson, 1974), have used these worms as
indicators of the trophic state of lakes. However, the comparison
between several regions of the same lake seems more successful than
the comparison between several lakes (Brinkhurst 1964).

     The present research, which attempts to describe the trophic
state of the nearshore region of Lake Geneva, is based on several
previous studies. Most of them were conducted in a very distinct
pollution gradient due to effluent of the Lausanne sewage treatment
plant (Lang 1978b). To judge the value of worm species as indicators of pollution and eutrophication level, we have classically
used their relative abundance in relation to the distance from the
source of pollution. However, species qualified as typically
oligotrophic in several papers reviewed by Milbrink (1973) were
found in very polluted situations. Moreover, no clear relationships appeared between the chemical pollution of the sediment and
the worm communities.

     For this reason, the value as trophic level indicators of worm
species used here has been established in a preceding paper (Lang
and Lang-Dobler 1979). In that study, we related quantitatively
the presence of every worm species to measured concentrations of
different pollutants in the sediment. For a given species, the
mean concentrations of several pollutants in all samples where that
species is present is used to define its chemical environment with
regard to pollution level. The multivariate comparison of these
different chemical environments, each characteristic of one worm
species, enables species to be divided into three main groups.

Two extreme groups of species are singled out: (1) species colonizing sediment where pollutant concentrations are very low; and (2) those where pollutant concentrations are very high. According to other studies, predominance in a community of species included in group (1) implies oligotrophic conditions, and in group (2) eutrophic conditions. Species in a third group which colonize the sediment where pollutant concentrations are intermediate are classified as mesotrophic.

This approach gives a basis for the calculation of three indices of trophy, which can be used to quantify the trophic state of a given worm community by the proportion of oligotrophic, meso-trophic and eutrophic species in it. The density and the structure of worm communities plus these three indices are used to describe the trophic level of the littoral zone of Lake Geneva in 1977.

MATERIAL AND METHODS

Sampling

From June to September 1977, 360 sediment samples were taken from around Lake Geneva on 120 transects perpendicular to the coast; the distance between these transects was 1 km. In every transect, three samples were collected at 10, 25 and 40 m deep. Sediment samples were taken with a Shipek sampler: a 15 cm^2 core was inserted inside the sediment to collect worms. The remaining sediment was used for chemical and grain size analyses (Davaud et al. 1977).

The nearshore zone of Lake Geneva was divided into 17 areas corresponding broadly to its geographical subdivisions (Figure 1 A). The biological results of the 19 to 26 cores taken inside every area were pooled together. Consequently the biological analyses were performed on 17 pooled samples. As shown by Dauer and Simon (1975), analysis by entire transects rather than individual stations decreases along-shore heterogeneity. This method overcomes local peculiarities and gives a more general picture of the situation.

Biological Analyses

In 1977, 6015 worms were collected and 18 species identified (Table 1). In each area, immature individuals were attributed to the different species according to the abundance of mature indivi-duals of these species in the community.

Worm species were divided into three groups according to their value as trophic level indicators (Lang and Lang-Dobler 1979). To quantify this approach, three indices of trophy were calculated: (1) an index of oligotrophy by adding the relative abundance of typically oligotrophic species in the community of every area

Table 1. Species of tubificid and lumbriculid worms collected in 1977 and their value as indicator of trophic level according to Lang and Lang-Dobler (1979). Relative abundance of species in 360 samples is indicated.

Number	Code	Species (authors)	Value as indicator of trophic level	Relative abundance (%)
1	Sl	*Stylodrilus lemani* (Grube)		6.7
2	Pvel	*Peloscolex velutinus* (Grube)	oligotrophic	10.8
3	Sh	*Stylodrilus heringianus* (Claparède)		12.9
4	Pvej	*Potamothrix vejdovskyi* (Hrabe)		108.0
5	Ap	*Aulodrilus pluriseta* (Piguet)		28.7
6	Al	*Aulodrilus limnobius* (Bretscher)		8.9
7	Phe	*Potamothrix heuscheri* (Bretscher)		226.1
8	It	*Ilyodrilus templetoni* (Southern)	mesotrophic	18.1
9	Lp	*Limnodrilus produndicola* (Verrill)		9.9
10	Lu	*Limnodrilus udekemianus* (Claparède)		1.8
11	Lh	*Limnodrilus hoffmeisteri* (Claparède)		111.3
12	Psb	*Psammoryctides barbatus* (Grube)		14.4
13	Lc	*Limnodrilus claparedeanus* (Ratzel)		3.3
14	Pha	*Potamothrix hammoniensis* (Michaelson)	eutrophic	255.2
15	Tt	*Tubifex tubifex* (Müller)		126.5
16	Pf	*Peloscolex ferox* (Eisen)		36.5
17	Pba	*Potamothrix bavaricus* (Oschmann)	unknown	7.1
18	Pbe	*Potamothrix bedoti* (Piguet)		13.8

Species 1 and 3 are lumbriculids, the others tubificids.

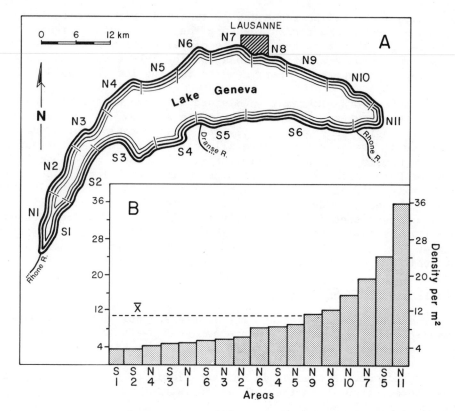

Figure 1. A: Location of 17 sampling areas along shore of Lake
Geneva. B: Mean density of worms (in thousands of
individuals per m^2).

(Table 1); (2) similarly an index of mesotrophy based on the
relative abundance of typically mesotrophic species and, (3) an
index of eutrophy. Therefore, the trophic state of every area of
Lake Geneva was characterized by the values of three indices (For
examples, see Figure 4). Factorial correspondence analysis
(Bertier and Bourouche 1975) was used to analyse relationships
between the 17 areas and the biological variables (worm species
or indices of trophy) which characterize them. Abundance of worm
species per area and more details on analyses will be given elsewhere (Lang and Lang-Dobler, submitted).

RESULTS

Worm Density

The mean density of worms increases where the allochthonous
inputs of organic matter are high (Lugrin 1974). The Rhône River,

the Dranse River and Lausanne sewage treatment plant are the major
sources of organic matter influencing the worm density (Figure 1B).
These results show that the currents disperse the input of the
Rhône River along the northern coast of the lake rather than on the
southern coast, where the worm density remains low in area S 6.

Therefore, the lake may be divided into two parts: (1) the
western region where allochthonous inputs are low and consequently
the worm density is depressed; (2) the eastern region, with the
exception of the area S 6, where allochthonous inputs are strong
and worm density increases. Similarly, in Bodensee and in several
lakes of North America, the worm density increases locally according
to the extent of allochthonous input (Zahner 1964, Cook and Johnson
1974). The extreme densities per core recorded in some localities
of Lake Geneva exceed approximately tenfold the mean values presented
here (Figure 1 B); they go beyond the highest values given in other
studies (op. cit). This result indicates that Lake Geneva is highly
eutrophic in some areas.

Structure of Worm Communities

The relationships between the 17 areas according to the worm
communities which inhabit them are analysed by factorial correspondence analysis. The areas with similar communities are grouped
together and the species characteristic of this group of communities
are singled out. A species is characteristic of a community, if it
contributes strongly to the factors explaining most the variance of
the data.

On this basis, four basic communities of worms are singled out
in Lake Geneva (Figure 2): (1) the oligotrophic species *Stylodrilus
heringianus* characterizes area S 6; (2) the western part of the lake
is colonized by a community defined by the mesotrophic *Potamothrix
heuscheri*; (3) areas directly subject to heavy organic inputs of
the Rhone and Dranse Rivers are occupied by a community characterized
by the eutrophic *Potamothrix hammoniensis*; near the mouth of rivers
Tubifex tubifex becomes more important; (4) the influence of the
Lausanne sewage treatment plant determines the presence of a community characterized by the eutrophic-polluted *Peloscolex ferox*.
However, Brinkhurst (1964) has also recorded an abundance of this
species in the polluted areas of some lakes.

These results allow a first classification of the littoral zone
of Lake Geneva according to eutrophication level. In 1977, the
largest part of the littoral was mesotrophic. As for density, the
eutrophic communities may be related to heavy organic input by
rivers or sewage treatment plants. In the only oligotrophic area,
special conditions (currents, steep bottom slope) decrease the
arrival and settlement of organic matter.

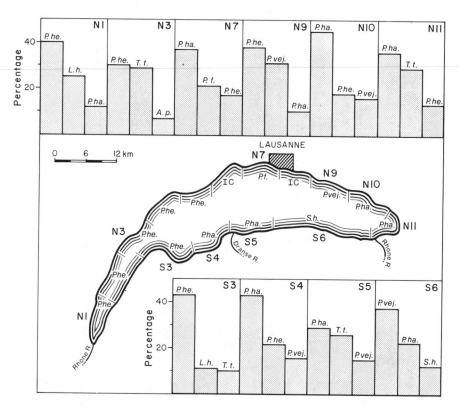

Figure 2.   Worm communities of Lake Geneva indicated with species
            characteristic of each community according to results
            of factorial correspondence analysis (Code for species,
            see Table 1). For some communities, relative abundance
            of three most abundant species is represented.
            IC = intermediate community (see text).

Indices of Trophy

        The structure of worm communities may also be expressed by
the proportion of oligotrophic, mesotrophic and eutrophic species.
Three indices of trophy were calculated to characterize the worm
community of every area according to the trophic level of the
sediment. The factorial correspondence analysis applied to these
results allows grouping of communities whose indices of trophy are
similar (Figure 3). The index of trophy nearest to a group of
communities characterizes their level of eutrophication. To provide
absolute reference communities, three fictitious communities are
entered into the analysis: (1) one typically oligotrophic community
in which 100% of species are oligotrophic (TO) (2) one typically
mesotrophic community, 100% mesotrophic species (TM) and (3) one
typically eutrophic community (TE). Distances on the factorial

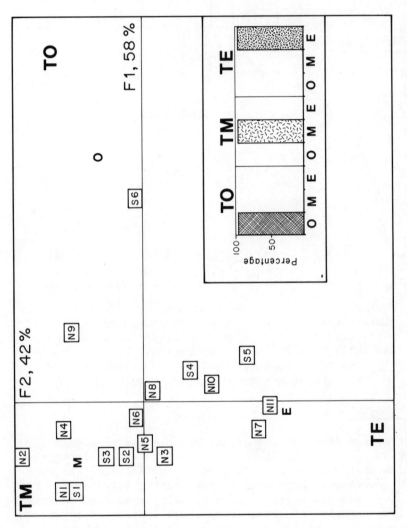

Figure 3. Plot of factors 1 and 2 of factorial correspondence analysis based on values of indices of oligotrophy (O), mesotrophy (M), and eutrophy (E) in each area of Lake Geneva. One typical oligotrophic community (TO), one typical mesotrophic community (TM), and one eutrophic community (TE) are represented.

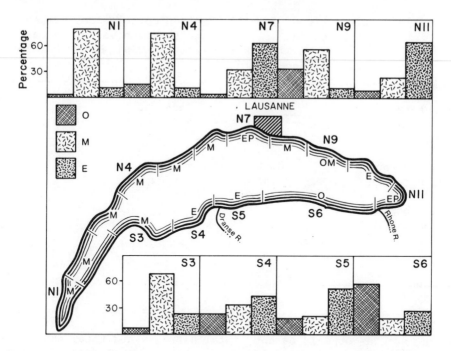

Figure 4. Trophic state of the littoral zone of Lake Geneva
           according to the factorial correspondence analysis
           based on indices of trophy (See Figure 3). O oligo-
           trophic area, M mesotrophic area, E eutrophic area,
           P polluted area. The values of the three indices of
           trophy in some areas are represented.
           N = north; S = south.

plot between these three fictitious and the 17 real communities
give an estimate of the trophic state of Lake Geneva in 1977. The
only oligotrophic community (S 6) is located as far from the other
real communities as from typically oligotrophic community (TO).
Consequently it may be called oligo-mesotrophic. Most of the
communities are rather mesotrophic (64,7%) and some are rather
eutrophic (Figure 4).

     The approach based on indices of trophy has the advantage of
providing absolute reference communities which allow comparison
of lakes independent of the species present.

Indices of Trophy and Sediment Chemistry

     To compare the biological approach of eutrophication provided
by the indices of trophy with sediment chemistry, the mean value
per area of several chemical variables recorded by Davaud et al.
(1977) were calculated. Some of these chemical variables, such
as the heavy metals, indicate the pollution level of the sediment;

Table 2. Significant (p ≤ .05) Pearson correlations between
          the indices of trophy and the mean values per area
          of the chemical parameters describing the sediment
          of Lake Geneva.

parameters	index of		
	oligotrophy	mesotrophy	eutrophy
organic carbon	− 0.52	+ 0.41	
total nitrogen	− 0.46	+ 0.49	
C : N ratio		− 0.52	+ 0.57
total phosphorus			+ 0.49
cadmium			+ 0.47
tin			+ 0.52
lead			+ 0.49
zinc			+ 0.44
chromium		− 0.42	+ 0.49
mercury		− 0.49	+ 0.52

others, such as organic carbon, total nitrogen and total phosphorus
indicate its level of eutrophication. There are significant correlations between sediment chemistry and the indices of trophy
(Table 2).

Total nitrogen and organic carbon in the sediment are positively correlated with the index of oligotrophy, and negatively
with the index of mesotrophy. These relationships may be explained
by the influence of oxygen on worm communities. As demonstrated by
Hargrave (1972) high content of organic carbon implies an increased
oxygen uptake by the sediment; i.e. decreased oxygen concentrations
at the water-sediment interface. Consequently the abundance of
oligotrophic worm species which are sensitive to oxygen deficit
decreases, whereas the importance of more resistant mesotrophic
species increases.

Total phosphorus and heavy metals are positively correlated
with the index of eutrophy. High concentrations of these elements
in the sediment indicate heavy allochthonous input. Similarly,
the values of the organic C: N ratio, which increase with the
importance of allochthonous input (Wetzel 1975), are positively
correlated with the index of eutrophy and negatively with the
index of mesotrophy. Consequently mesotrophic communities are
characteristic of areas subject mostly to autochthonous input
produced by algae and macrophytes, whereas eutrophic communities
prevail in the presence of strong allochthonous input.

These chemical results confirm the influence of organic matter and of allochthonous versus authochthonous inputs on the values of the indices of trophy. At the level of worm species, increased organic matter in sediment means more food and less oxygen available. Worm communities are structured by these two antagonistic factors.

DISCUSSION

The goal of the present study is the evaluation of the trophic state of the littoral zone of Lake Geneva during 1977. The structure of tubificid and lumbriculid communities was used as a descriptive tool to define eutrophication progress. The littoral zone of the lake was mostly mesotrophic, except in areas where high allochthonous input caused predominantly eutrophic conditions. The only oligotrophic area which may be considered as a refuge is the one where external organic input is suppressed because of the low human population density and of the general pattern of currents (S. Bauer, pers. comm.); moreover, the very steep bottom slope prevents the settlement of organic matter (Hargrave and Kamp-Nielsen 1977). The quantity of organic matter which settles on the bottom and the subsequent oxygen concentrations at the level of the water-sediment interface appear as the main factors structuring the worm communities. Thus these are linked to the carbon cycle which is essential to understand the processes going on in the lakes (Wetzel and Richey 1978). According to planktonic primary production, Lake Geneva is mostly mesotrophic, but sometimes the recorded values are typical of an eutrophic lake (Pelletier 1977).

At the beginning of the century, *Stylodrilus heringianus* and *S. lemani*, which are cold stenothermal species adapted to stable environments (Cook and Johnson 1974), were abundant in Lake Geneva (Piguet and Bretscher 1913). In 1977, these species became scarce in the littoral (Table 1), and were eliminated from the deepest area of the lake (Lang 1978a), whereas some tubificid species adapted to fluctuating conditions were very abundant in the littoral. Similarly, diatom species living in very stable physicochemical conditions are less able to adapt to chemical stress than the species colonizing unstable environments (Fisher 1977). Compared to the results of Juget (1967) collected between the years 1958 and 1967 at a depth of 30 m, our data show that the proportion of oligotrophic species in the community has decreased since that time.

The approach used in our work may be applied to other lakes with other communities composed of different species, provided that the value of the species present in the newly studied lake as trophic level indicators are checked. As a matter of fact, this approach assumes that the structure of worm communities is determined mostly by abiotic factors, such as organic matter and oxygen, without reference to biotic factors. However, our results suggest that interspecific competition restricted the important populations of

Peloscolex ferox to polluted areas unsuitable for its potential
competitor *Potamothrix vejdovskyi*. Similarly, in the laboratory,
Van Haven (1975) has shown that worm species typical of polluted
locations choose to colonize sediment where there is more oxygen
than in their natural habitat; this suggests that biotic factors
influence their distribution.

In extremely oligotrophic and eutrophic states, worm communities are likely to be physically controlled by scarcity of food
(bacteria colonizing organic matter) or oxygen deficit (Sanders
1968). In intermediate trophic states, as in the mesotrophic
Lake Geneva, competition may become a key factor. But to be important, competition does not need to occur very often (MacArthur 1972).
This is the main reason why competition is so difficult to observe
in the field, the observer arriving when the exclusion of one
species is accomplished. Therefore, to achieve a dynamic biological detection of trophic level based on worm species, we must
be able to predict the outcome of competition between two worm
species according to the actual chemical environment as done for
diatom species by Kilham and Kilham (1978).

ACKNOWLEDGEMENTS

Dr. J.M. Jacquet helped to improve the manuscript by his
suggestions. Dr. M. Lejeune assisted with statistical methods.
Miss L. Treloar corrected the English. Miss L. Cuvit and M. Fellrath
provided technical assistance. Miss N. Produit typed the manuscript.
Samples were taken and chemical analyses of the sediments were per-
formed by the "Laboratoire de sedimentologie de l'Université de
Genève" (Prof. J.P. Vernet). The study was supported by the
"Commission internationale pour la protection des eaux du lac
Léman contre la pollution."

REFERENCES

Bertier, P. and J.M. Bourroche, 1975, Analyses des données multi-
     dimensionnelles. Presses Universitaires de France, Paris.
Brinkhurst, R.O., 1964, Observations on the biology of lake
     dwelling Tubificidae (Oligochaeta). Arch. Hydrobiol., 60:
     385-418.
Brinkhurst, R.O., 1974, The benthos of lakes. The Macmillan
     Press, London.
Cook, D.G. and M.G. Johnson, 1974, Benthic macroinvertebrates of
     the St. Lawrence Great Lakes. J.Fish.Res.Board Can., 31:
     763-782.
Dauer, P.M. and J.L. Simon, 1975, Lateral or along-shore distribu-
     tions of polychaetous annelid of an intertidal, sandy habitat.
     Mar. biol., 31:360-370.
Davaud, E., F. Rapin, and J.P. Vernet, 1977, Contamination des
     sédiments côtiers pour les métaux lourds. Rapports de la

commission internationale pour la protection des eaux du lac
Leman contre la pollution (Secretariat: 23, av. de Chailly,
1012 Lausanne, Switzerland).

Fisher, N.S., 1977, On the differential sensitivity of estuarine
and open-ocean diatoms to exotic chemical stress. Amer.Natur.,
981:871-895.

Hargrave, B.T., 1972, Aerobic decomposition of sediment and
detritus as a function of particle surface area and organic
content. Limnol. Oceanogr., 17, 4:583-596.

Hargrave, B.T. and L. Kamp-Nielsen, 1977, Accumulation of sedimentary organic matter at the base of steep bottom gradients.
Proc. Symp., Interactions between sediment and freshwater
(ed. H.L. Golterman). Pudoc, Wageningen and Dr. W. Junk,
The Hague.

Juget, G., 1967, La faune benthique du Leman:modalites et deter-
minisme ecologique du peuplement. These de doctorat, Univ.
de Lyon.

Kilham, S.A. and P. Kilham, 1978, Natural community bioassays:
Predictions of results based on nutrient physiology and
competition. Verh.Internat.Verein.Limnol., 20:68-74.

Lang, C., 1978a, Factorial correspondence analysis of oligochaeta
communities according to eutrophication level. Hydrobiologia,
57:241-247.

Lang, C., 1978b, Approche multivariable de la detection biologique
et chimique des pollutions dans le lac Léman (Suisse). Arch.
Hydrobiol., 83:158-178.

Lang, C., and B. Lang-Dobler, 1979, The chemical environment of
tubificid and lumbriculid worms according to the pollution
level of the sediment. Hydrobiologia, 65:273-282.

Lang, C. and B. Lang-Dobler, Submitted. Degré d'eutrophisation
et de pollution du littoral lémanique en 1977 évalué à partir
de la composition des communautés de vers (tubificidés et
lumbriculidés). Rapports de la Commission internationale
pour la protection des eaux du lac Léman contre la pollution.

Lugrin, M., 1974, Premieres donnees sur les composes organiques
dissous dans le bassin du lac Leman. These de doctorat 3e
cycle. Universite de Paris VI.

MacArthur, R.H., 1972, Geographical Ecology. Harper and Row,
New York.

Milbrink, G., 1973, On the use of indicator communities of Tubificidae and some Lumbriculidae in the assessment of water
pollution in Swedish Lakes. Zoon,1:125-139.

Pelletier, J.P., 1977, Evaluation de la production primaire ou
production organique. Rapports de la commission internationale pour la protection des eaux du lac Léman contre la
pollution.

Piguet, E. and K. Bretscher, 1913, Catalogue des invertébrés de
la Suisse, fascicule 7, Oligochètes. Muséum d'histoire
naturelle de Genève.

Sanders, H.L., 1968, Marine benthic diversity : a comparative
     study. Amer. Nat., 102:243-282.
Van Hoven, W., 1975, Aspects of the respiratory physiology and
     oxygen preference of four aquatic oligochaetes (Annelida).
     Zool. Africana,10:29-45.
Wetzel, R.G., 1975, Limnology. W.B. Saunders Company, Philadelphia.
Wetzel, R.G. and J.E. Richey, 1978, Analysis of five North American
     lake ecosystems VIII. Control mechanisms and regulation.
     Verh. Internat. Verein. Limnol., 20:605-608.
Zahner, R., 1964, Beziehungen swischen dem Auftreten von Tubifici-
den und der Zufuhr organischer Stoffe in Bodensee. Int. Rev. ges.
     Hydrobiol., 49, 3:417-454.

POLLUTION BIOLOGY - THE NORTH AMERICAN EXPERIENCE

R. O. Brinkhurst

Ocean Ecology Laboratory
Institute of Ocean Sciences
P.O. Box 6000, Sidney, British Columbia
Canada   V8L 4B2

ABSTRACT

Earliest references to tubificids in pollution biology in North America were related to the simple abundance of the group in grossly polluted situations. With the improvement in taxonomy in the decade of the sixties, it was possible to recognize species assemblages, especially in the St. Lawrence Great Lakes. The distribution of these associations has now been worked out in considerable detail, and consulting companies and government agencies now work with identified species. Very few traditional laboratory tolerance tests have been done, but a start has been made on the investigation of the activity of worms in recycling sediment contaminants such as metals.

Before beginning this brief review of the use of aquatic oligochaetes in pollution biology in North America I must apologize for seeming to impose upon my audience again. Unfortunately my colleague Mr. J. Hiltunen was prevented from undertaking this review at the last minute, though at the time of writing every effort is being made to see if he can at least be present. I shall try to avoid overlap with the papers on the St. Lawrence Great Lakes and metal studies.

Among the earliest published records of the distribution of tubificids in relation to pollution in the U.S.A. were accounts of

471

R. O. BRINKHURST

the benthos of Lytle Creek (Gaufin and Tarzwell, 1952), Western
Lake Erie (Wright 1955), Green Bay, Wisconsin (Surbur 1959) and
other such studies. These studies were handicapped by a lack of
appropriate taxonomy, but it was possible to use the percentage
of worms in the total benthos made up of worms as a pollution
indicator as suggested by Goodnight and Whitley (1961). Indeed,
Carr and Hiltunen (1965) repeated the earlier survey of Western
Lake Erie and were able to show the degeneration of that huge
shallow industrialized bay in the interval between Wright's survey
of 1929-30 and theirs of 1961 but the picture was greatly improved
when Hiltunen (1969) reworked the material and identified the
species present. Taxonomic studies began with the account by
Goodnight (1959) and with my own papers on the American fauna
starting in the early 1960's. These made possible the move from
the situation which existed at the time of my account of pollution
biology published in 1965 (in which all the citations, bar one,
were European) to the summary of a long list of detailed ecological
studies on the St. Lawrence Great Lakes by Cook and Johnson (1974).
This extensive literature supported the division of the St.
Lawrence Great Lakes oligochates into three assemblages viz:

1. pollution tolerant
   a) *T. tubifex* $\pm$ *L. hoffmeisteri* (with *P. multisetosus*
      in all but the most extreme situations).
   b) These plus other *Limnodrilus* species (especially
      L. cervix and L. maumeensis) often with *B. sowerbyi*
      in less severely affected sites, and strongly
      eutrophic areas.
2. eutrophic-mesotrophic types
   Aulodrilus spp., *Potamothrix* spp., *P. ferox (Tubifex
   ignotus?)*
3. oligotrophic species
   P. variegatus, P. superiorensis, Rhyacodrilus spp.,
   *T. kessleri americanus, Phallodrilus hallae, S. heringia-
   nus*

Some of the results of these studies are illustrated by
Spencer (1979) in this volume. This type of work is being continued of course as in the studies by Stimpson et al. (1975) and
Nalepa and Thomas (1976), probably with more recent accounts I
have overlooked. Work in other areas includes that of Maciarowski
et al. (1977) and the acid lakes of Ontario (Roff and Kwiatkowski,
1977) but unfortunately there has been little attempt to use recent
methods of classification and ordination although simple indices
have been proposed such as that by Hommiller and Scott (1977) based
on Hamilton (in Brinkhurst et al.,1968 - not etc. 1969 paper
cited in error by the authors). What has happened, though, is
that many surveys are done by government agencies and exist as
intramural technical reports (e.g. Watt 1973, Cook 1975, Loveridge,

and Cook, 1976) or are carried out by consulting companies and submitted in unpublished reports to the client.

Some of the research literature have begun to show progress in the more dynamic approaches. Whitten and Goodnight (1967, 1969) were into this work, examining the role of worms in sediment/water exchanges by their ability to concentrate phosphorus, calcium and strontium and to pass these ions on to fish. A simple model of this type was referred to by me in a summary of studies in Toronto Harbour carried out by A. Appleby, K.E. Chua, M.G. Johnson and M. Wavre (Brinkhurst, 1972 and Fig. 10). These studies led to papers on the relationship between the patchy distribution of worms seen in the field and the niche specificity into which they have evolved (Brinkhurst, 1974, Brinkhurst and Austin, 1979, Chua and Brinkhurst, 1973) and to other studies of flux rates such as that of Wood and Chua (1973). Studies on oligochaete burrowing activity and sediment/water exchanges were also published by Davis (1974a,b) and Davis et al., 1975 and recent work on the mobilization of heavy metals via oligochaetes has accelerated interest in this field.

Traditional tolerance tests have not, so far, figured in the literature to a significant extent, but exceptions are the studies by Whitten and Goodnight (1966), Whitley (1968), Whitley and Sikora (1970) and that by Naqvi (1973) on the toxicity of twenty three pesticides to *B. sowerbyi* and to crayfish fed on exposed worms. There is a recent paper on the effects of X-irradiation on the coelomic cells of a tubificid (Block and Goodnight, 1976).

I have deliberately chosen to indicate the philosophical approaches used in pollution studies in North America in order to minimise the very lengthy presentation that a survey of published work would entail. Several other communications in this volume will expand upon the themes suggested here.

I would like to close this brief communication by paying tribute to the work of one of my younger American colleagues who is not here today. Having worked in the St. Lawrence Great Lakes and then on Wisconsin Lakes, Dick Howmiller died before he could fully utilise his expertise in his new home in California. I dedicate this paper to the memory of a warm, intelligent human being.

REFERENCES

Block, E.M. and C.J. Goodnight, 1976, The effect of X-irradiation on the coelomic cells of the tubificid *Limnodrilus hoffmeisteri*. Trans. Amer. Microsc. Soc., 95:23-34.
Brinkhurst, R.O., 1965, The biology of the Tubificidae with

special reference to pollution. Proc. 3rd Seminar on Water
Quality Criteria, Cincinnati,1962:57-65.

Brinkhurst, 1972, The role of sludge worms in eutrophication.
E.P.A. Ecological Research Series R3-72-004. 68 pp.

Brinkhurst, 1974, Factors mediating interspecific aggregation of
tubificid oligochaetes. J. Fish. Res. Bd. Canada,31:460-462.

Brinkhurst, R.O., A.L. Hamilton and H.B. Herrington, 1968,
Components of the bottom fauna of the St. Lawrence Great
Lakes. Univ. Toronto Gt. Lakes Inst. P.R., 33:1-49.

Brinkhurst, R.O. and M.J. Austin, 1979, Assimilation by aquatic
oligochaetes. Int. Revue ges. Hydrobiol., 63:

Carr, J.F. and J.K. Hiltunen, 1965, Changes in the bottom fauna
of Western Lake Erie from 1930 to 1961. Limnol. Oceanogr.,
10:551-569.

Chua, K.E. and R.O. Brinkhurst, 1973, Evidence of interspecific
interactions in the respiration of tubificid oligochaetes.
J.Fish Res. Bd. Canada,30:617-622.

Cook, D.G., 1975, A preliminary report on the benthic macroinvertebrates of Lake Superior. Fisheries and Marine Service
Tech. Rept., 572, 44 pp.

Cook, D.G. and M.G. Johnson, 1974, Benthic macroinvertebrates of
the St. Lawrence Great Lakes. J. Fish. Res. Bd. Canada,31:
763-782.

Davis, R.B., 1974a, Tubificids alter profiles of redox potential
and pH in profundal lake sediment. Limnol. Oceanogr.,19:
342-346.

Davis, R.B., 1974b, Stratigraphic effects of tubificids in profundal lake sediments. Ibid., 19:466-488.

Davis, R.B., D.L. Thurlow and F.E. Brewster, 1975, Effects of
burrowing tubificid worms on the exchange of phosphorus
between lake sediment and overlying water. Verh. Internat.
Verein. Limnol., 19:382-394.

Gaufin, A.R. and C.M. Tarzwell, 1952, Aquatic invertebrates as
indicators of pollution. Public Health Reports,67:57-64.

Goodnight, C.J., 1959, "Oligochaeta" In Ward and Whipple,
Freshwater Biology. T.E. Edmondson (ed.) Wiley, New York.

Goodnight, C.J. and L.S. Whitley, 1961, Oligochaetes as indicators
of pollution. Proc. 15th Ind. Waste Conf. Purdue Univ. Ext.
Ser. 106:139.

Loveridge, C.C. and D.G. Cook, 1976, A preliminary report on the
benthic macroinvertebrates of Georgian Bay and North Channel.
Fisheries and Marine Service Tech. Rept., 610, 46 pp.

Maciorowski, A.F., E.F. Benfield, and A.C. Hendricks, 1977,
Species composition, distribution and abundance of oligochates in the Kanawka River, West Virginia. Hydrobiologia,
54: 81-91.

Nalepa, T.F. and N.A. Thomas, 1976, Distribution of macrobenthic
species in Lake Ontario in relation to sources of pollution
and sediment parameters. J. Great Lakes Res., 2:150-163.

Naqvi, S.M.Z., 1970, Toxicity of twenty-three insecticides to a
    tubificid worm *Branchiura sowerbyi* from the Mississippi
    Delta. J. Econ. Entom., 66:70-74.

Roff, J.C. and R.E. Kwiatkowski, 1977, Zooplankton and Zoobenthos
    communities of selected northern Ontario lakes of different
    audities. Can. J. Zool., 55:899-911.

Spencer, D.R., 1979, The aquatic oligochaeta of the St. Lawrence
    Great Lakes In R.O. Brinkhurst and D.G. Cook (eds.)
    "Aquatic Oligochaete Biology," Plenum Press, New York.

Stimpson, K.S., J.R. Brice, M.T. Barbour and P. Howe, 1975,
    Distribution and abundance of inshore oligochates in Lake
    Michigan. Trans. Amer. Micros. Soc., 94:384-394.

Surber, E.W., 1960, Biological problems in water pollution.
    Trans. 2nd. Seminar on Biological Problems in Water
    Pollution, Cincinnati, 1959. p. 263-266.

Watt, W.D., G.H. Harding, J. Caldwell and A. McMinn, 1973, Ecology
    of the St. John River Basin. VI. Oligochaetes as water
    pollution indicators. Unpublished Report, St. John River
    Basin Board (Dept. of Environment, Fisheries Service,
    R.D.B. Halifax).

Whitley, L.S., 1968, The resistance of tubificid worms to three
    common pollutants. Hydrobiologia,32:193-205.

Whitley, L.S. and R.A. Sikora, 1970, The effect of three common
    pollutants on the respiration rate of tubificid worms.
    J. Wat. Poll. Cont. Fed., 1970:57-66.

Whitten, B.K., 1969, The role of tubificid worms in the transfer
    of radioactive phosphorus in an aquatic ecosystem In:
    Symposium on Radioecology, Nelson and Evans (Ed.). Proc.
    2nd Nat. Symp. Ann Arbor 1967, 270-277.

Whitten, B.K. and C.J. Goodnight, 1966, Toxicity of some common
    insecticides to tubificids. J. Wat. Poll. Cont. Fed., 1966:
    277-235.

Whitten, B.K. and C.J. Goodnight, 1967, The accumulation of
    SR-89 and CA-45 by an aquatic oligochaete. Physiol. Zool.,
    40:371-385.

Wood, L.W. and K.E. Chua, 1973, Glucose flux at the sediment-
    water interface of Toronto Harbour, Ontario, with reference
    to pollution stress. Can. J. Microbiol., 19:413-420.

Wright, S., 1955, Limnological survey of western Lake Erie.
    U.S. Fish and Wildlife Serv. Spec. Sci. Rept. Fisheries,
    139: 341 pp.

HEAVY METAL STUDIES WITH OLIGOCHAETES

P.M. Chapman[1], L.M. Churchland[2],

P.A. Thomson[2], and E. Michnowsky[2]

[1]Biology Department
University of Victoria
P.O. Box 1700
Victoria, B.C.
Canada   V8W 2Y2

[2]Department of the Environment
Inland Waters Directorate
502-1001 West Pender Street
Vancouver, B.C.
Canada   V6E 2M9

ABSTRACT

     Current studies suggest that oligochaetes, initially
considered by many authors to be intolerant to heavy metals, are
among the most tolerant benthic invertebrates. Heavy metal
studies with oligochaetes have concentrated almost exclusively
on tubificids, in particular *Limnodrilus hoffmeisteri* Claparede
and *Tubifex tubifex* (Muller), and the literature on this subject
is reviewed. In a recent study on metal levels in sediments
and tubificids in the Fraser River, B.C. conducted over a year
with monthly sampling for nine metals (Cu, Zn, Pb, Fe, Mn, Ni,
Co, Cd and Hg), there was no seasonal variability in tissue
metal levels. Tissue metal levels can only be determined
accurately if preservatives are not used and a correction is
made for gut sediment metal levels. Accumulation rates of metals
by tubificids in the Fraser River and in other areas are variable
and may reflect local differences in biologically available metal

levels; the actual route of accumulation (e.g. from sediment, bacteria or interstitial water) has not been determined.

INTRODUCTION

Oligochaetes have long been considered intolerant to heavy metals (Aston, 1973) and it has been suggested that their population reduction or absence might indicate heavy metal pollution (Hynes, 1960; Brinkhurst and Jamieson, 1971). However, the premise of oligochaete intolerance to heavy metals is based solely on a few studies involving copper and there is limited evidence that oligochaetes are sensitive to other heavy metals.

One of the early experiments on which the idea of oligochaete metal intolerance is founded is that of Jones (1938), in which the tubificid *Tubifex tubifex* was killed within hours in the laboratory by combined concentrations in water of copper and lead equivalent to over 1,000 mg 1^{-1}. A more recent laboratory experiment by Learner and Edwards (1963) showed that the naidids *Nais communis*, *Nais variabilis* and *Nais elinguis* were killed within six hours in hard or soft water by 1 mg 1^{-1} $CuSO_4$. A field study supporting the results of these laboratory experiments was carried out by Butcher (1946) in the river Churnet. He noted that populations of Tubificidae as well as populations of *Chironomus, Asellus,* leeches, molluscs and algae were eliminated by copper from effluent (receiving water concentrations of 0.12–1.2 mg 1^{-1}).

The first laboratory study on oligochaete tolerance to heavy metals other than copper was that of Whitley (1967), who used lead and zinc. He found that *T. tubifex* and *Limnodrilus hoffmesteri* had LD50s for lead that varied from 27.5 mg 1^{-1} at pH 8.5 to 49.0 mg 1^{-1} at pH 7.5. The LD_{50} for zinc at pH 7.5 was 46 mg 1^{-1}, however, zinc concentrations below 10 mg 1^{-1} were very toxic at a pH of 6.5 and 8.5. The results suggested that tubificids were more resistant than fish to lead and zinc in solution.

Similar tolerance experiments have been carried out by Brkovic-Popovic and Popovic (1977a), who studied the survival of *T. tubifex* exposed to solutions of copper, cadmium, mercury, zinc, chromium and nickel. They noted that, with the exception of mercury, toxicity depended on hardness and alkalinity. The 48 hour LD50s in mg 1^{-1} were: Cu, 0.006–0.89; Zn, 0.11–60.2; Cr, 0.06–4.57; Ni, 0.08–61.4; Cd, 0.03–0.72; Hg, 0.06–0.10.

The results of tolerance studies by Whitley (1967) and Brkovic-Popovic and Popovic (1977a) suggest that tubificid oligochaetes are more sensitive to copper, cadmium and mercury in solution than to zinc, chromium, nickel and lead. However,

because tubificids are infaunal animals that are not directly
exposed to conditions in the water column (Chapman, 1979), their
tolerances to heavy metals must also be considered on the basis
of metal levels in sediments and interstitial water.

Recent field surveys of heavy metal levels in sediments and
tubificids and laboratory studies with metals in solution indicate
that tubificids are tolerant of certain metals. Funk et al. (1973)
noted that zinc concentrations of 1,000-7,000 µg g^{-1} dry weight
(as determined by nitric-perchloric digestion) did not affect the
distribution of unidentified oligochaetes. Wentsel et al. (1977a)
measured metal levels in Palestine Lake, Indiana by nitric-
perchloric digestion and found that *Limnodrilus* spp. survived
cadmium, zinc and chromium levels (in µg g^{-1} dry weight) of 970,
14000 and 2100 respectively. These levels eliminated the midge
Chironomus tentans and most of the rest of the benthos.

Thus tubificid oligochaetes can tolerate high levels of
certain heavy metals and, in the case of cadmium, can tolerate
higher levels in sediments than in solution. The significance
of this tolerance is that tubificids (and possibly other oligochaetes) can be used in field surveys of heavy metals both as
monitoring tools (Chapman et al., in press) and possibly as
indicators of heavy metals pollution.

METHODS

Levels of nine heavy metals (Cu, Zn, Pb, Fe, Mn, Ni, Co, Cd
and Hg) in tubificids and in the sediments from which they were
collected were measured monthly from October 1977 to August 1978
at two stations in the Fraser River estuary, British Columbia
(Figure 1). Sediments were sampled subtidally with a Ponar grab
sampler and care was taken to exclude sediment which had contacted
the sides of the grab. The sediment was handled with clean plastic
instruments and three sediment sub-samples were placed in nitric-
acid cleaned glass bottles for analysis of sediment metal content.
The remaining sediment was screened on site through a 0.5 mm mesh
stainless steel sieve using water from the collection site.
Tubificids were sorted from the screened sediments within a day
of collection using stainless steel forceps.

Tubificids were often covered with mucous secretions which
collected sediment and these secretions were therefore removed
before analysis. This was most easily done by wearing surgical
silicone rubber gloves (washed in deionized, distilled water)
and drawing the worms across this surface.

Following cleaning, tubificids were put through two washes
with deionized, distilled water. No attempt was made to remove

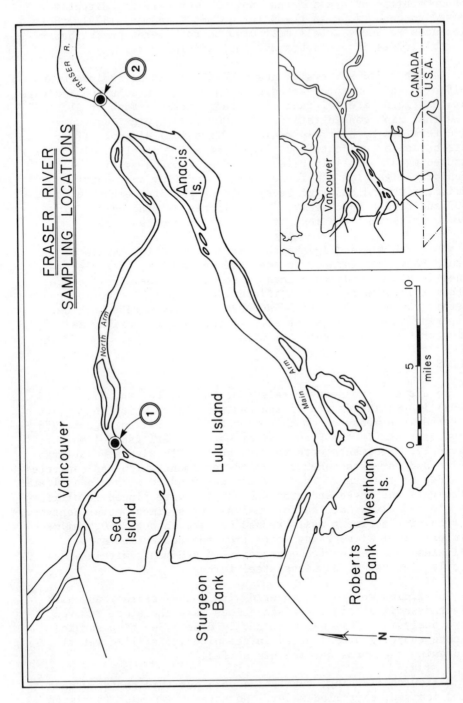

Figure 1.   Location on the Fraser River of sampling sites.

gut contents. When sufficient biomass had been collected, the worms were frozen at -20°C. For all metals except mercury, a total dry weight of at least 0.1 g was required for analysis; approximately the same amount was required for mercury alone. The sample for mercury analysis was air-dried after freezing and the sample for the other metals was freeze-dried.

'Sediments were analyzed for metals other than mercury by drying at 50°C, sieving through a 63 μm mesh stainless steel sieve and analyzing this sediment fraction by hydrogen peroxide-nitric acid digestion (4 ml 30% H_2O_2, 1 ml HNO_3 digested at medium heat for 2½ hours) and by using a Perkins Elmer 403 Atomic Absorption Spectrophotometer.

Tubificids were digested for all metals except mercury by a similar technique to the peroxide-nitric acid sediment digestion. The only difference in technique was in filtration, where sediment remaining from the digestion of tubificid tissue was collected on pre-weighed, acid-washed (0.2% HNO_3) 0.45 μm Sartorius membrane filters. The filters and sediment were dried to constant weight at 50°C and weighed. The weight of tissue sediment residue was multiplied by the heavy metal levels in the sediments from which the tubificids were collected and this value was subtracted from tissue metal levels to correct for gut sediment metal contents (cf. Bindra and Hall, 1977).

Mercury was measured by a wet digestion followed by a cold vapour technique (DOE Analytical Manual, 1974) using a 0.5 g sample of sediment which had been dried at 50°C then passed through a 63 μm mesh stainless steel sieve. Mercury tissue metal levels were not corrected for gut sediment metal contents, as the whole tissue sample was required for analysis.

Blanks were run for all samples; additionally, recovery rates were determined on bovine liver standards of known metal concentration. The salt and acid concentration of the blanks was matched as closely as possible with the sample.

During monthly sampling, tubificid species composition was determined by removing a representative sub-sample of 50 worms from each collection, preserving them in formalin and identifying them with Amman's lactophenol as the clearing agent. Age-groupings were also determined by visual examination of the genital segments.

RESULTS

The major objectives of the Fraser River study were to compare metal levels in sediments with metal levels in tubificids to determine bio-magnification and to make seasonal and geographi-

cal comparisons of metal levels in sediments and tubificids. In
order best to present these data for comparison, graphs have been
drawn for each metal (Figures 2-9) with the exception of cadmium.
Cadmium levels are not graphed because levels in sediments and in
tubificids were generally below detection limits (2.0 µg g^{-1} dry
weight).

Metal levels measured in tubificids were corrected for gut
metal contents in order to determine actual tissue metal levels.
The importance of making a gut sediment correction is illustrated
in Figures 2-8 in which both uncorrected and corrected tissue
metal levels are shown. In the case of mercury (Figure 9),
because a gut sediment correction was not done, only uncorrected
values are shown.

The mean values and 95% confidence limits for data on sediments and tubificids over the sampling period are shown in Table 1.
To determine whether significant differences existed between the
two stations, data from all collections were grouped and a one-way
anova comparison of means was run (Table 1).

There was no statistically significant (Table 1) difference
in sediment copper levels between the two stations. However,
copper levels in tubificids at station 2 were significantly
($P<0.05$) higher than at station 1 by an average of 65%. Levels
of copper in sediments and tubificids fluctuated, but there was
no discernable seasonal trend (Figure 2).

Zinc levels in the sediments were variable at station 2
whereas levels in the sediment at station 1 and in tubificids at
both stations were more consistent (Figure 3). Sediment zinc
levels were significantly higher at station 2 ($P<0.001$) by an
average of almost double the value at station 1 (Table 1), however
zinc levels in tubificids were not significantly different between
the two stations.

Lead levels in sediments were significantly higher ($P<0.001$)
and more variable at station 2 than at station 1 (Table 1). Lead
levels in tubificids were significantly higher ($P<0.05$) by an
average of three-fold at station 2 as compared to station 1. Lead
levels in tubificids at both stations were reasonably stable over
time (Figure 4).

Sediment iron levels were significantly higher ($P<0.05$) at
station 1 than at station 2 by an average of 9% (Table 1). There
was a great deal of variation within each collection and, consequently, there does not appear to be a seasonal pattern (Figure 5).
The levels of iron in tubificids fluctuated but there was no
obvious seasonal trend and levels were not significantly different
between the two stations.

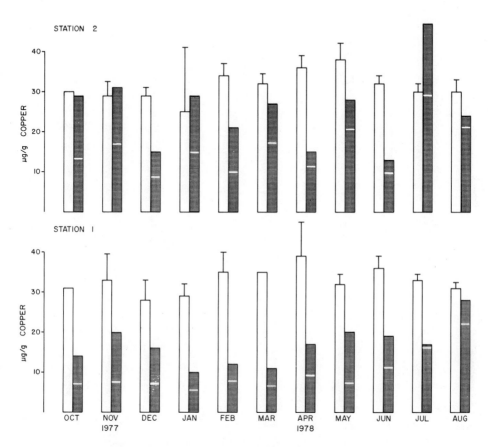

Figure 2.   Copper levels in sediments (clear bars) and tubificids
             (shaded bars). Mean values and 95% confidence limits
             (n=3) are given for sediment metal analyses with the
             exception of October, 1977, when n=2. Tubificid values
             are single analyses; the larger uncorrected values are
             indicated by a solid line and the smaller corrected
             tissue metal levels are indicated by a clear line. All
             values are in μg/g dry weight.

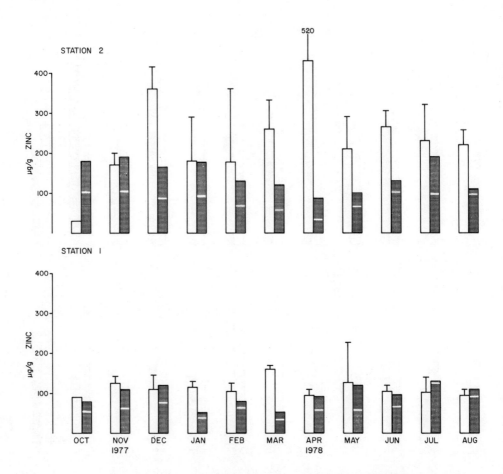

Figure 3.   Zinc levels in sediments (clear bars) and tubificids
             (shaded bars). Mean values and 95% confidence limits
             (n=3) are given for sediment metal analyses with the
             exception of October, 1977, when n=2. Tubificid values
             are single analyses; the larger uncorrected values are
             indicated by a solid line and the smaller corrected
             tissue metal levels are indicated by a clear line. All
             values are in µg/g dry weight.

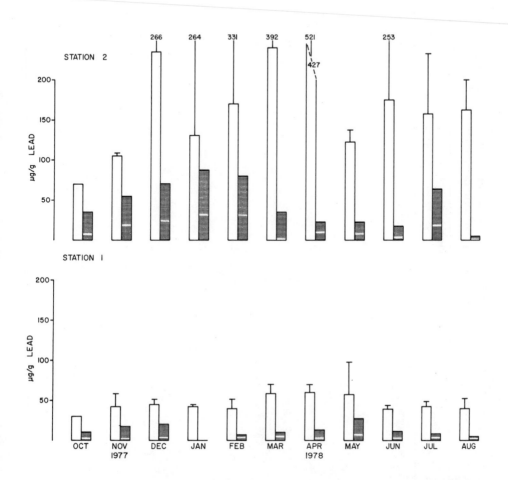

Figure 4. Lead levels in sediments (clear bars) and tubificids (shaded bars). Mean values and 95% confidence limits (n=3) are given for sediment metal analyses with the exception of October, 1977, when n=2. Tubificid values are single analyses; the larger uncorrected values are indicated by a solid line and the smaller corrected tissue metal levels are indicated by a clear line. All values are in µg/g dry weight.

486 P. M. CHAPMAN ET AL.

Table 1. Means (µg g^{-1} dry weight), 95% confidence limits and one-way anova comparison of the difference between means for metals in sediments and tubificids at stations 1 and 2.

METAL	SEDIMENTS STATION 1		STATION 2	TUBIFICIDS STATION 1		STATION 2
Cu	32.8± 1.1	n.s.	31.1± 1.4	9.5± 4.2	*	15.7± 3.1
Zn	113.4± 8.2	***	226.6± 34.9	65.1± 15.5	n.s.	83.5± 12.7
Pb	44.6± 3.9	***	162.1± 6.4	5.0± 1.3	*	15.1± 9.0
Fe	13694.3±708.6	*	12589.0±834.2	588.6± 459.8	n.s.	1171.6±584.9
Mn	314.9± 16.8	n.s.	335.7± 16.5	15.9± 13.9	n.s.	33.2± 14.2
Ni	32.7± 1.6	n.s.	31.4± 1.5	1.8± 1.3	n.s.	2.7± 1.8
Co	11.7± 0.7	n.s.	11.5± 1.0	2.7± 4.6	n.s.	1.8± 2.1
Cd	below detection limits			below detection limits		
Hg	0.065±0.003	***	0.108±0.022	0.11±0.01	**	0.43± 0.48

n.s. = not significant

  * = 0.05>P>0.01

 ** = 0.01>P>0.001

*** = P<0.001

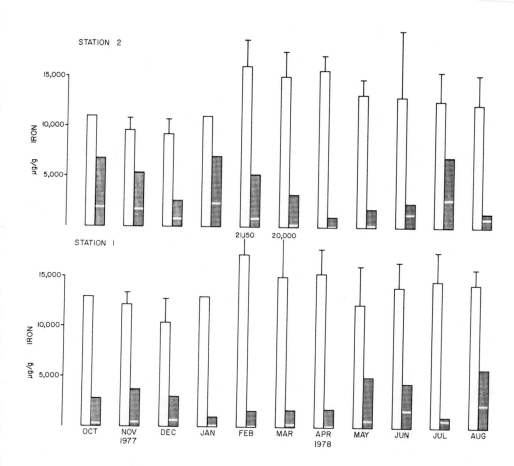

Figure 5.   Iron levels in sediments (clear bars) and tubificids
(shaded bars).   Mean values and 95% confidence limits
(n=3) are given for sediment metal analyses with the
exception of October, 1977, when n=2.   Tubificid values
are single analyses; the larger uncorrected values are
indicated by a solid line and the smaller corrected
tissue metal levels are indicated by a clear line.   All
values are in µg/g dry weight.

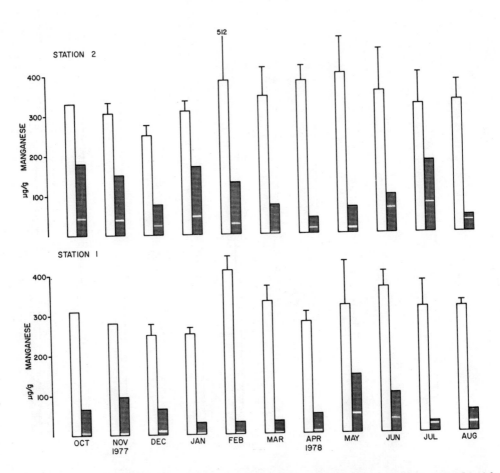

Figure 6.   Manganese levels in sediments (clear bars) and tubificids
(shaded bars).   Mean values and 95% confidence limits
are given for sediment metal analyses with the exception
of October, 1977, when n=2.   Tubificid values are single
analyses; the larger uncorrected values are indicated
by a solid line and the smaller corrected tissue metal
levels are indicated by a clear line.   All values are
in μg/g dry weight.

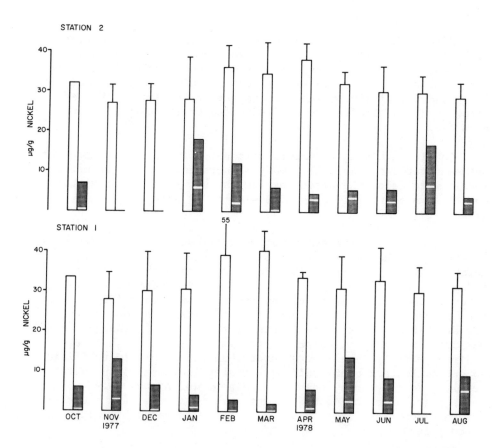

Figure 7. Nickel levels in sediments (clear bars) and tubificids (shaded bars). Mean values and 95% confidence limits are given for sediment metal analyses with the exception of October, 1977, when n=2. Tubificid values are single analyses; the larger uncorrected values are indicated by a solid line and the smaller corrected tissue metal levels are indicated by a clear line. All values are in µg/g dry weight.

Manganese levels in sediments were not significantly different between stations (Table 1) and there was considerable variation in monthly levels as in the case of iron (Figure 6). Manganese levels in tubificids were not significantly different between stations (Table 1).

Nickel levels in sediments were not significantly different between stations (Table 1) and, as was the case with iron and manganese, seasonal trends were not distinguishable (Figure 7). Measurements of nickel in tissue were not always precise due to the proximity of some values to the detection limits. Values below detection limits are not indicated on the bar graph in Figure 7.

Cobalt levels in sediments were not significantly different between stations (Table 1). Many measurements of tissue metal levels were below the detection limits, but the few measurable cobalt levels in tubificids were similar at both stations (Figure 8).

Sediment mercury levels were significantly higher (P<0.001) at station 2 than at station 1 by an average of 60% (Table 1). Mercury levels in tubificids were greater than the levels in sediments and, in the case of station 2, fluctuations in tissue mercury levels were greater than those in sediments (Figure 9). Mean concentrations of mercury in tubificids were four times higher (P<0.01) at station 2 than at station 1. Although these measurements were not corrected for gut sediment metal levels, sediment mercury levels were lower than the levels in tubificids so gut contents probably did not greatly increase tissue metal levels. For instance, mean gut sediment weight was 14.9% which means that, based on mean sediment mercury levels, gut sediment contents comprised less than 10% of the measured tubificid mercury levels.

From the measurements of tissue and sediment metal levels, accumulation patterns were determined by calculating the concentration factor (CF) which is the ratio of the mean concentration of metal in tubificids to that in sediment. Concentration of trace metals are normally used without regard to chemical state or uptake routes (Jenne and Luoma, 1977). Concentration factors (Table 2) were less than 1 for all metals except mercury, indicating that none of these metals are concentrated above sediment levels. Total mercury levels were 4.0 times as high in tubificids from station 2 as in sediments and were 1.7 times as high in tubificids from station 1 as in sediments.

There was no evidence of seasonal variation in species composition or in age class in tubificids identified from the

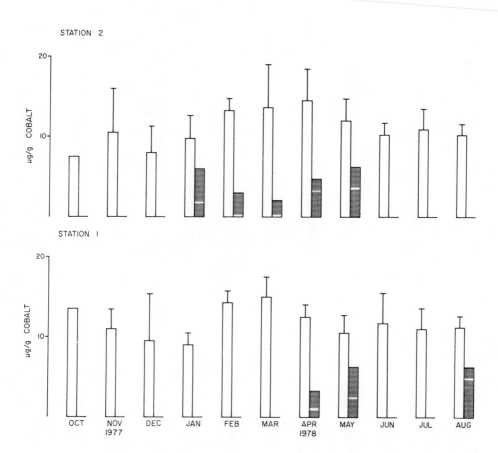

Figure 8.   Cobalt levels in sediments (clear bars) and tubificids
            (shaded bars). Mean values and 95% confidence limits
            are given for sediment metal analyses with the exception
            of October, 1977, when n=2. Tubificid values are single
            analyses; the larger uncorrected values are indicated
            by a solid line and the smaller corrected tissue metal
            levels are indicated by a clear line. All values are
            in μg/g dry weight.

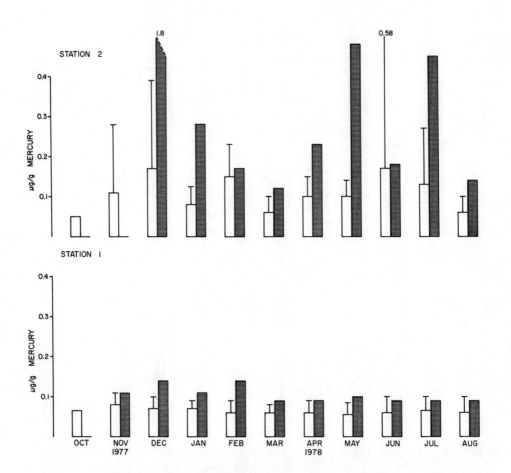

Figure 9.   Mercury levels in sediments (clear bars) and tubificids
(shaded bars).   Mean values and 95% confidence limits
are given for sediment metal analyses with the exception
of October, 1977, when n=2.   Tubificid values are single
analyses.   All values are in µg/g dry weight.

Fraser River collections. Two species, *L. hoffmeisteri* and
T. tubifex, dominated both study areas and comprised a mean of
77% and 22% of the collections, respectively (Figure 10).

Table 2.   Concentration factors for metals in tubificids. These
            values are ratios of metal concentrations in tubificids
            to those in sediments and are determined using the
            mean values from Table 1.

METAL	STATION 1	STATION 2
Cu	0.3	0.5
Zn	0.6	0.4
Pb	0.1	0.1
Fe	< 0.1	0.1
Mn	< 0.1	0.1
Ni	0.1	0.1
Co	0.2	0.2
Hg	1.7	4.0

DISCUSSION

Comparison with other studies

     Heavy metal levels in tubificids from the Fraser River are
compared to those reported in other published studies in Table 3.
The comparison is complicated by the fact that the other studies
used preservatives and different extraction techniques. Preservatives have been shown to increase measured tissue metal levels up
to a factor of three in some cases, and different extraction
methods may give different values for the same samples (Chapman
et al., in press). Another problem in cross-comparisons is that
of gut analysis. Although McNurney et al. (1977) found no statistical difference between lead levels in tubificids with full and
evacuated guts, gut sediment metal contents do affect measured
metal levels but, with the exception of Bindra and Hall (1977),
none of the other studies on metal levels in tubificids corrected
for gut sediment.

     The ratio of metals in tubificids and sediments is variable
depending on geographical area and may be a reflection of local
differences in biologically available metal levels. For instance,
Mathis and Cummings (1973) found that in the Illinois River, zinc,
lead, nickel, cobalt and cadmium were lower and copper was
slightly higher in tubificids than in sediments whereas in certain
areas of the lower Fraser River Bindra and Hall (1977) noted an
overall trend of copper, zinc, lead and manganese accumulation
above sediment levels. In South Africa (Greichus et al., 1977)
copper was accumulated above sediment levels while in Rhodesia

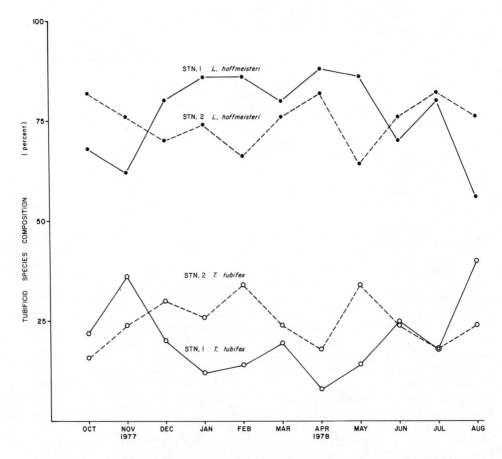

Figure 10.   Tubificid species composition at stations 1 and 2.
Values shown are percentages determined by identifying
a subsample of 50 worms.   The occasional presence of
species other than *L. hoffmeisteri* and *T. tubifex* is
not shown.

Table 3. Comparison of mean metal levels in oligochaetes in different studies and different geographical areas. Values are in µg/g dry weight unless otherwise indicated. Sediment metal levels (in µg/g dry weight) are indicated in parentheses. Asterisks indicate accumulation above sediment levels.

SPECIES	Cu	Zn	Pb	Fe	Mn	Ni	Co	Cd	Hg	REFERENCE	DIGESTION	COMMENTS
L. hoffmeisteri & T. tubifex	23* (19)	41 (81)	17 (28)			11 (27)	1.6 (6)	1.1 (2)		Mathis & Cummings, 1973	nitric-perchloric	Wet wt. Preserved in alcohol. Ill. River.
L. hoffmeisteri & T. tubifex	119*	439*	486*	13450	863*					Bindra & Hall, 1977	peroxide-nitric acid	Brunette River. (formalin)
	33* (17)	289* (62)	19 (39)	8900 (22400)	800* (610)							Salmon River. (formalin)
	34 (53)	86.8 (150)	5.7 (28)									Ladner Sidechannel (formalin)
unidentified oligochaetes	21* (15)	41 (49)	5.1 (9.0)		15 (340)				0.01 (0.19)	Greichus et al., 1977	nitric-perchloric & total Hg	Voëlvlei Dam, S. Africa. (formalin)
B. sowerbyi & others	72*	130*	1.3		28			0.05	0.08	Greichus et al., 1978	nitric-perchloric & total Hg	Lake McIlwaine, Rhodesia. (formalin)
L. hoffmeisteri & T. tubifex	16 (31)	84 (227)	15 (162)	1172 (12589)	33 (336)	3 (31)	2 (11)		0.43* (0.11)	This study	peroxide-nitric acid & total Hg	Station 2 (no preservatives)
	10 (33)	65 (113)	5 (45)	589 (13694)	16 (315)	2 (33)	3 (12)		0.11* (0.07)			Station 1 (no preservatives)

(Greichus et al., 1978) both copper and zinc were accumulated above
sediment levels.

     In the present study, higher sediment metal concentrations of
zinc, lead, iron, manganese and mercury at station 2 than at
station 1 were reflected in higher tubificid tissue metal levels
at station 2 than at station 1. Nickel and cobalt levels were
similar in sediments and tubificids at both stations while cadmium
levels were below detection limits. However, in the case of
copper, despite slightly higher sediment levels at station 1 than
at station 2, copper levels in tubificids at station 2 were significantly higher than at station 1. This suggests that copper at
station 2 is more biologically available than at Station 1. Also,
of the nine metals measured, only mercury was accumulated by
tubificids above sediment levels. In contrast, despite high
sediment levels of mercury in Voelvlei Dam, South Africa (Greichus
et al., 1977) and in Lake McIlwaine, Rhodesia (Greichus et al.,
1978), mercury was not accumulated by oligochaetes.

Uptake routes

     One of the major advantages to measuring heavy metal levels
in oligochaetes (and in other benthic organisms) is the fact that
tissue metal levels indicate, as chemical analyses do not, when
and where heavy metals are biologically available. Bindra and
Hall (1977) make the point that: "trace metals in benthic organisms
are not always highest in areas of the highest total trace metal
concentrations". They relate biological availability to a number
of factors: the geochemical phases of the metal, organic content
of sediments, Eh, pH and sulphide levels. Similarly, Hall (1976)
has suggested that heavy metals in sediments are mainly available
to organisms through the organic fraction.

     There are three possible uptake routes for metals in tubifi-
cids: by ingestion of bacteria which have concentrated metals in
sediments, by ingestion of sediment particles containing metals,
and by uptake from solution. Patrick and Loutit (1976) demonstrated
passage of metals from bacteria to tubificids (*Limnodrilus* sp. and
Tubifex sp.) for chromium, copper, manganese, iron, lead and zinc
and noted upper concentration values in the worms (in $\mu g \ g^{-1}$ dry
weight) of 30, 621, 25, 1944, 568 and 868 respectively. However,
accumulation can also proceed directly from solution. Dean (1974)
noted that unidentified tubificids accumulated ^{65}Zn from water
but not from sediment. D'Angelo and Signonle (1974) found uptake
of mercury by *T. tubifex* from freshwater was via the gut at low
concentrations ($0.1 \ mg \ 1^{-1}$) and by diffusion across exposed body
surfaces at higher concentrations ($0.3 \ mg \ 1^{-1}$). The latter
concentration probably represents a fatal dose (Brkovic-Popovic
and Popovic, 1977a).

During field studies, this gap in our knowledge of metals uptake by oligochaetes can be partly resolved by measurements of metal concentrations in interstitial water and in the water above the sediments. Metal levels in interstitial water at station 2 (G. Geesey, personal communication) and in the water above the sediments at both stations were three orders of magnitude less than 'in sediments, making water an unlikely route for uptake. Therefore, accumulation of metals in tubificids in the present study was probably by ingestion of bacteria and/or sediments. However, exact uptake routes have yet to be determined.

Moreover, it has recently been suggested by Bindra and Hall (1978) that oligochaetes can regulate tissue metal levels. These investigators attempted to determine the reason for differential bioaccumulation of metals. They did this by exposing a mixture of tubificids and naidids to sediments contaminated with copper, iron, lead and zinc. Experiments lasted six weeks and tissue metal levels were measured after 1, 2, 4 and 6 weeks. Concentrations of all five metals decreased after 1 week, increased to a peak value after 4 weeks and then showed a decrease after 6 weeks. The idea that oligochaetes can regulate uptake of heavy metals complicates the interpretation of bioaccumulation data.

Seasonal variations in tissue metal levels

There is no evidence in the present study of seasonal variation in tubificid metal levels at either station. However, distinct seasonal changes in metal levels have been observed in other organisms, usually in connection with biological cycles. For example, seasonal changes in copper levels in the whelk *Busycon canaliculatum* are associated with an annual cycle of behavioral and physiological change (Betzer and Pilson, 1975). Seasonal fluctuations in metal levels have been reported for scallops (Bryan, 1973), barnacles (Ireland, 1974) and amphipods (Zauke, 1977) with high metal concentrations occuring in winter or spring and low levels in summer or autumn. This variation is usually ascribed to renewed growth in the organisms. The lack of temporal variation in tissue metal levels in tubificids means that representative tissue metal levels can be determined without regard for sampling season, and suggests that variations in tissue metal levels are due to differences in biologically available sediment metal levels.

Passage of metals up the food chain

It has been suggested that tubificids may be an important food source for bottom feeding fish (Milbrink, 1973). Patrick and Loutit (1978) have shown that the metals chromium, copper, manganese, iron, lead and zinc can be passed up the food chain from

tubificids to fish feeding on tubificids. In the Fraser River,
Northcote et al. (1975) measured levels of nine heavy metals in
fish tissues and noted that only mercury was present in significant
amounts, averaging 0.67 μg g $^{-1}$ wet weight in two species of bottom-
feeding fish. Although it is tempting to use Patrick and Loutit's
(1978) study on bioaccumulation of metals from tubificids to fish
to partially explain high mercury levels in bottom-feeding fish
in the Fraser River, the role of tubificids in this context is
uncertain. For instance, Jernelov and Lann (1971) have suggested
that less than 25% of mercury accumulated by bottom-feeding fish
is transferred to them from tubificids and chironomids.

        Oligochaetes may also be responsible for the passage of
metals up the food chain by means of the release of metals from
contaminated sediments. Tubificids have been shown to raise
introduced pollen to the surface of sediments from depths of up
to 15 cm (Davis, 1974). Boddington et al. (1979) have shown that
L. hoffmeisteri causes a significant loss of inorganic and methyl
radio-labelled mercury from sediments via the resuspension of
fine material into the water column. In the Fraser River, sediment
metal levels measured in the present study were not unusually
high, thus bioturbation may not be a significant means for the
release of metals into the water column at stations 1 and 2.
However, in more polluted areas, bioturbation may make large
amounts of metals available for uptake by fish and other organisms.

Sub-lethal effects

        One facet of heavy metal pollution that has received little
attention is the sub-lethal effects of heavy metals on oligochaetes.
Sub-lethal effects are of interest both as regards the health of
natural populations and as a possible sensitive test of heavy
metal toxicities. In the case of benthic chironomid larvae, it
has been shown (Wentsel et al., 1977b, c) that these animals avoid
sediments containing high levels of heavy metals and that populations living in heavy metal contaminated sediments suffer
various impairments to growth and reproductive capabilities.
However, with the exception of two respiration studies and the
suggestion that high sediment mercury levels may affect setal
formation in Potamothrix hammoniensis (G. Milbrink, personal
communication), there is no information on the possible sublethal
effects of heavy metals on oligochaetes.

        Whitley (1967) suggested, in his experiments on the tolerance
of L. hoffmeisteri and T. tubifex to zinc and lead, that death
resulted from precipitation of the worms mucous coating, which
inhibited epidermal respiration. The effects of heavy metals on
the respiration rate of these two tubificids was investigated by
Whitley and Sikora (1970) with lead and nickel. In the case of

lead, increased concentrations resulted in a decrease in the
respiration rate, however, in the case of nickel there was no
clear relationship between metal concentration and respiration
rate. Similar experiments were carried out by Brkovic-Popovic
and Popovic (1977b) on the respiration rate of *T. tubifex* exposed
to solutions of copper, cadmium, mercury, zinc, chromium and
Nickel. These investigators noted a variety of different responses
depending on metal and dosage; however, at the acute lethal range
of concentration, copper, cadmium and mercury inhibited respiration while zinc, chromium and nickel stimulated respiration. The
results obtained using nickel seem to be contradictory.

Levels of sediment heavy metal levels measured in the present
Fraser River study were not high and tubificids collected from
stations 1 and 2 appeared to be healthy.

Species composition

Two final points concerning heavy metal studies with oligochaetes are the influence of species composition on metals uptake
and, conversely, the influence of heavy metals in sediments on
species composition. In the present study the relative proportions
of the two dominant tubificid species, *L. hoffmeisteri* and *T. tubifex*, were similar both seasonally and at the two sampling stations.
Thus differences between tissue metal levels at the two stations
cannot be ascribed to variations in species composition and there
is no evidence that sediment metal levels influence species composition.

However, it would be surprising if different species of
oligochaetes did not have different tolerances and abilities
to accumulate metals. Most heavy metal studies have been done
with tubificids, particularly the two pollution tolerant species
L. hoffmeisteri and *T. tubifex*. Studies on metals tolerances
need to be done with other common species of tubificids and with
other oligochaete families. The possibility of interspecific
differences makes it especially important that oligochaetes be
properly and completely identified in all heavy metal studies, but
an examination of the literature shows that this has rarely been
done. As a result, it is both difficult to compare different
studies and impossible to determine whether differences in heavy
metal tolerances exist. One of the main priorities in future
heavy metal studies with oligochaetes is that of determining, by
means of toxicity studies and sub-lethal effects, whether oligochaetes can be used as biotic indicators of heavy metal pollution.

ACKNOWLEDGMENTS

We thank Dr. T.G. Northcote, of the Institue of Animal
Resource Ecology, University of British Columbia, for supplying

laboratory facilities. We also thank Dr. W.E. Erlebach for his
critical review of the manuscript.

REFERENCES

Aston, R.J., 1973, Tubificids and water quality: a review.
    Environ. Pollut., 5:1-10.
Betzer, S.B. and M.E.Q. Pilson, 1975, Copper uptake and excretion
    by Busycon canaliculatum L. Biol. Bull., 148:1-15.
Bindra, K.S. and K.J. Hall, 1977, Geochemical partitioning of
    trace metals in sediments and factors affecting bioaccumula-
    tion in benthic organisms. Unpublished report prepared for
    NRC, Ottawa (contract # 032-1082/6073). 59pp.
Bindra, K.S. and K.J. Hall, 1978, Bioaccumulation of selected
    trace metals by benthic invertebrates in laboratory bioassays.
    Unpublished report prepared for NRC, Ottawa (contract
    # 032-1082/6073). 25pp.
Boddington, M.J., D.R. Miller and A.S.W. deGreitas, 1979, The
    effects of benthic invertebrates on the clearance of mercury
    from contaminated sediments. Limnol. Oceanogr., (submitted).
Brinkhurst, R.O. and B.G.M. Jamieson, 1971, Aquatic Oligochaeta
    of the World. U. of Toronto Press: Toronto. 860pp.
Brkovic-Popovic, I. and M. Popovic, 1977a, Effects of heavy metals
    on survival and respiration rates of tubificid worms: Part I-
    Effects on survival. Environ. Pollut., 13:65-72.
Brkovic-Popovic, I. and M. Popovic, 1977b, Effects of heavy metals
    on survival and respiration rates of tubificid worms:
    Part II - Effects on respiration rate. Environ. Pollut., 13:
    93-98.
Bryan, G.W., 1973, The occurrence and seasonal variation of trace
    metals in the scallops Pecten maximus and Chlamys opercularis.
    J. Mar. Biol. Ass. U.K., 53:145-166.
Butcher, R.W., 1946, The biological detection of pollution.
    J. Proc. Inst. Sewage Purif., 2:92-97.
Chapman, P.M., 1979, Seasonal movements of subtidal benthic
    communities in a salt wedge estuary as related to intersti-
    tial salinities. PhD. Thesis: U. of Victoria. 221pp.
Chapman, P.M., L.M. Churchland, P.A. Thomson and E. Michnowsky,
    1979, Tubificid oligochaetes as monitors of heavy metal
    pollution. Proc. Fifth Aquatic Toxicity Workshop: Hamilton,
    Ontario. P.T.S. Wong et al. (eds). Fish and Mar. Service
    Tech. Rept., 862:278-294.
D'Angelo, A.M. and G. Signonle, 1974, Investigations on mercury
    localization in a freshwater oligochaete (Tubifex tubifex).
    Igiene Mod., 66:286-291
Davis, R.B., 1974, Stratigraphic effects of tubificids in pro-
    fundal lake sediments. Limnol. Oceanogr., 19:466-488.
Dean, J.M., 1974, The accumulation of ^{65}Zn and other radionuclides
    by tubificid worms. Hydrobiologia, 45:33-38.
Department of Environment, 1974, Analytical Methods Manual. Inland
    Waters Directorate: Ottawa.

Funk, W.H., R.W. Rabe, R. Filby and J.I. Parker, 1973, An inte-
    grated study on the impact of metallic trace element pollu-
    tion in the Couer D'Alene-Spokane Rivers-Lake Drainage
    System. NTIS-PB 222-946.
Greichus, Y.A., A. Greichus, B.D. Ammman, J. Call, D.C.D. Hamman
    and R.M. Pott, 1977, Insecticides, polychlorinated biphenyls
    and metals in African Lake Ecosystems. I. Hartbeesport Dam,
    Transvaal and Vöelvlei Dam, Cape Province, Republic of
    South Africa. Arch. Environ. Contam. Toxicol., 6:371-383.
Greichus, Y.A., A. Greichus, B.D. Amman and J. Hopcraft, 1978,
    Insecticides, polychlorinated biphenyls and metals in African
    Lake Ecosystems. II. Lake McIlwaine, Rhodesia. Bull.
    Environ. Contam. Toxicol., 20:444-453.
Hall, T., 1976, A simple method for determining the concentration
    of heavy metals in the organic component of sediments. Mem.
    Ist. Ital. Idrobiol.,33:177-184.
Hynes, H.B.N., 1960, The Biology of Polluted Waters. University
    Press: Liverpool.
Ireland, M.P., 1974, Variations in the zinc, copper, manganese
    and lead content of Balanus balanoides in Cardigan Bay,
    Wales. Environ. Pollut., 7:65-75.
Jenne, E.A. and S.N. Luoma, 1977, Forms of trace elements in soils,
    sediments and associated waters: An overview of their deter-
    mination and biological availability, pp. 110-143. In:
    Biological Implications of Metals in the Environment, Tech.
    Info. Center, ERDA.
Jernelov, A. and H. Lann. 1971, Mercury accumulation in food
    chains. Oikos,22:403-406.
Jones, J.R.E., 1938, Antagonism between two heavy metals in their
    toxic action on freshwater animals. Proc. Zool. Soc. Lond.,
    A:481-497.
Learner, M.A. and R.W. Edwards, 1963, The toxicity of some
    substances to Nais (Oligochaeta). Proc. Soc. Water Treat.
    Exam.,12:161-168.
Mathis, J. and T.T. Cummings, 1973, Selected metals in sediments,
    water and biota in the Illinois River. Jour. Water Poll.
    Control Fed., 45:1573-1583.
McNurney, J.M., R.W. Larimore and M.J. Wetzel, 1977, Distribution
    of lead in sediments and fauna of small midwestern stream,
    pp. 167-177. In: Biological Implications of Metals in
    the Environment, Tech. Info. Center, ERDA.
Milbrink, G., 1973, On the vertical distribution of oligochaetes
    in lake sediments. Inst. Fresh Water Res., Drottningholm 53:
    34-50.
Northcote, T.G., N.T. Johnston and K. Tsumura, 1975, Trace metal
    concentrations in Lower Fraser River fishes. Westwater
    Tech. Report #7. Westwater Res. Center: U. of British
    Columbia. 41pp.
Patrick, E.M. and M. Loutit, 1976, Passage of metals in effluent
    through bacteria to higher organisms. Water Res., 10:333-335.

Patrick, E.M. and M. Loutit, 1978, Passage of metals to freshwater
     fish from their food. Water Res., 12:395-398.
Wentsel, R., A. McIntosh and V. Anderson, 1977a, Sediment conta-
     mination and benthic macroinvertebrate distribution in a
     metal-impacted lake. Environ. Pollut., 14:187-192.
Wentsel, R., A. McIntosh, W.P. McCafferty, G. Atchison and
     V. Anderson, 1977b, Avoidance response of midge larvae
     (Chironomus tentans) to sediments containing heavy metals.
     Hydrobiologia, 55:171-176.
Wentsel, R., A. McIntosh and G. Atchison, 1977c, Sublethal effects
     of heavy metal contaminated sediment on midge larvae (Chiro-
     nomous tentans) Hydrobiologia, 56:153-157.
Whitley, L.S., 1967, The resistance of tubificid worms to three
     common pollutants. Hydrobiologia, 32:193-205.
Whitley, L.S. and R.A. Sikora, 1970, The effects of three common
     pollutants on the respiration rate of tubificid worms. Jour.
     Water Poll. Control Fed., 42:57-66.
Zauke, G.P., 1977, Mercury in benthic invertebrates of the Elbe
     Estuary. Helgolander wiss. Meeresunters, 29:358-374.

THE RELATIONSHIP OF SAPROBIAL CONDITIONS

TO MASSIVE POPULATIONS OF TUBIFICIDS

H. Caspers

Institut for Hydrobiologie und Fischereiwissenschaft
Universitat Hamburg
Zeisweg 9, D2000
Hamburg, W. Germany

ABSTRACT

The tubificids play an important role as indicator organisms in the saprobial system. Less important than the species or genera present is the total number of individuals. The more organic enrichment in a sediment, the greater the number of tubificids, coincident with the disappearance of other benthic groups. Especially in polysaprobial waters, the fauna is limited to "lawns" of tubificids, which are chiefly involved with the remineralization of the detritus.

Conditions required to support massive populations of tubificids include not only a large detritus supply, but also the availability of at least a minimum amount of oxygen in the water immediately overlying the sediment. To obtain oxygen, the worms attempt to create water circulation by a characteristic swaying body movement. That means that the population density in terms of individuals per unit area is not strictly coupled with the degree of saprobity. The number of tubificids in the hypolimnion of eutrophic lakes is often less than would be expected from the quantity of organic materials. Deposits of detritus along the edge of streams, however, can support much greater numbers of the worms, as long as sufficient water exchange supplies the minimum oxygen requirement (Caspers, 1964). Conditions in estuaries are quite different in that tidal circulation generally results in high oxygen concentrations. Usually coarse sediments are not a suitable habitat for tubificids, but river shoals often support massive populations because they are lentic enough to allow the sedimentation of much detritus.

Habitat requirements for massive population development was studied in the Hamburg region of the Elbe including the harbor basin (Caspers and Schulz, 1964). In this freshwater part of the Elbe, divided into a north and south branch around a large island, tidal movements of 2.5 meters occur twice daily. A large amount of organic sediment has been deposited in the side canals and harbor basins which have only occasionally be dredged. Originally, after the construction of the harbor basis, only a sandy-silt was present, providing a habitat for a dense population of bivalves (*Sphaerium*). Very quickly, the harbor area filled with detritus, eliminating the bivalves and bringing about a massive development of tubificid populations. Oxygen availability is related to the river current. The region of rapid flow has a relatively small number of individuals, related to the low rate of detritus deposition. In spite of the detritus-rich sediment, the back harbor regions also have relatively few tubificids since the water exchange is insufficient

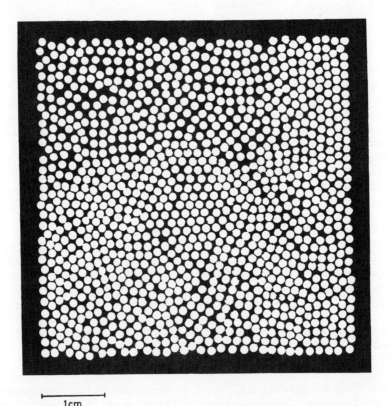

1cm

Figure 1. Density distribution of tubificids in Hamburg Harbor. The illustration is shown to scale with 2,211 worms (average diameter 0.73 mm) in an area of 1/360 m^2.

to provide a constant supply of oxygen. The open harbor regions, on the other hand, are optimal for tubificids because oxygen-rich waters overlie thick layers of detritus. Domestic sewage and fish processing wastes from the Altona Fishery Harbor provide an especially rich source of nutrients for the worms.

The maximum number of tubificids found was 800,000 m^{-2}. These were large individuals, reaching 15 cm in length, and belonged to several genera. The density of settlement poses a special question. The worms remain approximately vertical in the sediment. A model of this system was constructed. Holes representing worms of mean diameter 0.73 mm punched in a cardboard model of the sediment surface at the maximum observed tubificid density showed that in such a habitat with an overenrichment of nutrients and a satis-factory oxygen supply, physical space becomes the limiting factor. More individuals cannot exist because the posterior part of their bodies must extend into the water above the sediment to allow for respiration (Figure 1).

Whether or not young individuals can survive after they leave their cocoons under such crowded conditions is questionable. Such regions are very susceptible to rapid fluctuations: minor changes in the current can lead to movements of the sediments; as the tubificid populations become over-age, the worms die off, providing new room for settlement by young individuals.

A dense population of tubificids indicates polysaprobic conditions. Direct counts showed that in the presence of suffi-cient oxygen, the limit on the number of adult individuals is physical space. Normally, in limnic regions, polysaprobic conditions are generally coupled with a reduction of the oxygen quantity, which reduces the number of worms or totally eliminates them from the system.

REFERENCES

Caspers, H., 1964, Limitierenda Faktoren in übervölkerten Lebensraümen des Timnischen und marinen Benthos. Mitt. Hamburg Zool. Mus. Inst., 61 (Suppl.):1-13.
Caspers, H., and H. Schulz, 1964, Die biologischen Verhältnisse der Elbe bei Hamburg. Arch. Hydrobiol., 60:53-88.

POSTSCRIPT

(R.O. Brinkhurst)

This first symposium brought together many of the most active students of aquatic oligochaete biology, but inevitably several people were prevented from attending by problems of health, distance, finance and workload. Among those were three colleagues to whom greetings were sent in their absence, Dr. S. Hrabě (Brno), Dr. N. Sokolskaya (Moscow) and Mr. J. Hiltunen (Ann Arbor). Dr. D.H. Di Persia and Dr. M. Loden were eventually obliged to communicate their papers via Dr. W.J. Harman, and the editor had to attempt to present the work which Mr. Hiltunen would have covered had he been able to attend.

The proceedings opened with a consideration of limits to taxa in the freshwater and marine realms. Dr. W.J. Harman stressed the existence of the well-known "lump" versus "split" controversy, and mentioned the fact that our science has its subjective as well as objective components in decision making. The degree to which an author adheres to a classical type concept versus the need to accept ecologically induced variation (and, we might add, those due to reproductive stage) produces some of this continuing difference, but the dynamic view was given much support by the work of his student, Dr. M. Loden, who has been able to change naidid setae in a population from those of one species to those of another by changing the environment.

Another reason for differences in opinion as to the width of taxa is suggested here. Museum workers and primary taxonomists may approach the discipline from a very different viewpoint from the ecologist and applied biologist. The former may feel less obligation to work with second-hand collections, often poorly preserved in bulk, than his colleagues involved in large-scale surveys. They may also feel less need to place taxa discarded from revised genera than the author of a monograph who is obliged to put everything somewhere. It often seems the taxonomist is attracted to the study of difference, the ecologist working up his material begins by looking for similarity. Dr. Harman also empha-

sized the artificiality of the genus, and the important point that
a large proportion of monotypic genera in a family is inconvenient.
The concept of the need for large gaps between small genera, small
gaps between larger ones is worth some attention. Again the ecologist's primary complaint is the taxonomist's fascination for changing generic names.

Mr. C. Erséus revealed the great complexity of the emerging
marine tubificid fauna, where adaptations like the absence of the
gastrointestinal tract exceed anything seen in freshwater, but the
reduced diversity in setae, small size and complexity of the male
ducts make histological study even more of a necessity than for
the larger freshwater forms. Mr. Erséus appeals for the provision
of adequate type material for these species, and it is the absence
of good types that leads to so many problems (see Appendix and
Reynolds and Cook, 1976). A question to raise here is the thorny
issue of the single holotype when a species may be described from
whole animals (placement of pores) whole-mounts (setae, penis
sheaths) dissections (complex convoluted male duct, proportion of
parts) and sections (precise insertion of and number of prostate
glands and other details). Inadequate material may lead to the
omission of some of these preparations from the series, and for
scarce, poorly fixed specimens of moderately large worms this
author prefers the dissected genital segments plus head and tail
whole mount on the same slide. The dissection is carried out in
the mounting medium to prevent loss of tiny organs. The difficulty
encountered with this method has been the breakdown of the mounting
medium (which can, however, simply be replaced) and the continuous
clearing of the dissected material to the point at which it is no
longer of any great value. I recently used the stripped-off cuticle
of a large haplotaxid to establish the position of the otherwise
invisible genital pores, but this preparation became a uniform
glassy pink spot within weeks of obtaining it. New approaches, as
recommended by Mr. Erséus, will make the problem of types even more
difficult, as would be true for many of the taxa within *Capitella
capitata* recognized by Drs. F. and J. Grassle (and within species
discussed by M. Barbour, et al, and later by Dr. T.L. Poddubnaya).
The possibility of dividing ecologically complex, often opportunistic species such as *Capitella capitata, Limnodrilus hoffmeisteri,
Tubifex tubifex, Tubificoides gabriellae* and *Tubificoides pseudo-
gaster* (to name a few candidates) into ecospecies, subspecies or
some other subdivision will be hard to completely substantiate on
the basis of those anatomical features that need to be used to
identify ecological collections, so a decision to elevate these
finest divisions to specific rank should only be taken after considerable thought. We are inevitably dealing with continuous
variation right down to the level of the individual wherever our
perception is such that we can recognize it (human beings and our
pets and domestic animals).

One final point to be added here concerns the importance of error. In assembling large-scale synopses, similarities between species often emerge, and these may be identified as actual or potential synonyms. The apparent forward shift of reproductive organs observed in *Potamothrix bedoti*, for example, suggested allying it with the similar *Potamothrix bavaricus*. Several subsequent authors were able to demonstrate several other distinctive characters that can now be used to separate these taxa. Detailed ontogenetic studies of *Monopylephorus* need to be carried out to justify some of the suggested synonymies based on the appearance of median body-wall inversions at full maturity that carry the male and spermathecal pores inward from a superficial latero-ventral position. The synonomy of two species known from Lake Baikal and Lake Titicaca is suggested from their current descriptions but is contra-indicated by their distribution. The challenge (or error!) should stimulate those in possession of the necessary material or facts to publish them and clean up the problem. In a sense we only learn by making "mistakes"!

Two of the central problems in nomenclature not addressed in the symposium are the existence of a large number of old dormant names that have to be resurrected if new evidence establishes their true identity (i.e. the discovery of the type of *Limnodrilus profundicola* and the subsequent change of an otherwise well-established name, in this instance *Limnodrilus helveticus*) and the lack of control over non-refereed and poorly refereed publications that create new synonymies or extreme revisions. In any other field, bizarre publications can simply be ignored, but our nomenclatural procedures, for all their value, force one into an adversary role every time any kind of synopsis is attempted. The young feel the need to create order out of the "chaos" they perceive, and with time they become defenders of the overview the beginner often lacks. This postscript will end with a positive suggestion to aid in creating some degree of nomenclatural stability.

In the second section Dr. T. Timm began with a global survey of distribution patterns, and noted that, with increasing knowledge, there seem to be relatively few truly cosmopolitan species and those that are have been aided by human commerce. While distribution mechanisms are not yet well understood, some distributional histories are becoming clearer. The zoogeographic story, of course, has sometimes to adapt to changed appreciation of taxonomy, and in this regard it is worth noting a growing doubt about the presence of *Isochoetides* (or *Tubifex*) *newaensis* in North America as it has not been seen in recent surveys which have firmed up the earlier, somewhat broad superficial first efforts largely out of my own laboratory. Questions of relict status versus endemic speciation are raised in this group as in others. The cool-climate ancestry of many families may be related to glaciation survival, of course. The antiquity of the Haplotaxidae is accepted, and the views of

Dr. Timm are aligned with those of Jamieson (1978) in suggesting
that the Tubificina are not derivable from that family. Dr. Timm
also adheres to the classical view that oligochaetes are descended
from polychaetes, and so the multiple setae of the aquatic Tubificina are taken to be homologues of those polychaetes. The more
revolutionary view of the situation is that septate coelomates
with a few setae, derived from some type of sipunculid-like ancestor,
gave rise to something anatomically quite like a haplotoxid (the
whole argument appears in Clark, 1964). The odd setae of *Haplotaxis gordioides* are exceptional in the family, but may be an echo
of an earlier, less lumbricine setal arrangement in them. The
haplotaxid selected as a representative of the family by Jamieson
in his computer study of higher taxa happens to be the sole species
with its anterior-most male pores shifted back a segment to that
bearing the second pair, foreshadowing the trend to transposition
of male pores in the lumbricines. The family shows small traces of
its ancestry scattered among the species, making it hard to generalize about them. If they are ancestral to the oligochaetes and
similar to the ancestors of the polychaetes (and families such as
the Aeolosomatidae which differ profoundly from all other annelids)
then the multiple setae of the tubificines and the eye-spots of the
naidids would have to have evolved separately, presumably in response to a swimming mode of life at some point. A careful comparison of the setae of polychaetes and tubificines should be the
concrete result of this stimulating difference in views. Another
requirement is a computer analysis, such as that performed by
Jamieson, that actually includes representatives of the tubificines
as well as the lumbricines. It is worth noting, in passing, that
lumbricine setae do become multiplied in perichaetine forms, and
that, in a recently described phreodrilid, bifid setae change into
hair setae in sequential bundles. The profound lack of setal diversity in marine tubificids is noteworthy. The male ducts are often
complex, some have prostatic cells retained within the atrial muscle
layers (as in Phreodrilids) which may be a primitive condition. The
two or three bifid setae of marine species might not, then, be a
derived state but the beginning of a much greater setal diversity
in limnic species. It should be noted in this connection that the
Clitellinae has been dissolved in publications appearing after Dr.
Timm prepared his manuscript.

Drs. D.H. Di Persia, D.R. Spencer, M. Ladle and G.J. Bird
presented summaries of recent work in Argentina, St. Lawrence
Great Lakes of North America and southern Britain. Problems over
species complexes such as *L. hoffmeisteri* were raised once again.

In production studies various approaches are used. Drs. T.L.
Poddubnaya and G. Bonomi (with G. DiCola) presented detailed
accounts of life histories and production, going into much greater
detail than the pilot studies by my laboratory and others in the
1960's. The life histories are much more complex than was earlier

believed, and regeneration of the reproductive organs seems to
occur more than once in these relatively long-lived benthic species.
Production varies according to expected parameters such as temperature and food supply. Dr. V. Standen discussed the terrestrial
forms, and was critical of the use of production to biomass ratios,
partly because of the variable interpretation of them. R.O. Brinkhurst then outlined the work done on co-habiting species by a series
of workers in Canada, employing the physiological approach, measuring respiration, growth, ingestion and egestion. The most striking
finding here was that naturally co-existing species interact positively. The energy flow through a system is greatly increased by
mutual conditioning of the sediments by the species, suggesting that
monoculture experiments may underestimate production.

In habitat studies the emphasis was largely on environmental
factors involved in determining species distribution and abundance
of freshwater, groundwater, estuaries and the sea. The role of
tubificids, in particular, in relation to man and other users was
also emphasized through the relative importance of freshwater and
estuarine species as bioperturbers, fish food and oxygen users
(Drs. A. Grigelis, J. Juget, R. Diaz, I. Birtwell and D.R. Arthur).

Dr. P. Chapman indicated the importance of measuring interstitial variables rather than water-column parameters in the study
of benthic species. Dr. P. McCall summarized a body of recent
knowledge on the effect of worms on the geology and geochemistry
of sediments, where the possibility that the worms assist in the
recovery and recycling of materials that would otherwise remain
locked up in the sedimentary sink is being seriously considered.

Dr. O.W. Giere reported on tolerance and preference tests
using marine species, providing an ecophysiological basis to field
presumptions as to habitat preferences. In this connection the
absence of laboratory physiological tests on species held to occupy
different positions along a trophic gradient is surprising. The
fact that unifactorial tolerance tests reveal an adaptive capacity
wider than that suggested by field studies is not unusual, but is
less apparent in meiofaunal species. The maintenance of laboratory
gradients enabled preference tests to be performed, and synergistic
effects were demonstrated. Tolerance capacity in warm climates is
closely aligned with field distribution, far less so in temperate
areas where biotic factors may be of greater significance. Preference responses are of significance for the endobenthos, but again
the monofactorial approach is of limited value. Dr. Giere presumes,
as we all do, that marine species are limnogenic invaders into the
sea and it remains to be seen if that view changes with our greatly
increased knowledge of the phallodriline tubificids in particular.

   A minor issue in this section is the emphasis on the enchytraeids and tubificids and the lack of studies on other families.
Ecological work on the naidids is scarce (though some recent work
was not touched on at the meeting). One of the reasons for this
has been the recognition of the value of these two families in
terrestrial and aquatic soil studies and in pollution detection.
These subjects were addressed in the last series of papers, which
address the use of oligochaetes in pollution studies. Dr. K.
Kasprzak covered rural problems, both terrestrial and aquatic, and
stressed the need to measure the role of these organisms in energy
transformation and material circulation. Dr. G. Milbrink reported
on the European situation with special reference to Sweden. His
finding of elaborately modified setae in worms exposed to high
levels of mercury gives dramatic proof of environmental effects on
taxonomic characters. Drs. Lang and Lang-Dobler reported on the
trophic classification in Lake Geneva, producing a taxonomic
sequence which somewhat echoes that in the St. Lawrence Great Lakes
with some allowance for ecologically equivalent species in the two
regions. They did not apparently end up with the *Tubifex tubifex-
Limnodrilus hoffmeisteri* assemblage of polluted, rather than eutrophic, waters but this situation surely prevails in the most extreme
European situations. This author, substituting for Dr. J. Hiltunen,
briefly renewed the philosophical approaches to the use of benthos
in North American pollution studies without attempting an encyclo-
pedic coverage of the many reports and publications available.
Dr. P. Chapman and his co-workers discussed work in relation to
heavy metals and oligochaetes in general and specifically in the
Fraser River near Vancouver, British Columbia. Contrary to earlier
suggestions that worms were readily poisoned by metals, many seem
to be tolerant and to accumulate them. Their metal content stays
stable throughout the year, but the route of bioaccumulation is
not yet understood. Variation between species, the importance of
this accumulation to the food chain or the retrieval of metals
from the sedimentary sink, and the use of worms as indicators in
this field needs further attention.

   Dr. H. Caspers discussed those situations in which tubificids
totally dominate a grossly polluted urban environment of the type
that gave sludge-worms their name.

   At this first symposium many of the principal actors met each
other for the first time. Useful contacts were made or confirmed,
and much information was shared. Further symposia were thought
to be worth promoting, and so it is hoped that a similar group
will assemble in Pallanza, Italy in 1982, and possibly in Hamburg
three years later. While the normal format for exchange of views
seem appropriate for ecological work, there may be a need for some
sort of working session in that area. More clearly needed is some
forum for taxonomists to present to their colleagues the results
of group deliberation of some outstanding problems. I would like

to propose the creation of committees, one for each family, who
would seek opinions regarding the status of certain nomenclatural
problems. The replies could then be presented at special committee
meetings during the workshop. If any concensus emerges, it could
be reported in the proceedings of the symposium as a guide for
authors though it would not, of course, have any legal authority.
It might be that debate of such issues could shorten the time
elapsing between reviews that enable nomenclatural changes to be
made. If joint publication of brief notices in the scientific
literature are required this should not be too hard to achieve.
Hence, I wish to challenge the participants to respond to the
following recommendations to be formally addressed to the second
symposium.

1. That a small panel of taxonomists be created for each
   family, made up of at least one attendee of the second
   symposium plus other corresponding members willing to
   serve.
2. That the panel concerned with the Tubificidae seek a
   concensus, which could be incorporated into the proceedings of the second symposium on the following:
   a) the status of *Isochaeta virulenta*
   b) the generic position of *Limnodrilus newaensis*
      Mich.
   c) the status of *Monopylephorus rubroniveus* Lev.
   d) the status of *Peloscolex variegatus* Leidy
      sensu Brinkhurst.
3. That the panel concerned with the Lumbriculidae seek a
   concensus, which could be incorporated into the proceedings of the second symposium, on the following:
   a) the status of *Thinodrilus* Smith
                     Mesoporodrilus Smith
                     Sutroa Eisen
                     Metalamprodrilus Izossimov
                     Bythonomus Grube
4. That the panels concerned with other families determine
   the status of major, outstanding nomenclatural debates
   for inclusion in the proceedings of the second symposium.

These recommendations will be circulated by mail to a wide
audience prior to publication of these proceedings, and if the
response is positive the committees could be struck and some
progress achieved prior to the second symposium.

Most of these outstanding problems have been covered in the
papers of three recent authors, myself, Dr. S. Hrabě and Dr. C.
Holmquist. The list is naturally biased to those problems I can
readily identify, but which seem susceptible to solution without
the need for new anatomical evidence.

References additional to those
cited in the Appendix

Clark, R.B., 1964, Dynamics of Metazoan Evolution. Clarendon
    Press, Oxford.
Jamieson, B.G.M., 1978, Phylogenetic and phenetic systematics of
    the Opisthoporous Oligochaeta (Annelida, Clitellata).
    Evolutionary Theory, 3:195-233.
Reynolds, J.W. and D.G. Cook, 1976, Nomenclatura Oligochaetologica.
    UNB Press, Fredericton.

APPENDIX

Synonomies and Equivalent Names in Current Use Employed
Within This Volume.

1. Those names cited by various authors in the form used by an
earlier reference but which may be generally accepted to be
synonyms:

        Cited Name   Generally accepted current
                                              name

Cited Name	Generally accepted current name
Tubifex templetoni	*Ilyodrilus templetoni*
Ilyodrilus moldaviensis	*Potamothrix moldaviensis*
Ilyodrilus vejdovskyi	*Potamothrix vejdovskyi*
(*Potamothrix bedoti*	No longer a synonym of *P. bavaricus*, a valid species)
Psammoryctides curvisetosus	No longer in that genus, part of the *Peloscolex* complex but not placeable in the latest revision.
(*Aktedrilus monospermathecus*	No longer transfered to *Phallodrilus*)
Limnodrilus aurostriata	*Limnodrilus hoffmeisteri*
Limnodrilus helveticus	*Limnodrilus profundicola*
Pelodrilus	*Haplotaxis*
Phreoryctes gordioides	*Haplotaxis gordioides*

2. Names in common use as equivalent alternatives (see Appendix
3 for some clarification).

As used in this volume	Alternate also used here or elsewhere in recent literature
Limnodrilus spiralis	*Limnodrilus hoffmeisteri*
Limnodrilus newaensis	*Tubifex newaensis*
	Isochaetides newaensis
	Isochaeta newaensis
Monopylephorus rubroniveus	*Rhizodrilus pilosus* et al.
Monopylephorus irroratus	*Rhizodrilus irroratus* et al.

Alexandrovia	*Telmatodrilus*
Lumbriculus variegatus	
inconstans	*Thinodrilus inconstans*
Eclipidrilus lacustris	*Mesoporodrilus lacustris*
Peloscolex	*Tubificoides, Haber, Oriento-*
	drilus
	Spirosperma, Embolocephalus,
	Peloscolex, Edukemius

3.   Some causes of the examples of nomenclatural instability
identified in Appendixes 1-2. (*The editor's opinions are shown
in Italics in parenthesis where these refer to particular taxa*).

A.   Decisions Involving Type Species of Genera

    The reasons for several of the instabilities that have arisen
in the nomenclature of the Tubificidae in particular are that the
type specimens of the type species of certain genera are either
not available or are disputable.

A-1. The first case is easily disposed of. The types of *Isochaeta
virulenta* Pointner have been examined by Prof. S. Hrabě, and it
is his belief that the specimen is an immature *Limnodrilus* species.
As it has not been found in subsequent studies it seems best to
adopt his suggested alternative, and name the genus *Isochaetides*
with *baicalensis* as the generic type. This should hopefully find
general agreement.

    This does not, unfortunately, close the issue as Prof. Hrabě
restricts the genus to species lacking hair setae, and adds to
the earlier Brinkhurst assemblage *newaensis (which has male
reproductive organs of the Tubifex type in my opinion) and pseudo-
gaster (now seen as a member of the marine genus Tubificoides,
though still a problem)* as well as *Siolidrilus adetus, Limnodrilus
neotropicus* and *Limnodrilus lastockini*.

    The species excluded from this genus because they possess
hair and pectinate setae is *dojranensis*, placed in *Tubifex* by
Prof. Hrabě and in *Haber* by Dr. C. Holmquist (*despite the totally
characteristic long thin atrium with the small prostate gland,
genital setae and sheathless penis which accord with the genus
Isochaetides*). *Peloscolex nomurai* was first put in *Isochaeta*,
then returned to *Peloscolex* by this author, but Dr. C. Holmquist
places it "incertae sedis" (? = *Orientodrilus*). *Peloscolex freyi*
is suggested as a candidate for *Isochaetides* by Dr. C. Holmquist,
but is left "incertae sedis" (see below).

    The point at issue here should be resolvable, which is that,
while *Limnodrilus* is odd among the Tubificidae in having a consistent lack of hair and pectinate setae in the dorsal bundles in all

species, other good genera have some species with, others without (having only bifid dorsal setae) or even species divided into subspecies on this single characteristic (i.e. *Tubifex tubifex*). *(There seems no special reason for excluding species from the genus on this basis alone when other characteristics are more clearly aligned with those of this assemblage rather than any other).* The most important species affected here would be *newaensis,* a much studied European form, whose presence in North America must now be regarded as highly dubious as it has not been seen in a series of recent surveys.

A-2. The *Monopylephorus/Rhizodrilus* debate revolves partly around types, partly on the extension of specific limits. Dr. S. Hrabĕ considers *Monopylephorus rubroniveus* Levinsen to be unidentifiable, the subsequent re-description of new material by another author being discounted. Another school accepts the re-description in the absence of types of the earlier taxon. This seems to be purely a matter of opinion, but many recent authorities use *Monopylephorus*. The extension of the two main species *irroratus* and *rubroniveus* into two series of closely similar species can only be resolved by a study of the ontogeny of the male system, when it will become clear whether or not the male and spermathecal pores can open either separately or via a (?transitory) median bursa within a single population, or whether simple-pointed dorsal posterior setae can be regularly present in some populations, absent in others. This question is susceptible to testing, and the author hopes to be able to exploit a local population to this end. The extent to which a species may be protandrous may affect the reliance placed upon the presence of sperm in the spermathecae as evidence of complete maturation of the male system.

A-3. The genus *Peloscolex* is a special conundrum. As it was originally defined as either a genus or a subgenus of *Tubifex* (*a genus itself seemingly based as much on a lack of those characteristics that were used to define other tubificine genera as on positive characters of its own*) on the basis of the ornamentation of the body wall, and as this ornamentation made whole-mounts of limited value in determining internal organs, the taxon clearly became relatively unweildy as well as composed of a variety of separable taxa once the structure of the male ducts became better known. With the description of more and more marine forms, the distinctive *Tubificoides* assemblage drops out quite readily, with the exception that *benedeni* (or *benedii*) is placed in the separate genus *Edukemius* by Dr. C. Holmquist (*the degree of separation here being one of those splitter/lumper type of disagreements which are again matters of opinion, depending upon the focus on similarity or differences*).

The absence of type specimens of the type species *P. variegatus*
Leidy created an even worse problem. Perhaps taking an unconscious
leaf out of the book of W. Michaelsen (who elected to use the
obscure, then American, genus name *Ilyodrilus* for the European
species now separated off as the genus *Potamothrix*) this author
aligned some newly discovered American material with the poorly-
described taxon *P. variegatus* Leidy. Dr. C. Holmquist is of the
opinion that the new material properly belongs to a distinct
species (*P. confusus*) and that *P. multisetosus* is identical with
variegatus of Leidy, this suggesting a re-alignment of names that
would affect a reasonably large volume of ecological literature
(as evidenced in this volume where the names are used in their
pre-Holmquist sense). *(This author claims to be able to defend
the selection of the original neotype on the basis of anatomical
and ecological evidence).*

One point at issue here, as elsewhere in the genus, is the
identity of some short hair-like setae as homologues of the more
normally bifid or pectinate sigmoid setae rather than as simply
short hair setae. *(While the issue is not totally clear to me
in Dr. C. Holmquist's writing, the context suggests she uses the
term pectinate for bifid dorsal setae, and may regard thin simple-
pointed dorsals as all being hair setae - we perhaps need a term
such as the naidid "needle seta" to make clear our meaning here).*

The same author divides *Peloscolex* into eight genera but
leaves thirty-five taxa essentially homeless, and does not accept
the majority of geographical records not backed up by museum
specimens. *(This author accepts the separation of the marine
Tubificoides and Edukemius, suspects that the speciosus assemblage
plus the recently described turqini may well be separable as the
suggested genus Haber, that multisetosus and its very close ally
moszynskii are probably separable at the generic level, but under
a new name, and not Peloscolex because that is still useable for
P. variegatus sensu mihi and will probably be held to accomodate
Embolocephalus, Spirosperma, Orientodrilus and even Baikalodrilus
despite the odd shape of the worm).*

B.   Non-Equivalent Ranking

B-1. One of the more interesting causes of the finer division of
Peloscolex suggested by Dr. C. Holmquist than that contemplated
by the author is that quite fine histological differences revealed
by sections are used in preference to the coarser-grained detail
obtainable from dissections (see the following discussion on
methods). The usual lump/split problem does not matter too much
where the subdivisions are acceptable to two authors but at different rankings i.e. *Telmatodrilus (Alexandrovia) ringulatus* and
Alexandrovia ringulatus are recognisably the same taxon so long
as the sub-generic name is used (*and resolution of that difference*

*may occur as the question of the various types of agglomerated
sperm in the spermathecae is clarified by the discovery of more
material and more species - again susceptible to analytical study).*

B-2. The changes in the nomenclature of the Lumbriculidae noted
are again referable to another study by Dr. C. Holmquist, all of
these originating from a detailed follow-up of a single Alaskan
collection. Here again, the major differences are due to a
preference for fine divisions, so that most of the genera of Dr.
D.G. Cook end up as two or more "closely related" genera of Dr.
C. Holmquist, and as such cannot really be more than opinion on
either side unless the controversy leads to a clarification of
of homology versus analogy.

C.   Some Pertinent Literature Relating to the Previous Debate

Brinkhurst, R.O., 1979, On the types in the genus *Peloscolex*
     Leidy (Oligochaeta, Tubificidae). Proc. Biol. Soc. Wash.,
     (in press).
Brinkhurst, R.O. and B.G.M. Jamieson, 1971, Aquatic Oligochaeta
     of the World. Oliver and Boyd, Edinburgh, 860 pp.
Holmquist, C., 1974, On *Alexandrovia onegensis* Hrabĕ from Alaska,
     with a revision of the Telmatodrilinae (Oligochaeta, Tubifi-
     cidae). Zool. Jb. Syst., 101:249-268.
Holmquist, C., 1976, Lumbriculids (Oligochaeta) of Northern
     Alaska and Northwest Canada. Zool. Jb. Syst., 103:377-431.
Holmquist, C., 1978, Revision of the genus *Peloscolex* (Oligochaeta,
     Tubificidae) 1. Zool. Scripta, 7:187-208.
Holmquist, C., 1979, Revision of the genus *Peloscolex* (Oligochaeta,
     Tubificidae) 2. Zool. Scripta, 8:37-60.
Hrabĕ, S., 1966, New or insufficiently known species of the family
     Tubificidae. Spisy prir. Fac. Univ. Brne, 470:57-76.
Hrabĕ, S., 1967, Two new species of the family Tubificidae from
     the Black Sea with remarks about various species of the
     subfamily Tubificinae. Spisy prir. Fac. Univ. Brne, 485:
     331-356.
Loden, M.S., 1978, A revision of the genus *Psammoryctides* (Oligo-
     chaeta, Tubificidae) in North America. Proc. Biol. Soc.
     Wash., 91:74-84.
Spencer, D.R., 1978, Oligochaeta of Cayuga Lake, New York, with
     a description of *Potamothrix bavaricus* and *P. bedoti*. Trans.
     Amer. Microsc. Soc., 97:139-147.

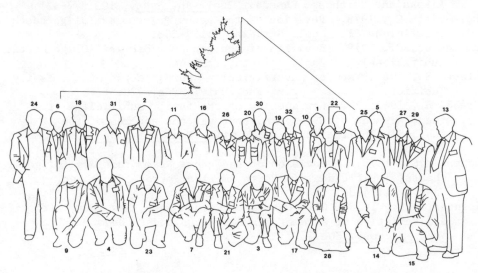

PARTICIPANTS

Numbers indicate identity on the group photograph.

1. H.R. Baker, Canada

2. M.T. Barbour, USA

3. I.K. Birtwell, Canada

4. G. Bonomi, Italy

5. R.O. Brinkhurst, Canada

6. H. Caspers, W. Germany

7. P.M. Chapman, Canada

8. L. Churchland, Canada

9. K.A. Coates, Canada

10. D.G. Cook, Canada

11. R. Diaz, USA

12. D.H. Di Persia, Argentina
    (in absentia)

13. M. Dzwillo, W. Germany

14. C. Erseus, Sweden

15. O.W. Giere, W. Germany

16. J. Grassle, USA

17. A. Grigelis, USSR

18. W.J. Harman, USA

19. J. Juget, France

20. K. Kasprzak, Poland

21. M. Ladle, UK

22. C. Lang and

    B. Lang-Dobler, Switzerland

23. S. Locy, USA

24. P. McCall, USA

25. G. Milbrink, Sweden

26. T.L. Poddubnaya, USSR

27. D. Spencer, USA

28. V. Standen, UK

29. T. Timm, USSR

30. M. Wetzel, USA

31. S. Whitley, USA

By correspondence: M.S. Loden, USA
Observer: 32. S. Piper, USA